ENGINEERING CHEMISTRY

ENGINEERING CHEMISTRY

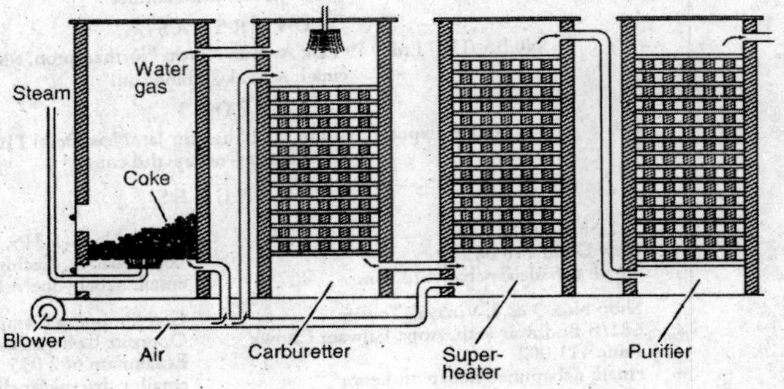

Bidhan Chandra Ray

MSc (Gold-medalist), PhD; Professor,
Department of Chemistry, Jadavpur University, Kolkata

Satyanarayan Das

MSc, PhD; Reader, Department of Chemistry (Retd), Tamralipta Mahavidyalaya,
Tamluk, Purba Medinipur; Ex-Reader in Chemistry, Surendranath Evening College, Kolkata

Suchandra Biswas

MSc, PhD; Assistant Professor, Department of Chemistry,
Bankim Sardar College, Tangrakhali, South 24 Parganas

New Central Book Agency (P) Ltd

LONDON

HYDERABAD ERNAKULAM BHUBANESWAR

NEW DELHI KOLKATA PUNE GUWAHATI

NCBA

REGD OFFICE

8/1 Chintamoni Das Lane, Kolkata 700 009, India
email: ncbapvtltd@eth.net

OVERSEAS

NCBA (UK) Ltd, 149 Park Avenue North, Northampton, NN3 2HY, UK
email: ncbauk@yahoo.com

EXPORT

NCBA Exports Pvt. Ltd, 212 Shahpur Jat, New Delhi 110 049
email: ncbaexp@ncbapvtltd.com

BRANCHES

212 Shahpur Jat
New Delhi 110 049
email: ncbadel@ncbapvtltd.com

Shop Nos. 3 & 4, Vinayak Towers
681/B Budhwar Peth, Appa Balwant Chowk
Pune 411 002
email: ncbapune@ncbapvtltd.com

Shop No. 15, Bharati Towers, Ground Floor
Forest Park, **Bhubaneswar** 751 001
email: ncbabub@ncbapvtltd.com

House No. 3-1-315, 1st Floor
Nimboliadda, Kachiguda, **Hyderabad** 500 027
email: ncbahydb@ncbapvtltd.com

GSS Shopping Complex, 1st Floor
Opposite College Post Office, Convent Road
Ernakulam 682 035
email: ncbaernk@ncbapvtltd.com

Radhamayee Complex (Second Floor)
Opp. SBI, Panbazar, Jaswanta Road
Guwahati 781 001
email: ncbaguwahati@ncbapvt.ltd.com

ENGINEERING CHEMISTRY • Bidhan Chandra Ray, Satya Narayan Das and Suchandra Biswas

First Published: July 2008
Reprinted: June 2014

PUBLISHER AND TYPESETTER
New Central Book Agency (P) Ltd
8/1 Chintamoni Das Lane, Kolkata 700 009

PRINTER
New Central Book Agency (P) Ltd
Web-Offset Division, Dhulagarh, Sankrail, Howrah

COVER DESIGNER
Soumen Paul

TECHNICAL EDITOR
Dipan Roy

PROJECT TEAM
Suman Dutta, Mithu Karmakar and Prabhat Jas

ISBN 978 81 7381 583 6

Price ₹ 285.00

CONTENTS

PREFACE

Engineering Chemistry, meant for the beginners, is a sincere attempt to make the engineering students understand the fundamentals of chemistry in a lucid manner. Engineering students—burdened by their main disciplines—often find chemistry a 'necessary evil', to be surmounted somehow. This book, we hope, will help the students not only in this, but will also make them interested in the subject.

It has been our endeavour to make the book as user-friendly for the target readers as possible. The book has been written in accordance with the IUPAC nomenclature and convention. We have provided solved problems with each chapter and also solved West Bengal University of Technology Question Papers of the last three years.

We are thankful to (Prof) Dr Dulal Chandra Mukherjee and Dr Sreedhar Gupta of Heritage Institute of Technology; Sri Prabir Kumar Panda, Assistant Professor, Siliguri Institute of Technology, WB; and Dr Jaya Mukherjee of Jadavpur University for their timely help and cooperation. They had gone through the manuscript and given us their valuable suggestions. We are indebted to them.

We have tried to make each of the chapters as exhaustive as possible. But we are fully aware of our limitations. Hence, some errors might have crept herein. Suggestions and criticisms from our learned colleagues and also from inquisitive students for the betterment of this book will be thankfully received for future upgradation.

Kolkata
Rathayatra
4 July 2008

Bidhan Chandra Ray
Satya Narayan Das
Suchandra Biswas

1

ATOMS AND MOLECULES

ATOMS

The *atoma* or *atom* derived from the two Greek words –A (not) temnein (to cut), i.e., something that is not divisible.

(a) Hindu philosophers of the ancient time thought matters around us were not continuous but consisted of small particles.

(b) About 5th century B.C., Greek atomists Leucippus and his student Democritus thought that materials were made up of ultimately small indivisible units. This idea was suppressed and not came into light for several hundred years due to the objection of Greek philosopher Aristotle (384–322 B.C.). All these ideas were only suppositions, no scientific basis was there.

(c) John Dalton's (an English Mathematician and Chemist in 1808) concept of atom was based on successful explanation of some experimental observations such as the laws of chemical combinations, etc. Thus, Dalton's idea is a landmark of the development of modern science.

This atomic concept of Dalton cannot explain the Gay Lussac's laws of gaseous volume; why do the elements have different masses and why do the elements have different chemical activities? Later, discovery of electron, proton, neutron, isotopes and isobars seemed to have proved the failure of the Dalton's idea, but still his ideas formed the basis of modern thinking.

(d) Avogadro introduced the idea of the existence of molecules to explain the Gay Lussac's laws of gaseous volume.

(e) Prout (1815) gave the most important idea about the composition of matter. Hydrogen is the simplest and lightest element and all other elements are composed of the hydrogen atoms. So, the atomic weights of other elements will be simple multiple of the atomic weight of hydrogen (whole number rule). This is not actually met with, as atomic weights of elements are not whole number and hence discarded at that time. This idea gets momentum in present years due to the development of the conception of matter-energy relationship and nuclear reactions.

FUNDAMENTAL PARTICLES

(i) Stable particles (existence for a few seconds to a few minutes)

Electron, e^{-1} (unit of electricity and named by Stonny in 1881)

In 1897 Prof. J. J. Thomson (a British physicist) discovered electron by passing high voltage electricity (50 000 volts) through a highly evacuated (0.01 mm of Hg) discharge tube when cathode rays are obtained and he showed that the rays consist of negatively charged very light particle, called electron. Experimental determination of specific charge, i.e., charge to mass ratio and charge was $e/m = 1.76 \times 10^7$emu/g = 5.27 × 10^{17}esu/gm and charge, $e = 4.77 \times 10^{-10}$ esu or 1.602×10^{-19} coulomb and hence, mass of electron is 9.1×10^{-28}g.

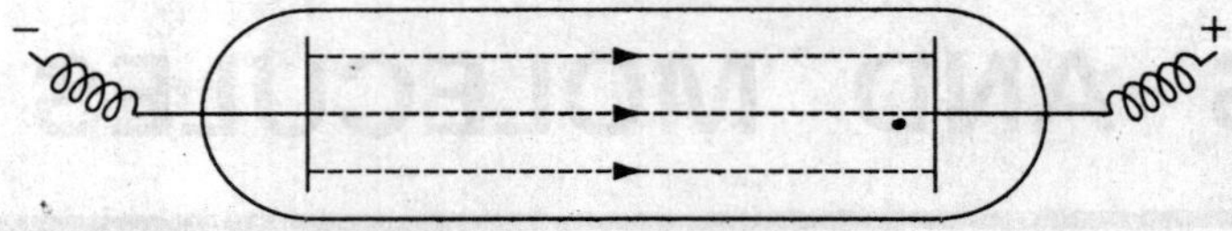

Fig. 1.1 : Thomsons experiment on Cathode ray.

Proton ($_1^1H$, $_1^1P$)

Matter as a whole being neutral, scientist thought that as a counterpart of electron there must be some positive particle in the atom. In 1886 Goldstein used a perforated cathode at the centre of the discharge tube when he found positive rays emerged out in the tube but travel away from the anode. He found that these positive rays are dependent on the nature of residual gas in the tube after evacuation and charge to mass ratio, e/m, is maximum when the tube was initially filled with hydrogen. He called this positive particle as proton. The charge of proton is equal to that of electron with positive sign, and mass is about 1836 times the mass of electron and 1.007277 amu or 1.67×10^{-24}g.

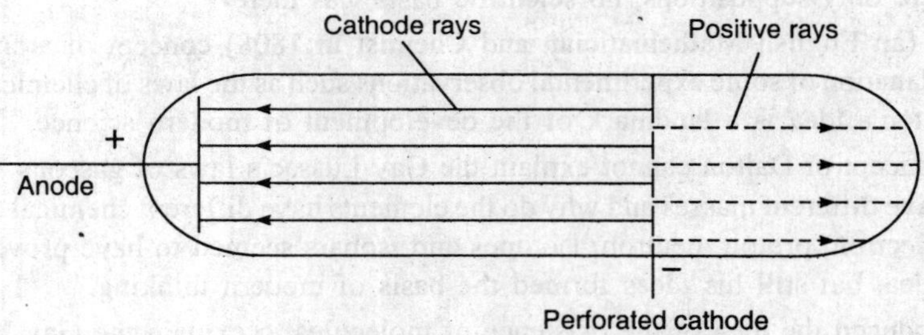

Fig. 1.2 : Discovery of proton with perforated cathode.

Neutron (James Chadwick, 1932) $_1^1n$

In 1920 Rutherford postulated the existence of neutral particle in atom. In 1930 Bothe and Backer allowed á particle to fall on beryllium when high penetrating radiation, first assumed to be ã ray, was obtained. In 1932, F. Joliot and I. Curie proved these to be neutral particles with mass equal to that of proton. These are called neutron.

(ii) Unstable particles (existence from 10^{-16} sec to 10^{-6} sec)

Positron (e^+ or β^+, average life 10^{-9} sec)

In 1932, C. D. Anderson discovered the existence of these particles on analyzing the particles emitted when high-energy cosmic ray impinges on metals. These particles have mass equal to that of electron but with unit positive charge.

Meson (π, μ, k)

This Greek word means the particle of mass in between that of electron and proton. Dr. H. J. Bhaba named the particles as meson in 1939. Analysis of the particles emitted when cosmic ray impinges on metal proved the existence of these particles. Mesons are of three types:

(i) π-meson or pion π^+, π^-, π^0 , mass approximately 270 m_e,

(ii) μ-meson or muon; μ^+ , μ^-, mass about 206 m_e

(iii) k-meson or kaon k^+, k^- , $2k^0$ and mass is 966 m_e.

Neutrino and anti-neutrino (ν and $\overline{\nu}$)

These are hypothetical particles. Their existence was supposed to explain the mass balance and energy balance during different nuclear reactions. Besides these, unstable particles like anti-proton, anti-neutron, etc are also obtained.

ATOMIC MODELS

Thomson's water melon model

After the discovery of electron and proton, Thomson pictured atom as watermelon where positive charge is distributed throughout the atom and negative charge electrons are held in it as seeds. Or it can be represented like an ornament box, where cotton pad is the positive charge and the valuable gems are the electrons.

Rutherford's nuclear model or solar model (1911)

When α-rays are allowed to strike on 0.0004 mm gold foil, 1 in 20 000, α particles are scattered at an angle greater than 90°. From this observation he concluded that all the positive charges and almost all the masses of the atom is located at a central position within the atom, called nucleus, and the electrons revolve round it with high speed just like the solar system. The rest of the atom is void.

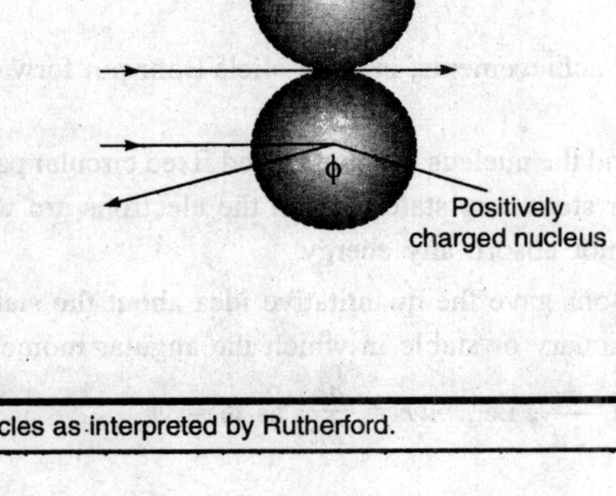

Fig. 1.3 : Scattering of α-particles as interpreted by Rutherford.

Demerits: When the charged particles revolve round the centrally located high charge, the revolving charge will radiate energy and the radiation spectra will be continuous. But the fact is that this spectrum is discontinuous. It means the electron will radiate energy in some position and does not in other position.

The discontinuous nature of the spectra is observed in hydrogen spectra. Hydrogen atom on excitation yields a series of spectral lines. These series are named after their discoverers – Lyman, Balmer, Paschen, Brackett and Pfund – the first one is in the ultra violet region; second one is in the visible region and the others are in the infrared region. Balmer arbitrarily gave the wavelength, λ of the second series. The general form of the equation for these series is $\frac{1}{\lambda} = R_H\left(\frac{1}{n_1^2} - \frac{1}{n_2^2}\right)$, where $n_2 > n_1$ and R_H is called Rydberg constant, whose arbitrary value is 109677.75 cm^{-} for hydrogen atom. For Lyman series $n_1 = 1$ and $n_2 = 2, 3, 4, \cdots$, etc. For Balmer series $n_1 = 2$ and $n_2 = 3, 4, 5, \cdots$, etc.

Bohr model (1913)

To explain the discontinuous nature of spectra, Bohr proposed some theory describing the structure of an atom that based on the following:

(i) Planck's quantum theory

When solids are heated, they emit electromagnetic radiation in the form of wave. The amount of radiation at a certain temperature depends on its wavelength. Classical physics assumed that atoms and molecules could emit or absorb any arbitrary amount of energy. Planck was of the opinion that atoms or molecules cannot absorb or radiate any amount of energy but only in discrete quantity like small packages or bundlés, called quantum. The emitted or absorbed energy $E = h\nu$ or, $E = nh\nu$, where n is principal quantum number, ν is the frequency of radiation and h is the Planck's constant, ($h = 6.625 \times 10^{-27}$ erg-sec or 6.625×10^{-34} joule-sec). So, energy emitted or absorbed will be *h í*, 2*h í*, 6*h í*, and so on but not any fractional number.

(ii) Photoelectric effect (Albert Einstein in 1905)

When metal surface is exposed to light, electrons are emitted from the surface; hence, called photoelectric effect. For this, a minimum amount of energy, called threshold energy, is required. The number of emitted electrons depends on the intensity, i.e., brightness of the light but the energy of emitted electrons are not dependent on it. Einstein explained the phenomenon in a new outlook. He considered beam of light as a stream of particles, called *photon*. When the kinetic energy of the photons is less than the binding energy of electrons in the metal surface, no electron will be emitted. When K.E. = B.E., electrons are just knocked out and are free to be ejected. If K.E. > B.E., then electrons are emitted having some kinetic energy. So, using Planck's equation, $E = h\nu$ = K.E. + B.E. and kinetic energy of the emitted electron will be K.E. = $h\nu$ – B.E.

Based on these two scientific achievements, in 1913 Niels Bohr put forward three postulates regarding the structure of the atom.

1. The electrons revolve round the nucleus in some closed fixed circular path with some definite energy. These are called orbits or stationary states. When the electrons are within these stationary states, they will neither radiate nor absorb any energy.
2. In his second postulate Bohr gave the quantitative idea about the stationary states. He stated that only those orbits are stationary or stable in which the angular momentum of the electron will be some integral multiple of $\frac{h}{2\pi}$, i.e., $mvr = \frac{nh}{2\pi}$ as $\omega = vr$.

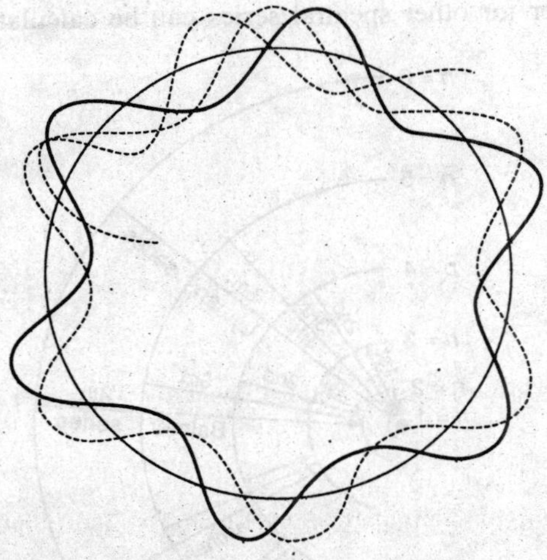

Fig. 1.4 : A Bohr orbit in terms of electron-waves. As the waves become smaller, the circular Bohr orbit is obtained as a limiting case. For an orbit or radius, r, $2\pi r = n\lambda$, where n is an integer and λ is the wavelength. Substituting $\lambda = h/mv$, one gets $mvr = nh/2\pi$.

3. In the third postulate he gave the idea about the origin of spectra. Although electrons are stable in stationary states, if its transition takes place from a higher energy state E_2 to a lower energy state E_1, the excess energy will be radiated and will give rise to atomic spectra.

   Hence, $\Delta E = E_2 - E_1 = h\nu = hc/\lambda = hc\bar{\nu}$, where $\bar{\nu} = \frac{1}{\lambda}$ or $\nu = \Delta E/h$, where c is the velocity of light and $\bar{\nu}$ is the wave number. On the basis of these postulates, the following calculations can be done:

   Radius of the nth orbit, $r = \dfrac{n^2h^2}{4\pi^2 mze^2}$.

   Energy of the electron in nth orbit, $E = -\dfrac{2\pi^2 z^2 e^4 m}{n^2h^2}$.

   Velocity of the electron in nth orbit, $v = \dfrac{2\pi ze^2}{nh}$.

   Here n = principal quantum number, h = Planck's constant
   m = mass of electron, z = atomic number of the atom (z =1 for Hydrogen)
   e = charge of the electron

Interpretation of hydrogen spectra

When electron transition takes place from higher energy state, E_2, to lower energy state, E_1, the radiated energy will give rise to emission spectra. If electron transition occurs from second energy state to the first energy state it will give rise to Lyman series and its wave number can be calculated as

$$E_2 - E_1 = \frac{2\pi^2 z^2 e^4 m}{h^2}\left(\frac{1}{1^2} - \frac{1}{2^2}\right) = hc\bar{\nu}$$

or, $\bar{\nu} = \dfrac{1}{\lambda} = \dfrac{2\pi^2 z^2 e^4 m}{h^3 c}\left(\dfrac{1}{1^2} - \dfrac{1}{2^2}\right) = R_H\left(1 - \dfrac{1}{2^2}\right)$ here $R_H = \dfrac{2\pi^2 z^2 e^4 m}{h^3 c}$. R_H is called the Rydberg constant for hydrogen.

Similarly, the wave number for other spectral series can be calculated.

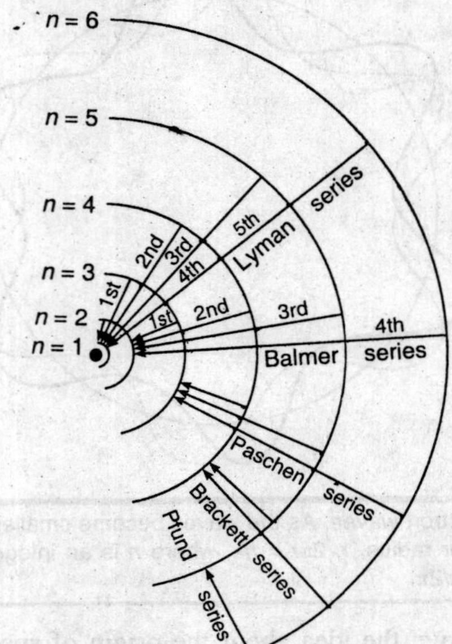

Fig. 1.5 : Transitions leading to the five spectral series of the hydrogen atom.

NUMERICAL

Problem 1. Electromagnetic radiation of wavelength 242 nm is just sufficient to ionize sodium atom. Calculate the ionization energy of the atom in kJ/mole.

Solution: Wavelength λ = 242 nm = 242×10^{-9} m

$$E = \frac{hc}{\lambda} = \frac{6\cdot 62 \times 10^{-34}\,\text{Js} \times 3 \times 10^{8}\,\text{m}}{242 \times 10^{-9}\,\text{m}} = 8\cdot 21 \times 10^{-19}\ \text{J/atom}$$

$= 8{\cdot}21 \times 10^{-19}$ J/atom $\times\ 6{\cdot}02 \times 10^{23}$ (Avogadro's number)

$= 4{\cdot}94 \times 10^{5}$ J/mole $= 4{\cdot}94 \times 10^{2}$ kJ/mole

Problem 2. Calculate the K.E. of the electron which has a wavelength of 4.8 pm.

Solution: $\lambda = h/p = h/mv$ $\quad v = \frac{h}{m\lambda} = \frac{6\cdot 63 \times 10^{-34}\,\text{kg m}^2\text{s}^{-1}}{9\cdot 11 \times 10^{-31}\,\text{kg} \times 4\cdot 8 \times 10^{-12}\,\text{m}} = 1\cdot 516 \times 10^{8}\,\text{ms}^{-1}.$

$$\therefore \text{K.E.} = \tfrac{1}{2}mv^2 = \tfrac{1}{2} \times 9\cdot 11 \times 10^{-31}\,\text{kg} \times \left(1\cdot 516 \times 10^{8}\,\text{ms}^{-1}\right)^2 = 1\cdot 047 \times 10^{-14}\ \text{kg m}^2\text{s}^{-2}.$$

Problem 3. If the electron of an unexcited hydrogen atom is excited by absorption of 1.76×10^{-18} J per atom, what would be the wavelength of radiation emitted when the atom returns to the ground state?

Solution: $\Delta E = h\nu = hc/\lambda$

$\lambda = hc/\Delta E$

$$\therefore\ \lambda = \frac{6\cdot 6 \times 10^{-34} \times 3 \times 10^{8}}{1\cdot 76 \times 10^{-18}} = 1\cdot 125 \times 10^{-7}\,\text{m}.$$

Demerits of Bohr model

1. Bohr calculated the energy of an electron (as a particle) revolving around the nucleus in a circular orbit. But according to Heisenberg's uncertainty principle, the position and velocity cannot be calculated accurately.
2. When spectral lines are viewed through spectroscope, each line appears to consist of more closely spaced lines called fine structures. This indicates more energy states are there within the same stationary states.
3. Bohr treated electron as particle, but modern concept accepts its wave particle duality.
4. Bohr's theory is applicable to only one electron system like H, He^{+1}, H_2^+, Li^{+2}, Be^{+3}, but not to multi-electron systems.

Sommerfeld's model (1915)

Azimuthal quantum number

To explain the limitations of Bohr model and to explain the fine structures of the hydrogen spectrum, Sommerfeld modified Bohr's concept from two different angles:

(i) By analogy of the planetary motion in the solar system, the electronic orbits are not circular but elliptical with the nucleus at one of its focus. So, during revolution along this elliptical orbit, sometimes the electron will be at the closest and sometimes at the furthest distance from the nucleus, i.e., the distance varies, consequently the velocity of the electron will vary.

(ii) If velocity varies its relative mass and hence, energy will also vary. The variation of the mass is given by the equation $m = \dfrac{m_0}{\sqrt{1-\dfrac{v^2}{c^2}}}$, where m_0 is the mass at rest and m is the relative mass when in motion. The motion of the electron in elliptical path can be described by two quantum numbers the radial quantum number n_r and azimuthal quantum number k (l according to wave mechanics) such that $n = n_r + k$ and

$$\frac{n}{k} = \frac{a}{b} = \frac{\text{length of semi major axis}}{\text{length of semi minor axis}}$$

and ellipticity of the orbit is determined by the relative values of n_r and k.

When $n = 4$, k has values 1, 2, 3 and 4 (according to wave mechanics, $l = 0, 1, 2, 3$).

According to Sommerfeld, for different values of n and l, the spliting of orbital can be shown as in Fig 1.6(a).

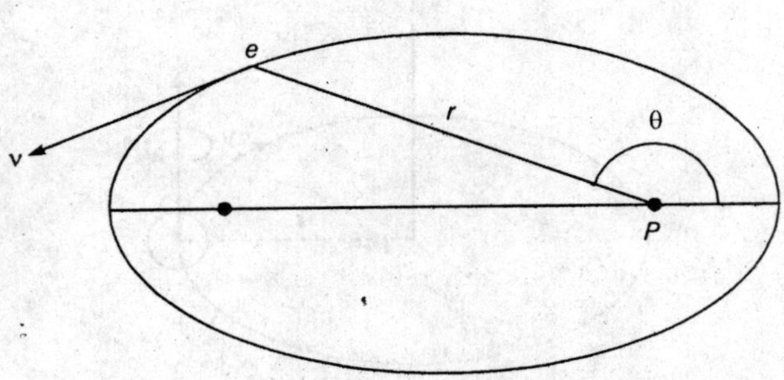

Fig. 1.6 : Revolution of electrón along eliptical path.

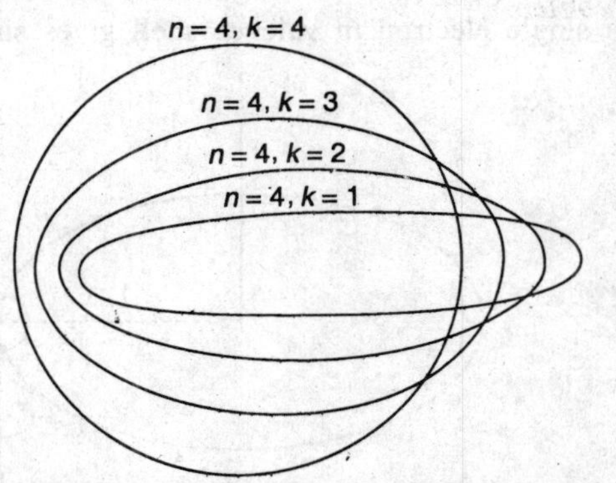

Fig. 1.6a : The different orbitals in $n = 4$ quantum state.

Zeemann effect (Magnetic quantum number m_l)

When atomic spectra is taken by placing the substance in high intensity magnetic field, each fine structure in the atomic spectra again splits to still more finer structures, called *hyperfine structures*. To explain these hyperfine structures, Zeemann introduced another quantum number, called *magnetic quantum number* m_l that indicates the orientation of the electronic orbits in space aligned with the applied field. Hence, the electronic orbits are not two dimensional, but three-dimensional. The values of m_l depends on the value of l (azimuthal quantum number) such that $m_l = \pm l$ passing through zero; thus total $2l + 1$ states are possible.

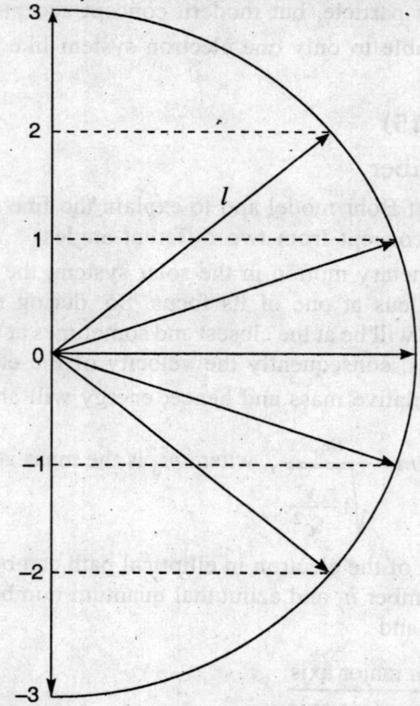

Fig. 1.7 : Space quantization in a magnetic field.

Spin quantum number m_s (it indicates the rotation of electrons on its own axis)

Compton (1921) first gave the idea about the spinning of electrons on its own axis. Sodium having a single electron in valence shell gives single yellow D-line. Observation under high resolving power

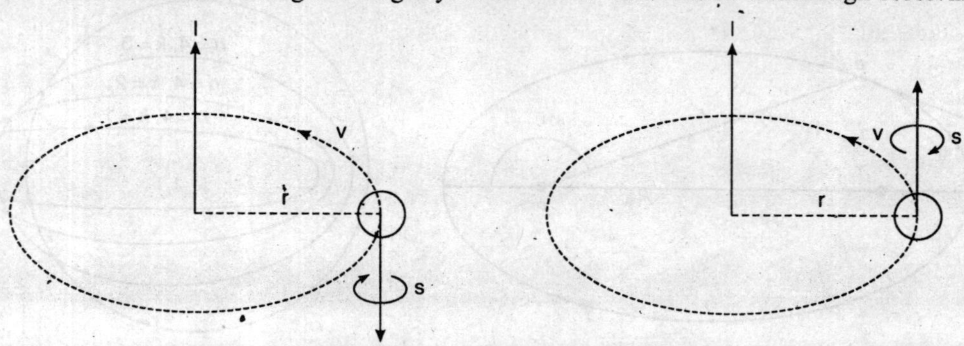

Fig. 1.8 : Orientation of spin momentum with reference to orbital angular momentum.

spectroscope shows that single D-line is actually a doublet D_1 and D_2 having space of 6Å between them. This can be explained by assuming spinning of electron on its own axis together with orbital motion round the nucleus. Spin momentum of the electron is indicated by another quantum number m_s; it has value $+\frac{1}{2}$ or $-\frac{1}{2}$.

The number of maximum electrons that can hold any orbital can be understood considering Aufbau principle, Pauli's exclusion principle and Hund's rule.

Aufbau principle

According to this principle, electrons will occupy orbital of lowest energy available. The order of these orbitals are $1s > 2s > 2p > 3s > 3p > 4s > 3d > 4p > 5s$ and so on.

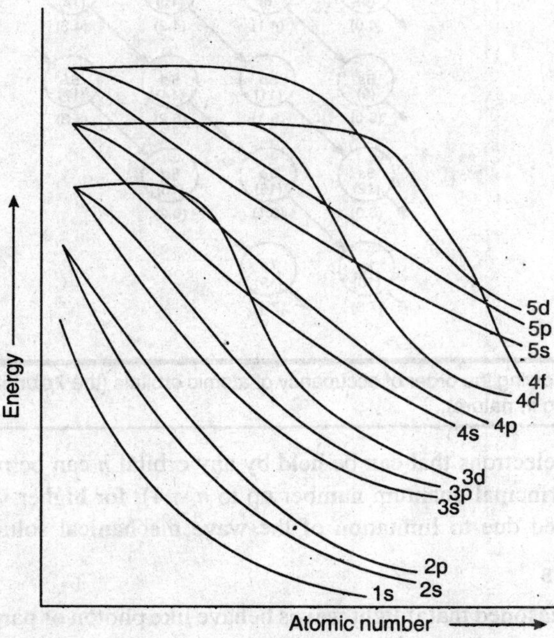

Fig. 1.9 : Nature of variation of orbital energy with atomic number.

Pauli's exclusion principle (1925)

According to this principle, no two electrons of the same atom will have all four quantum numbers n, l, m_l, and m_s identical. If two electrons are in the same quantum state i.e., if n, l, and m_l is identical then they must vary in m_s. Hence, they will have opposite spin.

Hund's rule

This rule states for degenerate states, i.e., states having equal energy, the electron will occupy singly i.e., their spins will remain parallel. Hence, half-filled or full-filled orbitals are comparatively more stable than any other electronic arrangement.

So, the different energy states and their sub-states capable of accommodating maximum number of electrons can be represented as follows:

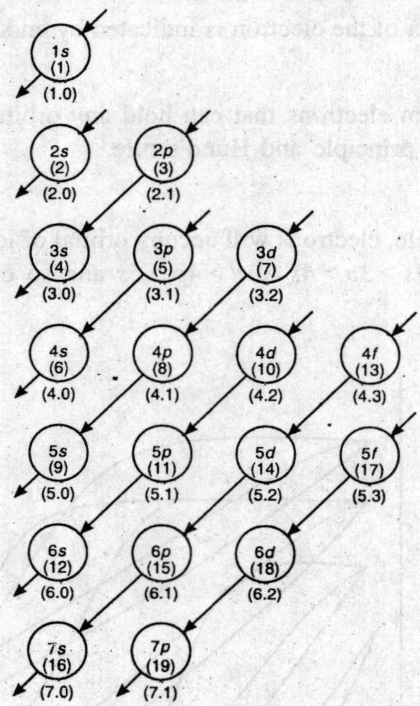

Fig. 1.10 : Mnemonic scheme showing the order of occupancy of atomic orbitals (the 7*p* orbitals are not occupied in any of the elements occurring in nature).

The maximum number of electrons that can be held by any orbital n can be represented by the general formula $2n^2$ (where n is the principal quantum number up to $n = 4$), for higher values of n, the maximum number is 32 (This is achieved due to limitation of the wave mechanical solutions).

Dual nature of electrons

In 1924 Louis de Broglie reasoned that if light waves behave like photon or particles, then small particles like electron can possess wave properties and in Bohr's model electrons bound to the nucleus behave like standing or stationary waves. If so then the wavelengths of the standing waves must fit the circumference of the orbit.

So $2\pi r$ (circumference) $= n\,\lambda$ (λ is the wavelength of electron wave, r is the radius of the orbit and n is integers like, 1, 2, 3, etc.).

de Broglie gave a relation between the particle property momentum and wave property, i.e., wave length λ of electron as $\lambda = \dfrac{h}{mv}$ [We know $E = mc^2 = h\nu = \dfrac{hc}{\lambda}$].

This relation holds well for micro-particle but not for macro bodies.

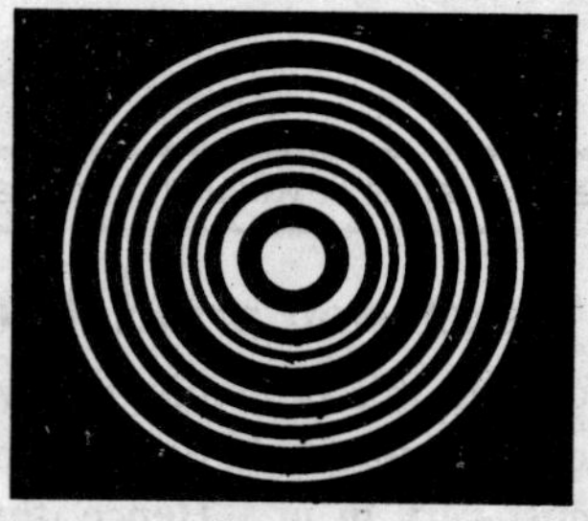

Fig. 1.11 : Electron beam diffractium pattern.

C. P. Davisson and L.H. Germer (1927) allowed a stream of electron beam to fall on nickel crystal at angles predicted by Bragg's equation. They got a diffraction pattern having alternate bright and dark concentric circles. This definitely proves the wave nature of electron.

Again, if a beam of electron is allowed to strike a fluorescent screen (e.g., ZnS screen) a spot of light is produced on the screen, which is localized. It does not spread out from the spot. This shows the particle nature of electron.

Heisenberg's Uncertainty principle

For macro particles one can measure the velocity, position, etc. of a particle and predict their path by Newtonian mechanics. But for micro particles like electron it cannot be done accurately. Werner Heisenberg (1927) stated the fact in terms of uncertainty principle. He stated that it was impossible to determine both position and momentum of micro particles simultaneously with perfect accuracy. According to him, in doing such measurements, if there be indeterminacy of momentum Δp, and indeterminacy in the measurement of position Δx, then $\Delta x \Delta p \geq h/4\pi$. Photon is used to locate moving electron. This photon strikes the electron, and the reflected photon is viewed through microscope. In doing so, the uncertainty in the location of position is $\pm\lambda$. If shorter wavelength, i.e., high frequency light is used, the position will be more accurate but due to collision, momentum of the electron will change. Thus, the two quantities cannot be measured with accuracy simultaneously.

Development of quantum mechanics or wave mechanics

As the moving electron has both wave nature and particle nature, considering these, E. Schrödinger and W. Heisinberg independently developed two methods of which Schrödinger's method is widely accepted and called the *wave mechanics*. He derived the equation for hydrogen atom as

$$\frac{\delta^2\psi}{\delta x^2}+\frac{\delta^2\psi}{\delta y^2}+\frac{\delta^2\psi}{\delta z^2}+\frac{8\pi^2 m}{h^2}(E-V)\psi=0,$$

where x, y and z are the space coordinates, m is the mass of the electron, h is the Planck's constant, E is the total energy of electron, V is the potential energy of electron, ψ is the amplitude of the electron wave, called *wave function*. $\frac{\delta^2\psi}{\delta x^2}$ is second derivative of ψ with respect to x-coordinate and so on. The solution of this equation shows that only certain solutions are permissible; each solution is called function and corresponds to definite energy states, called *orbital of the electron*. These orbitals have definite energy. Ψ is a state function and represents the amplitude of the wave and ψ^2 is the charge density and probability of finding the electron with definite energy within certain space.

This equation can be solved by transferring the Cartesian coordinate to spherical coordinates r, θ and Φ as represented in the figure.

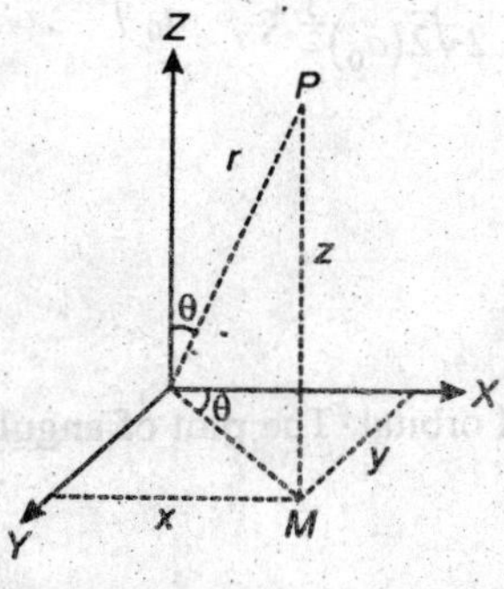

Fig. 1.12 : Relationship between Cartesian coordinates, x, y, z and the spherical polar coordinates $r\,\theta$, ϕ ($x = r\sin\theta\cos\phi$; $y = r\sin\theta\sin\phi$; $z = r\cos\theta$).

Separation of the ψ function as function of r, θ, Φ and solution of each part, the ø equation takes the form in terms of n, l, and m_l as

$$\Psi = \frac{1}{\sqrt{2\pi}} e^{-m\phi i} \times \sqrt{\frac{2l+1}{2} \times \frac{(l-m)!}{(l+m)!}} (1-z^2)^{\frac{m}{2}} \frac{\partial^{l+m}}{\partial z^{l+m}} (z^2-1)^l$$

$$\times \sqrt{\left(\frac{2z}{na_0}\right)^3 \frac{(n-l-1)!}{2n\{(n+l)!\}^3}} e^{\frac{-Zr}{na_0}} \left(\frac{2zr}{na_0}\right)^l \times \frac{\partial^{2l+1}}{\partial \rho^{2l+1}} \left[e^{\rho} \frac{\partial^{n+1}}{\partial \rho^{n+1}} (\rho^{n+1} e^{-\rho}) \right]$$

Here, $\cos\theta = z$, $\rho = 2\alpha r$ and

$$\frac{8\pi^2 \mu E}{h^2} = -\alpha^2, a_0 = \frac{h^2}{4\pi^2 me^2}, \therefore \lambda = \frac{4\pi^2 \mu Ze^2}{h^2}.$$

RADIAL DISTRIBUTION FUNCTION AND SHAPES OF ORBITALS

The wave functions for electrons of different orbitals obtained by putting the value of n, l and m are as follows:

$$\Psi_{100} = \frac{1}{\sqrt{\pi}} \left(\frac{z}{a_0}\right)^{\frac{3}{2}} e^{\frac{-zr}{a_0}} . \rho \qquad \Psi_{200} = \frac{1}{4\sqrt{2\pi}} \left(\frac{z}{a_0}\right)^{\frac{3}{2}} \left(2 - \frac{zr}{a_0}\right) e^{\frac{-zr}{2a_0}}$$

$$\Psi_{210} = \frac{1}{4\sqrt{2\pi}} \left(\frac{z}{a_0}\right)^{\frac{5}{2}} e^{\frac{-zr}{2a_0}} r \cos\theta$$

$$\Psi_{21+1} = \frac{1}{4\sqrt{2\pi}} \left(\frac{z}{a_0}\right)^{\frac{5}{2}} e^{\frac{-zr}{2a_0}} r \sin\theta \cos\phi \quad \Psi_{21+l} = \frac{1}{4\sqrt{2\pi}} \left(\frac{z}{a_0}\right)^{\frac{5}{2}} e^{\frac{-zr}{2a_0}} r \sin\theta \sin\phi.$$

The radial part of these wave functions is

$$R_{100} = \frac{2}{(a_0)^{\frac{3}{2}}} e^{\frac{-zr}{a_0}} \qquad R_{200} = \frac{1}{2\sqrt{2}(a_0)^{\frac{3}{2}}} \left(2 - \frac{r}{a_0}\right) e^{\frac{-zr}{2a_0}}$$

$$R_{210} = \frac{1}{2\sqrt{6}(a_0)^{\frac{5}{2}}} \times r \times e^{\frac{-zr}{2a_0}}.$$

Shapes of the orbitals

The wave function of an electron is called orbital. The plot of angular functions, θ and Φ, determines the geometry of the orbital.

s orbital

Ψ_{100} and Ψ_{200} are spherical in space and termed as 1*s* and 2*s* orbitals. Their projection on plane paper is circular and the charge distribution is symmetrical.

p-orbital

These orbital has principal quantum number, n, as an integer and $l = 1$, and $m_l = 0 = p_z$, $m_l = +1 = p_x$, $m_l = -1 = p_y$ orbital. Thus, for three possible values for m, there are three p-orbital and their shapes are dumb-bell type with two lobes. The two lobes are symmetrical; one lobe is marked (+)ve and the other (−)ve symmetry. Thus, the p-electrons tend to concentrate along certain preferred directions, which are mutually perpendicular to each other.

The diagrammatic representations of these orbitals are as follows:

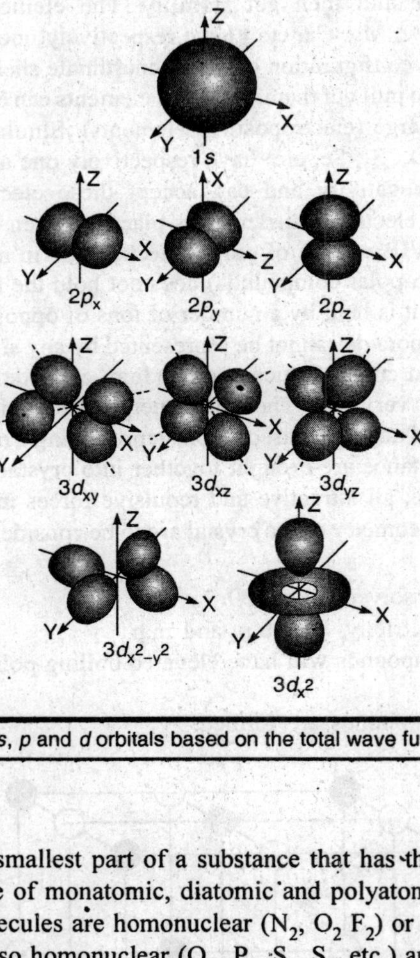

Fig. 1.13 : Conventional shapes for s, p and d orbitals based on the total wave function $\Psi_{n,l,m}$.

MOLECULES

As we know, molecules are the smallest part of a substance that has the properties of the substance and has free existence. Molecules are of monatomic, diatomic and polyatomic. Monatomic elements are He, Ne, Na, C, etc, the diatomic molecules are homonuclear (N_2, O_2 F_2) or heteronuclear (NO, CO, HF, HCl, etc., Polyatomic molecules are also homonuclear (O_3, P_4, S_4, S_8, etc.) and heteronuclear (NH_3, PH_3, CO_2, etc.). In other words, diatomic or polyatomic molecules are formed due to chemical combination between atoms of same or different types. The different elements have different chemical activity that depends on the electronic arrangements in the valence shell. It has been found that the monatomic inert gas molecules have octet or doublet (He) configuration in the valence shell and hence magnetic moments of the valence shell electrons are zero, i.e., all electrons are paired and are stable. It is the inherent property of the elements to attain stable configuration, octet or doublet structure. Although there are some stable compounds, e.g., BH_3, PCl_3, PCl_5, SF_6, H_2^+, $CuCl_2$, etc., which have either short of or extended octet or underlying 18, electron arrangement. But in all these compounds during combination between atoms, electrons are paired. Hence, it is called that "the rule of two is more fundamental than the rule of eight." Again, some stable molecules are also known, which have unpaired electron, e.g., NO, NO_2, etc. Let us discuss the reasons and methods how the combination between them occurs. We shall confine our discussion in the formation of diatomic molecules only. In 1916, Kossel attempted to explain the formation and nature of electrolytes, whereas Lewis attempted to explain the same for non-polar compounds.

Kossel theory

The electrolytes either in solution or in molten state dissociate into cation (+) and anion (–), i.e., some parts of heteronuclear diatomic molecules lose electron and attain (+)ve charge and the other accept it to attain (–)ve charge and then get stability. The elements of periodic table group IA (Li, Na, K, etc.) and group IIA (Be, Mg, Ca, etc.) have respectively one and two electrons excess in their valence shell beyond the inert gas configuration of their penultimate shell and hence less attracted by their nucleus, i.e., less energy required to pull out them. So, these elements can easily loose these valence electrons to attain stability and positive charge (electropositive elements). Similarly, elements of group VIIB (F_2, Cl_2, Br_2, I_2 etc) and group VIB (O_2, S_2, Se, etc) have respectively one and two electron short of the octet configuration, i.e., high electronegativity and can accept these electrons assuming negative charge (electronegative elements). Thus, electron transfer takes place between them. So, the compounds formed between IA or IIA with VIIB or VIB will be of purely electrostatic in nature and hence are electrovalent compounds. Any bond (unlike non-polar compounds) does not hold the ions of these compounds but each ion having electrical field around it is held by a number of ions of opposite charge by simple electrostatic force. Thus, the electrovalent compounds cannot be represented by any simple formula, but are aggregation of large number of opposite ion in crystal lattice, e.g., sodium chloride is not NaCl but Na_6Cl_6, CsBr as Cs_8Br_8, etc. So, the lattice energy is very high. The lattice energy of an ionic compound is the energy released when large numbers of positive and negative gaseous ions representing a molecule of the material, separated from each other by an infinite distance are brought together into crystal lattice which is characterized by definite inter-ionic distances. Here, all attractive and repulsive forces in all direction and from nearer to far remote ions depending on the geometry of the crystal are to be considered. The electrovalent compounds have the following properties:

(a) They are soluble in polar solvent like H_2O.
(b) The solutions conduct electricity, high b.p. and m.p.
(c) The solutions of these compounds will have elevated boiling points and depressed freezing points
(d) They form solid crystals.

Now let us see how these compounds are formed:

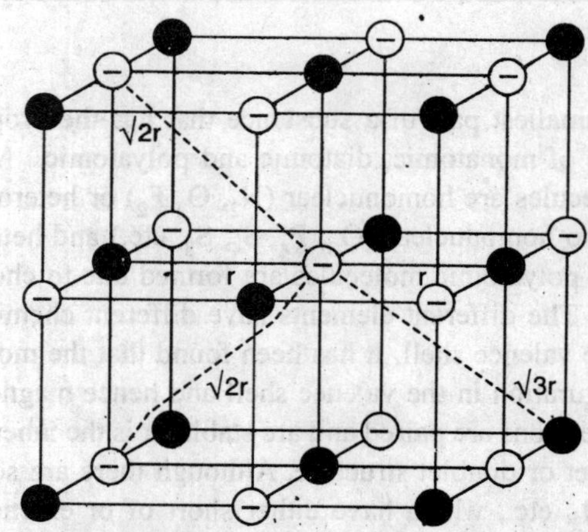

Fig. 1.14 : Unit cell of Na^+Cl^-.

(i) NaCl formation

$$\underset{281}{Na} \xrightarrow{-e^-} \underset{28}{Na^+} + e^-$$

$$\underset{287}{Cl + e^-} \longrightarrow \underset{288}{Cl^-}$$

So the compound is $Na^+ Cl^-$ and in crystalline form, it is a face centered cubic lattice with coordination number 6:6.

(ii) Formation of $MgCl_2$

$$\underset{282}{Mg - 2e^-} \longrightarrow \underset{28}{Mg^{+2}} \qquad \underset{287}{2Cl + 2e^-} \longrightarrow \underset{288}{2Cl^{-1}}$$

The compound is $Mg^{+2}Cl_2^-$. It has also crystalline structure.

(iii) Formation of AlF_3

$$\underset{283}{Al - 3e^-} \longrightarrow \underset{28}{Al^{+3}} \qquad \underset{27}{3F + 3e^-} \longrightarrow \underset{28}{3F^{-1}}$$

The compound is AlF_3.

BORN-HABER CYCLE

Using a thermodynamic cycle, known as *Born-Haber cycle*, the lattice energy is determined which is in good agreement with the theoretical value. The cycle gives correlation between the heat of formation H_f of a compound with electron affinity, *A*; ionization potential, *I*; heat of sublimation, *S*; dissociation energy, *D* and lattice energy, *U*. The construction of the cycle is based on the assumption that the formation of an ionic crystal may occur either by direct combination of elements or by the alternative method in which reactants are vaporized, then converted to gaseous ions and the gaseous ions are converted to products. The cycle may be constructed as follows:

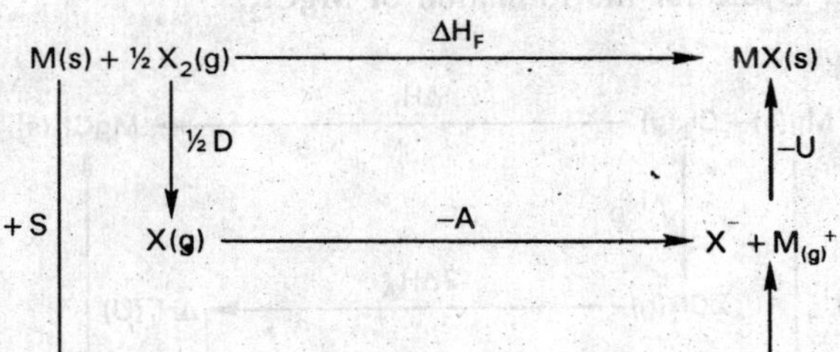

So $\quad -\Delta H_F = S + I + \frac{1}{2}D - A - U \quad$ or, $\quad U = \Delta H_F + S + I + \frac{1}{2}D - A.$

Problems on Born-Haber cycle

1. Calculate the lattice energy of RbCl from the following data:

ΔH_s of Rb = + 20.6 kcal mole^{-1}, ΔH_D of Cl_2 = 57.6 kcal mole^{-1}

ΔH_i of Rb = + 95.9 kcal mole^{-1}, ΔH_A of Cl_2 = –90.7 kcal mole^{-1}

ΔH_f of RbCl = –105 kcal mole^{-1}

Solution: We know $\Delta H_f = \Delta H_s + \Delta H_i + 1/2\ \Delta H_D + \Delta H_A + \Delta H_L(U)$

or, $-105 = 20.6 + 95.9 + 57.6/2 - 90.7 + \Delta H_L(U)$

or, $\Delta H_L(U) = -\ 159.6$ kcal mole^{-1}.

2. Calculate the lattice energy of MgS from the following data; energies are in kcal mole^{-1}. Given,

ΔH_s of Mg = 36.5, ΔH_D of S_8 = 133.2 × 8, ΔH_i of Mg = 520.6, ΔH_A of S = −72.4

ΔH_f of MgS = −82.2.

Solution: $\Delta H_f = \Delta H_s + \Delta H_i + 1/8\ \Delta H_D + \Delta H_A + \Delta H_L(U)$

or, $-82.2 = 36.5 + 520.6 + 133.2 - 72.4 + \Delta H_L(U)$

or, $\Delta H_L(U) = -700.1$ kcal mole^{-1}.

3. Calculate the electron affinity of Cl_2 from the following data:

$\Delta H_L(U)$ of KCl = −169.4 K cal mole^{-1}, ΔH_f of KCl = −104.2 kcal mole^{-1}

ΔH_s of K = + 21.5 kcal mole^{-1}, ΔH_i of K = + 100 kcal mole^{-1}

ΔH_D of Cl_2 = 57.8 kcal mole^{-1}

Solution: $\Delta H_f = \Delta H_s + \Delta H_i + 1/2\ \Delta H_D + \Delta H_A + \Delta H_L(U)$

or, $-104.2 = 21.5 + 100 + 28.9 + \Delta H_A - 169.4$

or, $\Delta H_A = -85.2$ kcal mole^{-1}

$= -3.7$eV (1eV = 23 kcal mole^{-1}).

4. Calculate the lattice energy of KI from the following data; energies are in K cal mole^{-1}.

ΔH_f of KI = −78.9 kcal/mole, ΔH_s of K = +21.7 kcal/mole,

ΔH_i of $K_{(g)}$ = 99.6 kcal/mole, ΔH_D of I_2 = 50.8 kcal/mole,

ΔH_A of I_2 = −76.6 kcal/mole.

Solution: $\Delta H_f = \Delta H_s + \Delta H_i + 1/2\Delta H_D + \Delta H_A + \Delta H_L$

$-78.9 = 21.7 + 99.6 + 25.4 - 76.6 + \Delta H_L$

$DH_L = -78.9 - 21.7 - 99.6 - 25.4 + 76.6$

$= 149.6$ kcal/mole.

5. Draw the Born-Haber Cycle for the formation of $MgCl_2$.

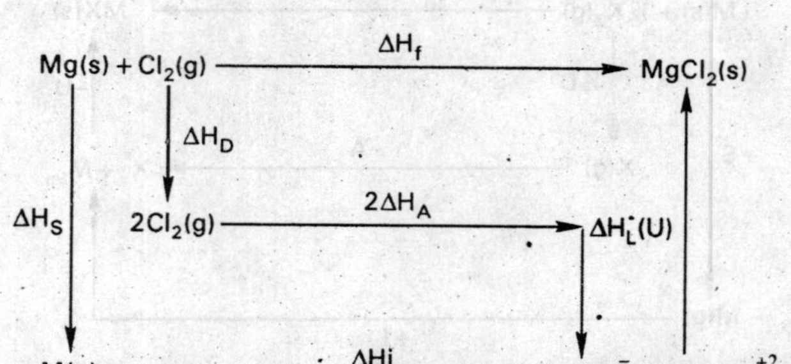

COVALENCE

Lewis concept of electron pairing in chemical bond formation, attainment of octet configuration is the main feature of covalent bond formation although the existence of one-electron bond (H_2^+) and three-electron covalent bond (in NO) are known. The method of electron pairing and their energy consideration was approached in three ways: (a) the valence bond (VB) or atomic orbital (AO) method, (b) the molecular orbital (MO) method and (c) hybridization method of which the third method is more acceptable.

Valence bond method

The essential features of valence bond approach is that

(i) Only the valence electrons of the combining atoms interact to get paired.

(ii) After bond formation each combining atom attains closed electronic configuration.

iii) The atomic orbitals retain their individuality.

(iv) Chemical bond is formed as a result of considerable degree of overlap of suitable orbitals and the paired electrons reside in the overlapping zone in between the combining atoms.

Thus the formation of homonuclear diatomic molecule like H_2, O_2, N_2, and heteronuclear molecules like H_2O, NH_3, etc. can be shown as under:

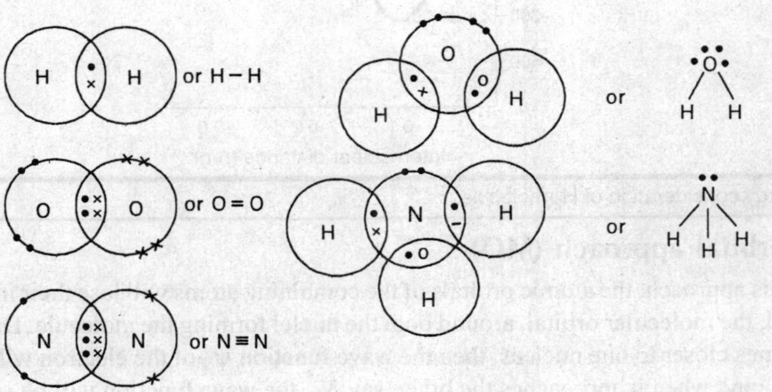

As both the combining atoms have equal electro negativity, there will be symmetrical charge distribution and the molecule formed is non-polar. If the combining atoms have different electronegativity, the paired electron will be slightly dragged towards the higher electronegative element, so, the electron charge over that atom will be higher and will attain $-\partial$ negative charge and the other will have $+\partial$ positive charge. Thus, gaseous HCl molecule can be represented as below:

$$\text{H}\bullet + .\ddot{\underset{..}{\text{Cl}}}: \longrightarrow \text{H}\bullet \; .\ddot{\underset{..}{\text{Cl}}}: \longrightarrow \overset{+\partial}{\text{H}} - \overset{-\partial}{\text{Cl}}$$

Thus HCl molecule is a polar molecule. CO, NO all will be polar with $-\partial$ charge over the oxygen atom. CO_2 is non-polar although each C-O bond is polar.

$$\overset{+\partial}{\text{C}} - \overset{-\partial}{\text{O}} \qquad \overset{+\partial}{\text{N}} - \overset{-\partial}{\text{O}} \qquad \overset{-\partial}{\text{O}} = \overset{+\partial}{\text{C}} = \overset{-\partial}{\text{O}}$$
$$\mu > 0 \qquad\qquad \mu > 0 \qquad\qquad \mu = 0$$

Formation of PCl_5, SF_6, etc. is the common examples where the essential feature of the valence bond approach is violated; phosphorous has 10 electrons and sulphur has 12 electrons in their valence shells.

Limitations

1. NO is an odd electron molecule but stable.
2. Formation of B_2H_6 (electron deficient molecule) cannot be explained.
3. According to this approach O_2 will be diamagnetic, but actually it is highly paramagnetic.

Theoretical approach of the formation of covalent bond shows that after the bond formation, electrons lose their individuality and resonance is the natural outcome to get the stability of the molecule. Electrons are paired either with symmetric spin E_s of lower energy or with unpaired spin E_A of higher energy.

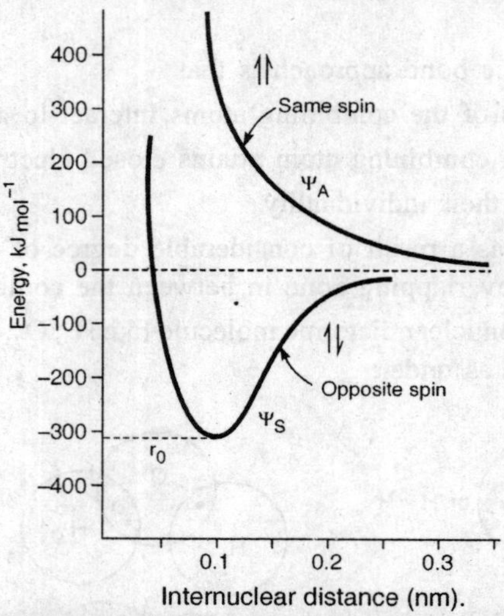

Fig. 1.15 : Energy consideration of H_2 molecule.

Molecular orbital approach (MO)

According to this approach, the atomic orbitals of the combining atoms will lose their individuality and form a common orbital, the molecular orbital, around both the nuclei forming the molecule. During revolution, when the electron comes closer to one nucleus, then the wave function ψ_A of the electron will like that of the closet nucleus, say N_A, and when it approaches the other, say N_B, the wave function will be ψ_B. The common wave function will be the combination of the two wave functions. $\psi_{MO} = \psi_A \pm \lambda \psi_B$.

Calculation shows that the combination of the two atomic orbital will give rise to two molecular orbitals, one of lower energy, $\psi_b = \psi_A + \lambda\, \psi_B$ or called bonding orbital and the other of higher energy, $\psi_a = \psi_A - \lambda\psi_B$ called anti-bonding orbital. The electrons now occupy these orbitals according to Aufbau principle. The formation of the MO's can be represented as under:

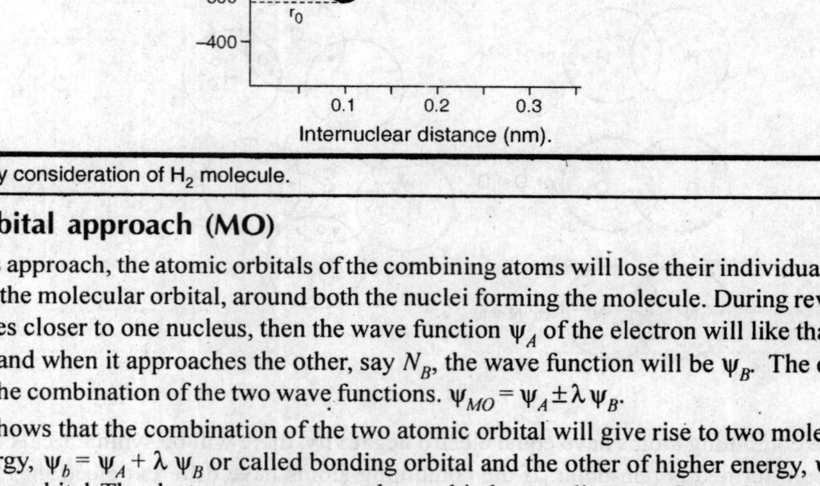

Fig. 1.16 : Addition of two 1*s* A.O.s in terms of (a) boundary surfaces of the A.O.s and (b) corresponding charge clouds. The sign of the wave function appears in (a) but not in (b), since density of charge cloud is proportional to Ψ. Combination of two 1*s* atomic orbitals having opposite sign of wave function in terms of (a) boundary surface of the atomic orbital and (b) charge density.

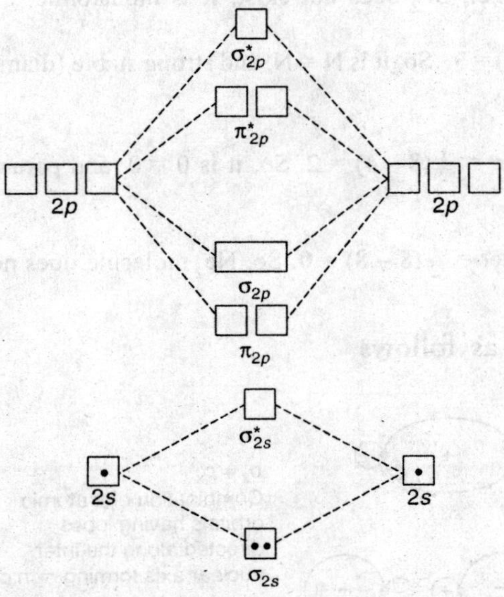

Fig. 1.17 : M.O. diagram for Li_2.

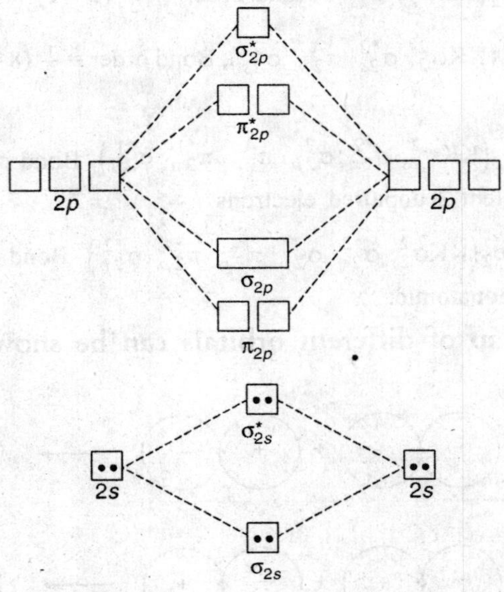

Fig. 1.17 : M.O. diagram for Be_2.

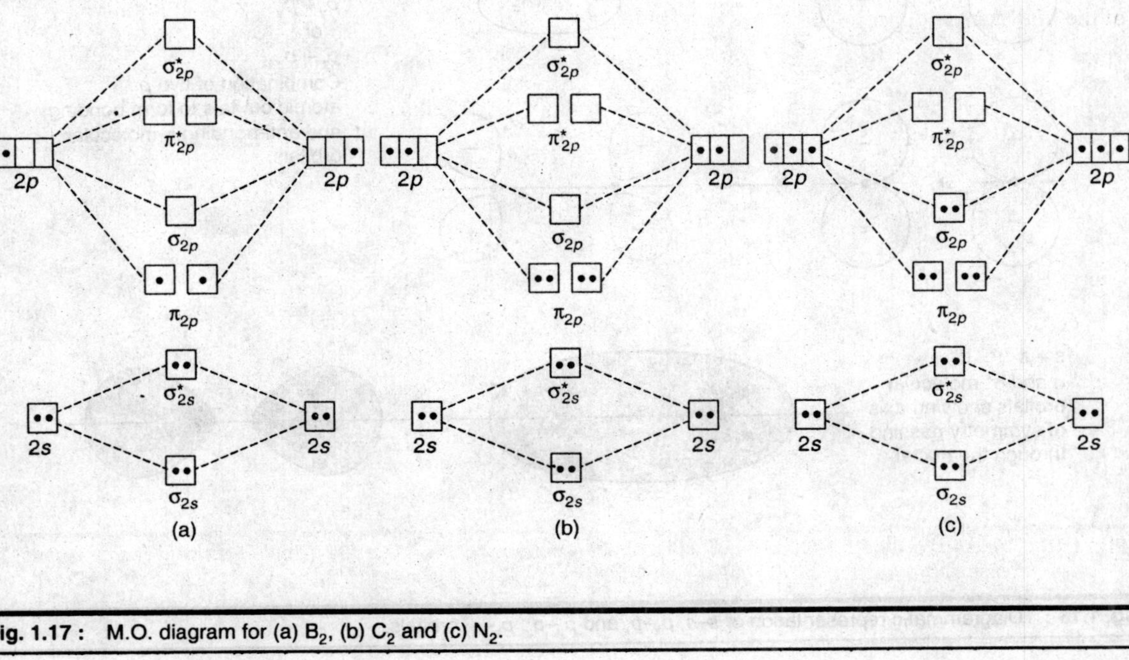

Fig. 1.17 : M.O. diagram for (a) B_2, (b) C_2 and (c) N_2.

Thus, the molecular orbital diagram of Be, N_2, O_2, F_2, Ne are

$Be_2\left(KK\sigma_{2s}^2\ \sigma_{2s}^{*2}\right)$, Bond order $= \frac{1}{2}\ (2-2) = 0$, i.e., Be_2 does not exist. It is monatomic.

$N_2\left(KK\sigma_{2s}^2\ \sigma_{2s}^{*2}\ \pi_{2p}^4\ \sigma_{2s}^2\right)$, Bond order $= \frac{1}{2}\ (8-2) = 3$. So, it is N ≡ N, and strong stable (diamagnetic) bond.

$O_2\left(KK\sigma_{2s}^2\ \sigma_{2s}^{*2}\ \sigma_{2p}^2\ \pi_{2p}^4\ \pi_{2p}^{*1}\ \sigma_{2p}^{*1}\right)$, Bond order $= \frac{1}{2}(8-4) = 2$. So, it is 0 = 0, and paramagnetic equivalent 2 unpaired electrons.

$Ne_2\left(KK\sigma_{2s}^2\ \sigma_{2s}^{*2}\ \sigma_{2p}^2\ \pi_{2p}^4\ \pi_{2p}^{*4}\ \sigma_{2p}^{*2}\right)$, Bond order $= \frac{1}{2}(8-8) = 0$. So, Ne_2 molecule does not exist. It is monatomic.

Overlap of different orbitals can be shown as follows

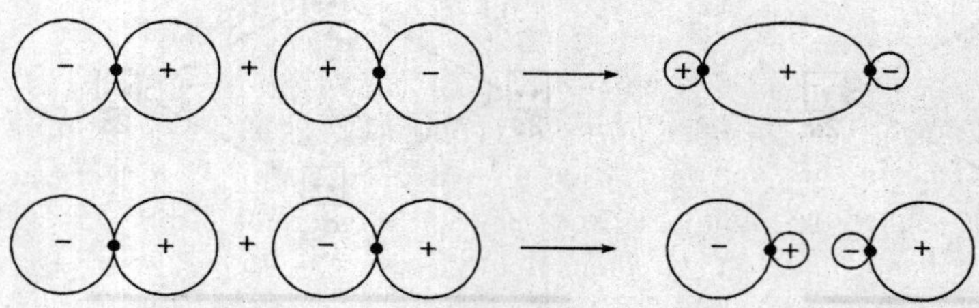

$p_x + p_x$
Combination of p atomic orbitals having lobes directed along the inter-nuclear axis forming σ-m.o.'s.

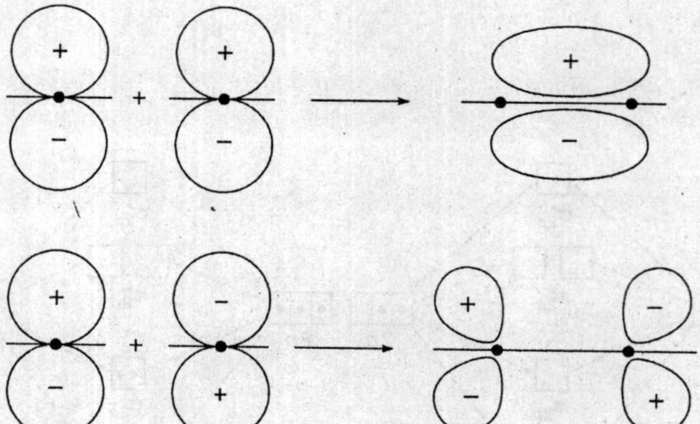

$p_y + p_y$
or
$p_z + p_z$
Combination of two p atomic orbitals to form bonding and anti-bonding π-molecular orbitals.

s + s
σ and σ* molecular orbitals showing axis of symmetry passing through the nuclei.

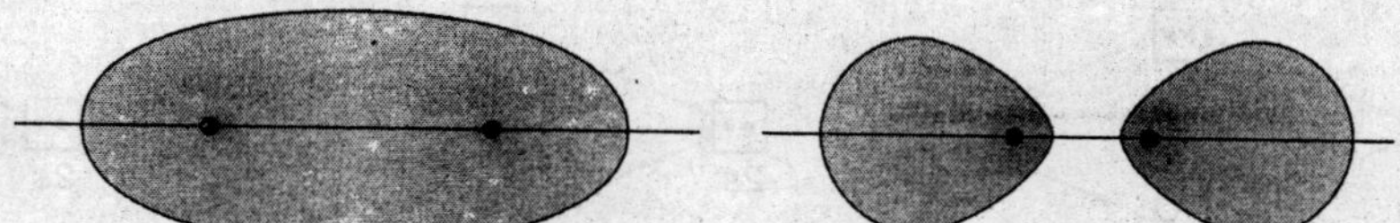

Fig. 1.18 : Diagrammatic representation of s–s, p_x–p_x and p_y–p_y, p_z–p_z orbitals.

Limitations

1. This concept cannot explain the inter-valence bond angles.
2. Relative strength of the bonds cannot be estimated.

Hybridization Concept

According to valence bond concept the bond angles of the compounds H_2O = 90°, bond strength √3 times of *s–s* bond, (relative strength of *s–s* = 1, *s–p* = √3 and *p–p* = 3). In NH_3, CH_4 also bond angle is 90°. Three-CH bond energy is √3 and one-CH bond energy is 1. in CH_4

But the experimental values of bond angles are H_2O = 104.5°, NH_3 107.3°. CH_4 = 109.28° and in each case bond strength is still higher. To explain the bond angles and relative strength of the bonds, Pauling proposed the hybridization concept.

According to this concept, in atoms or ions, when several outer electronic orbitals have more or less identical energies they can get mixed or combined to form corresponding number of equivalent bonding orbitals with greater electron density in particular directions different from the atomic orbitals from which they are constructed. This new orbital is called hybrid orbital. The relative strength of this hybrid orbital is greater than pure *s* or *p* atomic orbital and the bonds formed with the combining orbital through better overlap with the orbitals of other atoms along the direction. Thus, the molecule assumes certain geometrical configuration.

The structure of H_2O, NH_3, CH_4 molecules can be understood in this line.

H_2O: Oxygen has $2s^2\ 2p^4$ valence electrons. It undergoes sp^3 hybridization, two hybrid orbitals contain 2 paired of electrons and the other two has single unpaired electrons. These, then, overlap with the 1*s* orbital of hydrogen forming two single σ bonds. The two paired of non-bonding electrons repel each other, and so due to $l_p - l_p$ interaction the H-O-H angle decreases slightly and the value is 104.5° instead of 109°28′.

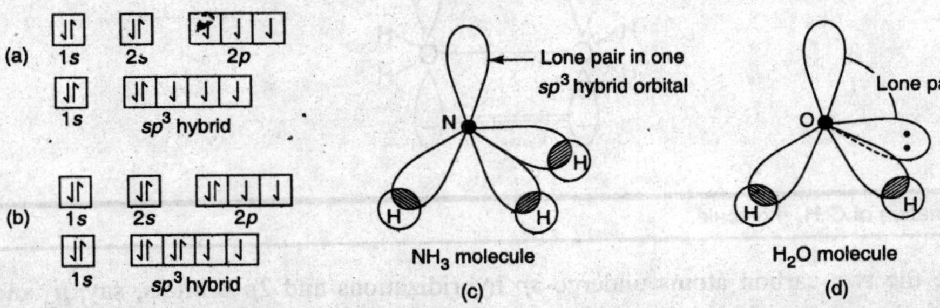

Fig. 1.19 : Hybrid orbitals of (a) nitrogen and (b) oxygen and the geometry of (c) amonia and (d) water molecule.

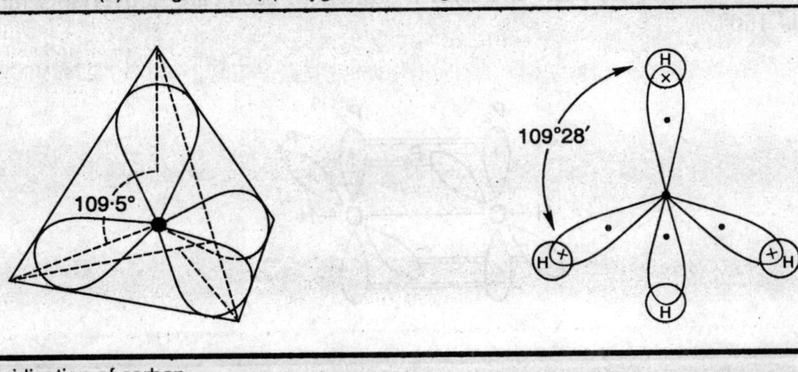

Fig. 1.20 : sp^3 hybridization of carbon.

Likewise CH_4 can be constructed as in Fig 1.20.

Actual geometry of a molecule primarily depends on the type of hybridization of orbitals of central atom, different interactions resulting from lone pair bond pair electrons, resonance and ionic character of the bond due to overlapping with the non-equivalent atom. Atoms of different electronegativity deform the molecule from actual geometry.

Formation of C_2H_6, C_2H_4, C_2H_2, can be shown.

C_2H_6: Each carbon has sp^3 hybridization. One of these hybrid orbitals of each carbon overlap to form C–C σ-bond, and others overlap with 1*s* orbital of hydrogen.

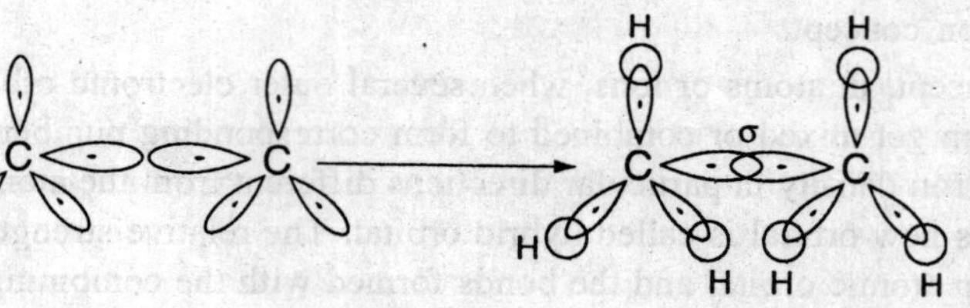

Fig. 1.21 : Formation C_2H_6 molecule.

C_2H_4: Each carbon atom undergoes as sp^2 hybridization and, say, p_2 remains unhybridized, that overlaps sideways to form π-bonds.

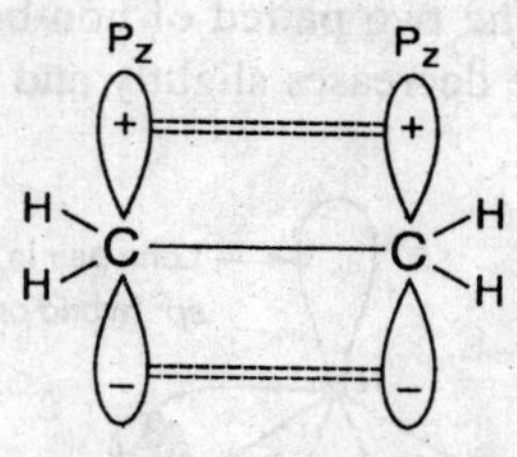

Fig. 1.22 : Formation of C_2H_4 molecule.

C_2H_2: Here the two carbon atoms undergo *sp* hybridizations and 2*p* orbitals, say, p_z and p_y remain unhybridized, they undergo sidewise overlap to form two π-bonds–one hybrid orbital of each carbon overlaps to form C–C σ-bond, the other two hybrid orbitals of each carbon overlap with 1*s* of hydrogen; the bond angle is 180°.

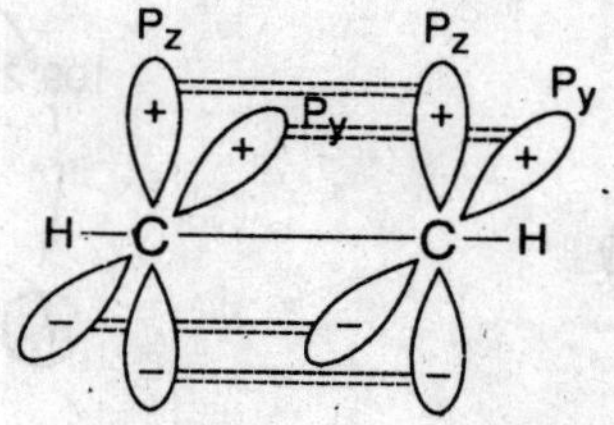

Fig. 1.23 : Formation of C_2H_2 molecule.

When *d* orbitals of transition metal ions are involved in hybridization the shape of the compounds are of special interest. The types of hybridization, their geometry and examples are given in the table.

dsp^2	Planar	$AuCl_4^-$, $PtCl_4^-$, $Pt(NH_3)_4^{+2}$
d^2sp^3	Octahedral	$Cr(CO)_6$, $[Co(NH_3)_6]^{+3}$, SF_6.
dsp^3	Trigonal bipyramid	$[Ni(CN)_5]^{-3}$, $Fe(CO)_5$. PCl_5

COORDINATE BOND OR DATIVE BOND

It is a special type of covalent bond where shared pair of electrons is contributed by one atom only, the other atom accepts the share without any contribution of electron, i.e., one sided sharing of electrons. Thus, one of the combining parts is electron pair donor and the other part is acceptor.

According to Lewis concept of acid and base, the acceptor part behaves as an acid and the donor part as base, so coordinate bond formation is a type of acid-base reaction. The donor part, i.e., base will have electronic configuration with lone pair and the acceptor will have two electrons short to have stable configuration. Once the coordinate bond formed, the electron pair will lose its individuality. It is denoted by

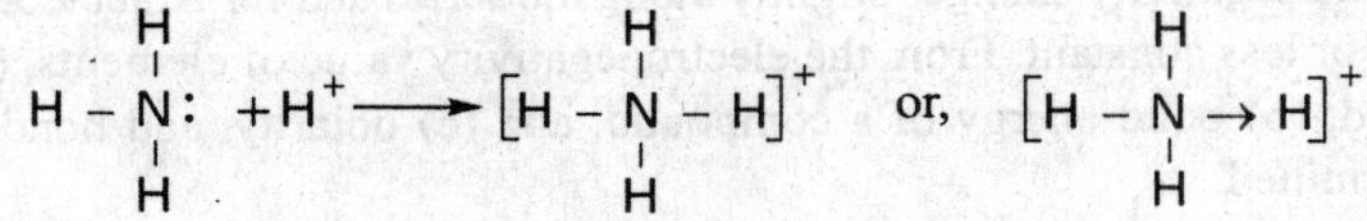

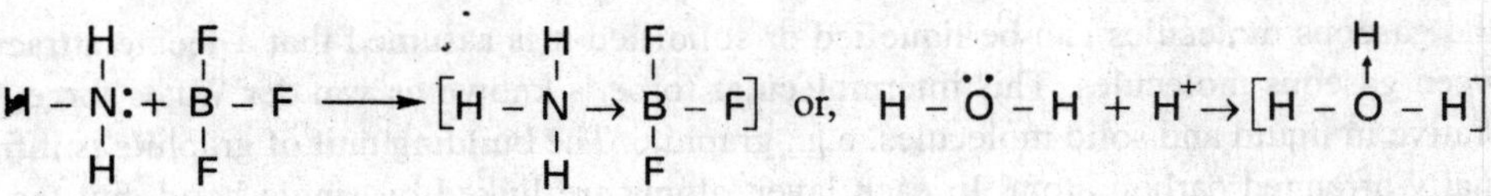

The transition metal complexes are formed where transition metal ions have incomplete *d*-shell (i.e., less than ten electrons) and capable of accepting pair of electrons. The combining ligands have lone pair of electrons that they donate to the metal ion forming coordinate bonds. So, complex formation is a type of acid-base reaction. The $[Co(NH_3)_6]Cl_3$ compound, the complex moiety is formed that can be represented in this way. Co^{3+} has $3d^6$ configuration and undergo d^2sp^3 hybridization and assume octahedral configuration. Each: NH_3 having a lone pair of electron donates it to Co^{3+} having six vacant hybrid orbitals that accommodate them.

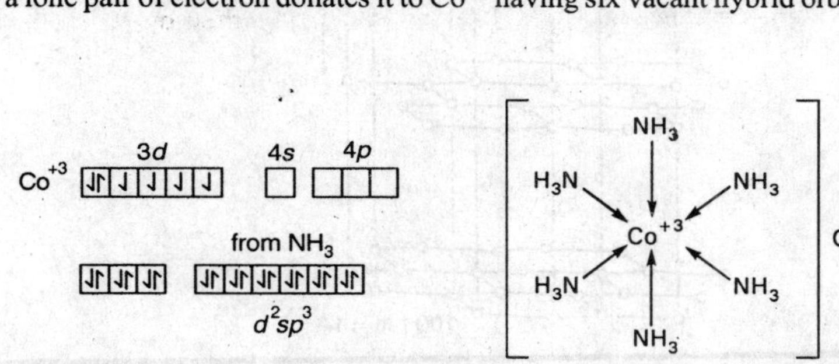

Fig. 1.24 : Formation of $[Co(NH_3)_6]Cl_3$ molecule.

ELECTRONEGATIVITY

Electronegativity is not the inherent property of an atom. As stated earlier, in hetero-diatomic molecules polar character develops. Out of the two atoms one will enjoy the shared electron pair more than the other. This develops due to the difference in affinity towards the bonded electron. This property is known as electronegativity.

Thus, electronegativity is the tendency of an element in a molecule to attract the bonded electron towards itself. Several methods have been developed to measure the electronegativity of elements of which Pauling's scale is more acceptable.

Pauling's Scale of electronegativety: If A_2 and B_2 molecules combine to form the compound A–B, then

$$\chi_A - \chi_B = 0 \cdot 28\sqrt{E_{A-B} - (E_{A-A} \times E_{B-B})^{\frac{1}{2}}},$$

where χ_A and χ_B are the electronegativity of A and B, E_{A-B} is the bond energy of A–B, E_{A-A} is that of A_2 and E_{B-B} is that of B_2.

Considering B_2 as H_2 and taking $E_H = 0$, electronegativity of other elements were calculated. Thus, introducing some correction, electronegativity of F = 4, O = 3.5, C = 2.5, B = 2.0 and so on.

Electronegativity of elements increases from left to right along a period, and decreases down the group. For transition elements, electronegativity changes slightly along the series and for f-block or inner transition elements, this value is more or less constant. From the electronegativity value of elements, (a) percent ionic character of a covalent bond, (b) bond energy of a compound, and (c) polarity, and bond moment of the covalent bond can be determined.

VAN DER WAALS OR INTERMOLECULAR FORCES

Since the gaseous molecules can be liquefied or solidified it is assumed that a feeble attractive forces exists between gaseous molecules. This intermolecular force is known as van der Waals force. This force is also operative in liquid and solid molecules, e.g., graphite. The building unit of graphite is infinite layers of hexagonally arranged carbon atom. In each layer, atoms are linked by single bond, but the layers are held by weak van der Waals forces at large distance of 0.335nm.

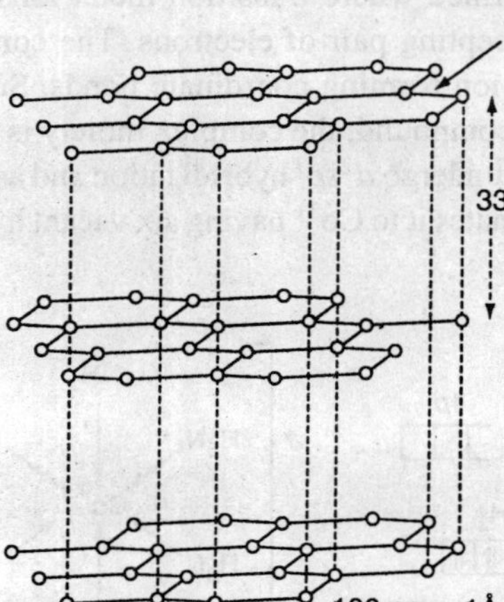

Fig. 1.25 : Structure of graphite.

The origin of van der Waals force can be considered as the force of attraction between the instantaneous fluctuating dipoles between the two adjacent molecules. Say two Cl_2 molecules are at their closet approach.

In one molecule at an instant, electron density concentrate at one end and becomes a dipole, likewise the other molecule may also behave in a same manner having opposite charges at the closet distance resulting in attraction. In the next moment, the charges of the dipole may change their position due to constant motion of electron and so, creates two new dipoles oppositely oriented. These fluctuating dipoles result in attraction.

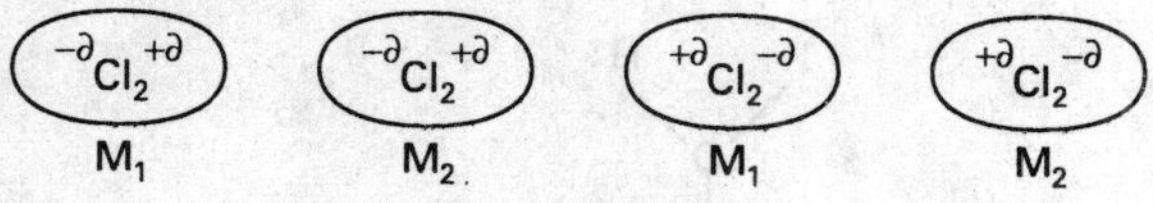

Fig. 1.26 : Fluctuating dipole in adjacent Cl_2 molecules resulting in van der Waals attraction.

This force is very weak in nature and is opposed by the repulsive force between electron clouds of the adjacent molecules and also between the nuclei of neighboring atoms, and of course attractive force is greater.

HYDROGEN BOND

This bond is formed in the hetero-diatomic when one element is strong electronegative (F, O, N, etc.) and of small size and the other atom is hydrogen. So, in such molecules, partial ionic character develops, e.g., HF molecule. The situation is $H^{+\partial} - F^{-\partial}$. When number of molecules is high the situation is

$$H^{+\partial} - F^{-\partial} \cdots H^{+\partial} - F^{-\partial} \cdots H^{+\partial} - F^{-\partial}.$$

Hydrogen atom in one molecule is bonded to F^- atom by covalent solid line and simultaneously bonded to F^- of the nearer molecule by weak electrostatic force. This weak force is termed as hydrogen bond (shown by dotted line). It has greater bond distance than the covalent bond. Apparently hydrogen atom here is bivalent. Hydrogen bond is of two types: (1) Intermolecular hydrogen bond, i.e., bonds between two same or different molecules, e.g., *p*-nitro phenol and (2) Intramolecular hydrogen bond, i.e., bond within the same molecule, e.g., *o*-nitro phenol (shown in the figure).

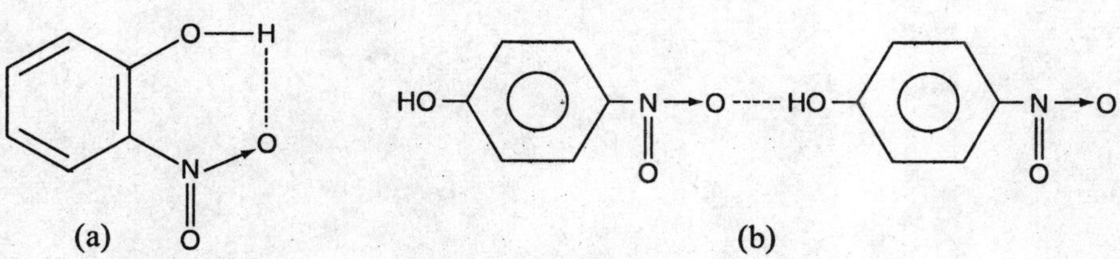

Fig. 1.27 : Hydrogen bond in (a) *o*-nitrophenol and (b) *p*-nitrophenol.

Intramolecular hydrogen bond forms a 5 membered or 6 membered ring. In water the hydrogen bond is formed in this way. In liquid water the bond formation takes place in long chain. But in ice it has open-cage type structure in which one oxygen atom in one molecule is tetrahedrally surrounded by other four oxygen atoms of other four water molecules leaving some empty space in the ice crystal. Hence, volume of ice becomes greater than the volume of water taken. 11cc water forms 12 cc ice, so ice floats on water with $\frac{1}{12}$th part above the water surface. When it melts, volume decreases as a result of continuous breaking down of the hydrogen bond. This breaking down becomes complete at 4°C. Hence water has the highest density at 4°C.

Hydrogen bond is very weak, its strength is about 1/10th of the covalent bond, and is electrostatic in nature.

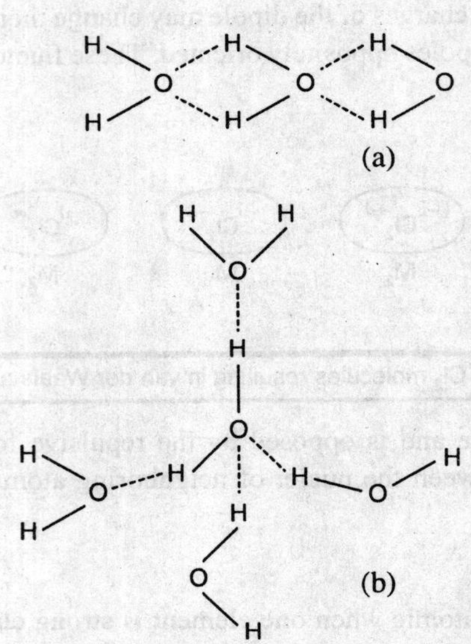

Fig. 1.28 : Formation of hydrogen bond (a) in water and (b) in ice.

The consequences of hydrogen bond formation result in the formation of associated molecules, increases the effective molecular weight of the compound. This causes the increase in the melting point and boiling point of the compound. That is why, being members of the same periodic group H_2S is gas but H_2O is liquid. Water boils at 100°C but due to dative and extensive hydrogen bond, H_2O_2 boils at 151°C.

Again, H_2S is acidic but H_2O is neutral. Alcohols of lower molecular weight like CH_3OH, C_2H_5OH, etc., form hydrogen bond with water and hence, soluble in water.

2

TRANSITION METAL CHEMISTRY

CLASSIFICATION OF ELEMENTS

Depending on the type of electron arrangements of the valence shell, elements are classified as follows:

(a) Normal or representative elements

The elements of different groups of the periodic table are called *normal* or *representative elements*. Group I and Group II elements are called *s*-block elements. These are strong electropositive elements. Group I (ns^1) elements are called the *alkali metals* and Group II (ns^2) elements are called *alkaline earth metals*. Group III and onwards are the *p*-block elements, group VIII (ns^2p^6) are the noble gases, group VII (ns^2p^5) are highly electronegative elements called *halogens*, group VI (ns^2p^4) are also electronegative, called *chalcogens*, group V (ns^2p^3) elements have both electropositive and electronegative characters to some extent, elements of group IV (ns^2p^2) are the amphoteric in character, group III (ns^2p^1) have peculiar characteristics, their compounds are known as electron deficient compound.

(b) Transition elements

The term transition means intermediate. The elements belonging to this class have both metallic and non-metallic characteristics, lower valent states are basic and higher valent states are acidic, e.g., Cr^{+2}, Cr^{+3} are basic but Cr^{+6} as CrO_4^{-2} is acidic. The elements have in general $(n-1)\ d^{1-10}\ ns^{2/1}$ electronic configuration, In other words, either the atom or any one of the oxidation state of the element will have incomplete *d*-shell. General feature of this class of elements will be discussed later. These elements occupy the position in the periodic table from group III onwards to group II in the sub-groups *A* or *B* depending on their electropositive character. In the modern extended periodic table, they occupy group IIIA to Group VIIIA and group IB and IIB.

(c) Inner transitional elements

These elements have incomplete *f*-shell, i.e., inner shell of the transition series $(n\text{-}2)\,f^{1-14}(n\text{-}1)\,d^{10}ns^2$. Incomplete $4f^{1-14}$ are called the *lanthanides* (the first member being Lanthanum) and incomplete $5f^{1-14}$ shell are called *Actinides*(the first element being Actinium). These elements together are called rare earths as they occur in very small amount in nature. U-92 occurs in nature and uranium onwards are artificial or synthetic elements, called *transuranic elements.*

(d) The missing elements

Technetium (Tc-43), Promethium (Pm-61), Astatine (At-85) and Francium (Fr-87), these elements are not found in nature as their half-lives are very short.

GENERAL CHARACTERISTICS OF THE TRANSITION ELEMENTS

As stated earlier the transition elements –as atom or any one of its oxidation state will have incomplete *d*-shell. Accordingly there are three transition series in the periodic table. The elements with their electronic configuration are shown in the table.

TABLE 2.01: *Electronic configuration of d-block elements*

3d (or first)		*4d (or second)*		*5d (or third)*	
Element (At. No.)	***Configuration***	***Element (At. No.)***	***Configuration***	***Element (At. No.)***	***Configuration***
Sc (21)	$3d^1$, $4s^2$	Y (39)	$4d^1$, $5s^2$	La (57)	$5d^1$, $6s^2$
Ti (22)	$3d^2$, $4s^2$	Zr (40)	$4d^2$, $5s^2$	Hf (72)	$5d^2$, $6s^2$
V (23)	$3d^3$, $4s^2$	Nb (41)	$4d^4$, $5s^1$	Ta (73)	$5d^3$, $6s^2$
Cr (24)	$3d^5$, $4s^1$	Mo (42)	$4d^5$, $5s^1$	W (74)	$5d^4$, $6s^2$
Mn (25)	$3d^5$, $4s^2$	Tc (43)	$4d^5$, $5s^2$	Re (75)	$5d^5$, $6s^2$
Fe (26)	$3d^6$, $4s^2$	Ru (44)	$4d^7$, $5s^1$	Os (76)	$5d^6$, $6s^2$
Co (27)	$3d^7$, $4s^2$	Rh (45)	$4d^8$, $5s^1$	Ir (77)	$5d^7$, $6s^2$
Ni (28)	$3d^8$, $4s^2$	Pd (46)	$4d^{10}$, $4s^0$	Pt (78)	$5d^9$, $6s^1$
Cu (29)	$3d^{10}$, $4s^1$	Ag (47)	$4d^{10}$, $5s^1$	Au (79)	$5d^{10}$, $6s^1$
Zn (30)	$3d^{10}$, $4s^2$	Cd (48)	$4d^{10}$, $5s^2$	Hg (80)	$5d^{10}$, $6s^2$

Due to involvement of $(n\text{–}1)$ *d*-electrons the characteristics of transition elements are somewhat different from that of representative elements. A few of these general properties are mentioned below.

(a) Atomic radius

The valence shell ns^2 is constant. So the atomic radii remain more or less constant, although slight variation is observed due to the screening effect of $(n\text{–}1)$ *d*-electrons. At the begimning of the series, $(n\text{–}1)$ *d*-electrons are less and screening effect is more with increased nuclear charge, but towards the end of the series, the screening effect is counter balanced by the opposing effect of increased nuclear charge. Thus, the atomic radii in pm:

Sc–144	Cr–117	Co–116	Zn–125
Ti–132	Mn–117	Ni–115	
V–122	Fe–117	Cu–117	

The increased value of Zn is due to higher repulsion between the added electrons in the same $(n\text{–}1)$ *d*-shell. Down the group the atomic radii increase.

(b) Ionic radius

For the same oxidation state of the metal ions, the ionic radii will decrease along the period as effective nuclear charge increases. For different oxidation states it will decrease with increasing oxidation states as effective nuclear charge increases. In general the ionic radii of the transition metal ions are smaller than that of *s*- and *p*-block elements

(c) Density

Since the atomic radii, rather atomic volumes are low compared to *s*- and *p*-block elements, the metal lattices are close packed and hence of high densities.

(d) Metallic character

The transition metals have hcp or ccp or bcc lattices together with metallic bond formed among themselves with the unpaired $(n-1)$ d-electrons present in them. The metallic bond formed is more extensive with more unpaired electrons. So, Mn, Cr, Mo, W, etc are very hard. Zn, Cd, Hg have no unpaired electrons and are soft (Hg is liquid). The metals are lustrous, malleable, ductile and good conductors of heat and electricity.

(e) Ionization energy or potential

It is the amount of energy needed to remove the most loosely bound electron from an isolated gaseous atom to infinity. As the increased nuclear charge and screening effect, the two opposite forces are operative in them, the ionization potential increases irregularly. The second ionization potential of Cr and Cu are exceptionally high as the removal of second electron occurs from the stable half filled and fulfilled d-shell.

(f) Oxidation states

All the metals exhibit variable oxidation states due to the involvement of ns and $(n-1)$ d-electrons, the energy difference between ns and $(n-1)$ d-shell is very low. The general oxidation state is +2 state and using d-electrons they can exhibit other oxidation states. The highest oxidation state is generally the sum total of ns and unpaired $(n-1)$ d-electrons. The different oxidation states with electronic configuration is listed in the table

TABLE 2.02: *Oxidation states of first transition series (or 3d-series)*

Elements	*At. No.*	*Configuration*	*Oxidation States*
Scandium (Sc)	21	[Ar] $4s^2$, $3d^1$	+2, +3
Titanium (Ti)	22	[Ar] $4s^2$, $3d^2$	(+2), +3, +4
Vanadium (V)	23	[Ar] $4s^2$, $3d^3$	+2, +3, +4, +5
Chromium (Cr)	24	[Ar] $4s^1$, $3d^5$	(+1), +2, +3, +4, (+5), +6
Manganese (Mn)	25	[Ar] $4s^2$, $3d^5$	+1, +2, (+3), +4, (+5), +(6), +7
Iron (Fe)	26	[Ar] $4s^2$, $3d^6$	(+1), +2, +3, (+4), (+5), (+6)
Cobalt (Co)	27	[Ar] $4s^2$, $3d^7$	(+1), +2, +3, (+4)
Nickel (Ni)	28	[Ar] $4s^2$, $3d^8$	(+1), +2, +3, (+4)
Copper (Cu)	29	[Ar] $4s^1$, $3d^{10}$	+1, +2, (+3)
Zinc (Zn)	30	[Ar] $4s^2$, $3d^{10}$	(+1), +2

If the element has more than one oxidation states the relative stability of the ions can be understood from their standard oxidation potential data.

(g) Paramagnetism

If an element or compound has unpaired electrons, then in high intensity magnetic field, the atom or compound is attracted and hence the weight of the compound increases. From the increased weight the number of unpaired electrons present in them can be calculated. On the other hand, if they have all electrons paired, then the atom or compound is repelled by the field, and hence weight of the compound decreases. These are called diamagnetic compound. Most of the transition metal compounds are paramagnetic in nature as the metal ions have $(n-1)$ d-unpaired electrons. These unpaired electrons have both orbital moment and spin moment due to their orbital motion and spin motion respectively. When these compounds are placed in high intensity magnetic field, the total spin moment and orbital moment overcome the magnetic moment induced by the applied field and hence are attracted, e.g., Cu^{+2} $(3d^9 4s^0)$ and Cu^{+1} $(3d^{10} 4s^0)$. Cu^{+2} is paramagnetic and Cu^{+1} is diamagnetic. Fe^{+2} and Fe^{+3} here $(3d^6 4s^0$ and $3d^5 4s^0)$ configuration. If the complexing ligand is weak (H_2O, F^-, etc) Fe^{+2} will have four unpaired electrons and Fe^{+3} will have five unpaired electrons. Hence, both are paramagnetic, but moment of Fe^{+3} will be higher. If the complexing ligand is strong (NH_3, CN^-, etc.) then quenching of 3*d*-orbital occurs. In Fe^{+2}, $3d^6$ occupy t_{2g}^6 all paired and diamagnetic, in Fe^{+3}, $3d^5$ occupy t_{2g}^5 with one unpaired electron, hence paramagnetic.

The different ions with weak ligand and their unpaired electrons can be known as

Sc^{+2} (d^1), Ti^{+2} (d^2), V^{+2} (d^3), Cr^{+2} (d^4), Mn^{+2} (d^5), Fe^{+2} (d^6), Co^{+2} (d^7), Ni^{+2} (d^8), Cu^{+2} (d^9), Zn^{+2} (d^{10}). So paramagnetic moment increases from Sc^{+2} to Mn^{+2} and then decreases and is zero in Zn^{+2}.

(h) Formation of coloured ions

Transition metal compounds either in solid state or in solution are coloured e.g., Cu^{+2}, i.e., $[Cu^{+2}(H_2O)_4]^{+2}$ is blue, but Cu^+Cl is white, Fe^{+2}(green), Fe^{+3} (yellowish brown), Cr^{+2} (blue), Cr^{+3} (green), Cr^{+6} (orange), Mn^{+2} (pale pink), Mn^{+3} (red), Mn^{+6} (green), Mn^{+7} (purple), etc. The different ions with number of unpaired electrons and their corresponding colour is given in the table

TABLE 2.03: *Colours of different hydrated transition metal ions*

Ion	*Outer configuration*	*No. of unpaired electrons*	*Colour*
Sc (III)	$3d^0$	Nil	White (colourless)
Ti (IV)	$3d^0$	Nil	White (colourless)
Ti (III)	$3d^1$	One	Purple
V (III)	$3d^2$	Two	Green
V (II)	$3d^3$	Three	Violet
Cr (III)	$3d^3$	Three	Violet
Mn (III)	$3d^4$	Four	Violet
Mn (II)	$3d^5$	Five	Pink
Fe (III)	$3d^5$	Five	Yellow
Fe (II)	$3d^6$	Four	Green
Co (II)	$3d^7$	Three	Pink
Ni (II)	$3d^8$	Two	Green
Cu (II)	$3d^9$	One	Blue
Cu (I)	$3d^{10}$	Nil	White (colourless)
Zn (II)	$3d^{10}$	Nil	White (colourless)

During compound formation, the ligands have the effect on the metal ion *d*-orbital that splits up to t_{2g} and e_g states. For hexa coordinated compounds, the unpaired electrons in the 3*d* shell absorb light energy and are excited, transition of the unpaired electrons takes place within t_{2g} to e_g with absorption and the complimentary colour reaches our eyes and we accordingly observe the colour.

(i) Catalytic property

The transition metals have variable oxidation states utilizing their incomplete 3*d*-shell. In these different oxidation states they can form some unstable intermediate products (transition compounds) and activate the reacting substances which then form the compounds, e.g., Fe in Haber's process acts as catalyst as follows:

$$Fe + \tfrac{1}{2}H_2 \rightarrow FeH \qquad Fe + \tfrac{1}{2}N_2 \rightarrow FeN$$

$$FeH + \tfrac{1}{2}H_2 \rightarrow FeH_2 \qquad FeN + FeH_3 \rightarrow 2Fe + NH_3$$

$$FeH_2 + \tfrac{1}{2}H_2 \rightarrow FeH_3$$

Similarly, V_2O_5 in contact process acts as

$$V_2O_5 + SO_2 \rightarrow V_2O_4 + SO_3 \qquad 2V_2O_4 + O_2 \rightarrow 2V_2O_5$$

(j) Non-stoichiometric compounds (Berthollide compounds)

Transition metals form some compounds particularly with O, S, Se, etc of group VI elements which do not follow the laws of chemical combination particularly the law of definite proportion, e.g., ferrous oxide is $Fe_{0.93}O$. Other compounds are $VH_{0.56}$, $TiO_{1.7\text{-}1.8}$, $WO_{2.86\text{-}2.92}$, etc. This variation occurs partly due to the variable oxidation states of the metal and partly due to the defects in the solid crystal. Knowing the exact formula, the percent of other oxidation states can be determined, e.g., in $Fe_{0.93}O$, number of Fe^{+2} present = 0.93, number of Fe^{+2} missing = 0.07. Each Fe^{+2} has two positive charges, so total positive charge missing = 0.07 × 2 = 0.14. This 0.14 positive charge is to be balanced by Fe^{+3}. Now one Fe^{+2} replaced by one Fe^{+3} increases one positive charge, so 0.14 positive charge will be compensated by $0.14Fe^{+3}$. So in the compound, 0.93 Fe^{+2} contains $0.14Fe^{+3}$.

So, % of $Fe^{+3} = \frac{0.14}{0.93} \times 100 = 15.05\%$.

Non-stoichiometric compounds often show semi-conductivity, fluorescence, etc.

(k) Interstitial compounds

Small atoms like H, B, C, N, etc can easily occupy the interstitial positions or void sites in the crystal lattice of the transition metals. This results in some special properties of the metal like extremely high melting point, extreme hardness, metallic conductivity, and metallic lustrous, chemical inertness, opacity, etc., e.g., steel, chrome steel, stainless steel (these are solid solution of metals). Alloy formation of the transition metals occurs in which one transition metal atom substitutes the lattice position of the crystal of other metal as they have more or less equal size.

(l) Complex formation

It is the inherent property of the transition metal cations to form coordinate bond with opposite, i.e., negative ions or neutral molecule containing at least one lone pair of electron that is available for the coordinate bond formation. Here central cation is the acceptor, i.e., Lewis acid and the surrounded ligands anion or neutral molecules are donor, i.e., Lewis base. These compounds contain the main complex moiety, or coordination sphere, that is not easily decomposable. Hence these constituting parts of the complex moiety cannot be detected, e.g., in $K_4[Fe(CN)_6]$ the complex moiety is $[Fe(CN)_6]^{-4}$ and the existence of Fe^{+2} and CN^- cannot be detected easily. The complex compounds are of three types:

(a) ***Cationic complex***: Here, the central metal ion is cation and surrounding ligands are neutral molecule, or the negative group such the complex moiety carries the positive charge. This charge is balanced by anions present outside the complex moiety and held by electrostatic force. e.g., $[Cu^{+2}(NH_3)_4]SO_4$, $[Co^{+3}(NH_3)_6]Cl_3$, $[Co^{+3}(NH_3)_5Cl]Cl_2$.

(b) ***Anionic complex***: Here, the central cation is bonded ~~to the negative anion~~ through coordinate bond forming the complex moiety. This carries the excess negative charge of the anions, thus forming anionic complex. The excess negative charges of the anions are balanced by other metal cation or by other cationic complex. e.g., $K_4.[Fe^{+2}(CN)_6]^{-4}$, $K[Ag^{+1}(CN)_2]^-$, $[Pt(NH_3)_4][PtCl_4]$, $[Co^{+3}(NH_3)_6][Co^{+3}(CN)_6]$, etc.

(c) ***Neutral complex***: In this class, the total charge of the metal cation is balanced by the anionic ligands such that the complex moiety carries no charge. e.g., $[Co^{+3}(NH_3)_3Cl_3]$, $[Ni^0(CO)_4{}^0]$, $[Fe^0(CO)_5]$.

The above classes are simple and the ligand occupies single coordination position.

Another class of complexes is known where the single ligand can occupy more than one coordination position. These are called chelate complex and will be discussed later. e.g:, dimethyl glyoxime complex of nickel. The different types of ligands are listed as follows:

TABLE 2.04: *Some of the typical ligands*

Formula (name)	*Formula (name)*
Neutral (monodentate): H_2O (*aqua*) **Note** : Until recently, the term *acqo* was used. NH_3 (*ammine*) Co (*carbonyl*) NO (*nitrosyl*) CS (*thiocarbonyl*) CH_3NH_2 (*methylamine*) C_5H_5N (*pyridine*) **Anion (monodentate):** Br^- (*bromo*) CN^- (*cyano*) Cl^- (*chloro*) F^- (*fluoro*) I^- (*iodo*) OH^- (*hydroxo*) NH_2^- (*amido*) NO_3^- (*nitrato*) NO_2^- (*nitro*) ONO^- (*nitrito*) CH_3COO^- (*acetato*) SCN^- (*thiocyanato*) NCS^- (*isothiocyanato*) O_2^- (*oxo*) O_2^{2-} (*peroxo*) SO_4^{2-} (*sulphato*) $S_2O_3^{2-}$ (*thiosulphato*) CO_3^{2-} (*carbonato*) NH_2^{2-} (*imido*)	**Neutral (bidentate):** $H_2N(CH_2)_2NH_2$ (ethylenediamine or en) **Neutral (tridentate):** 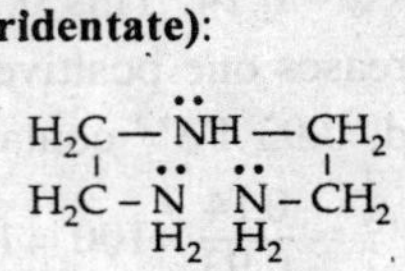 (*diethylenetriamine* or *dien*) **Neutral (tetradentate):** 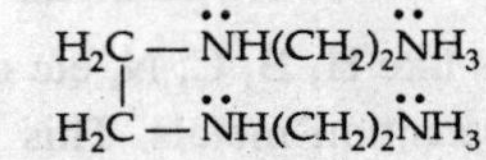 (*triethylenetetramine* or *trien*) **Anion (bidentate):** $^-OOC\ .\ COO^-$ (*oxalato*) 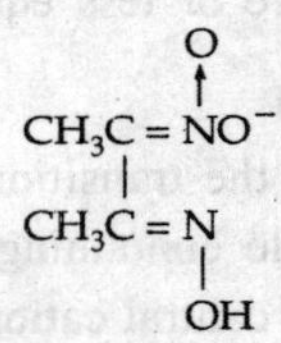 (*dimethylglyoximato*) **Anion (pentadentate):** 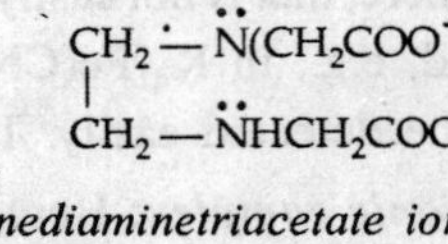 (*ethylenediaminetriacetate ion* or *edta*$^{3-}$) **Anion (hexadentate):** 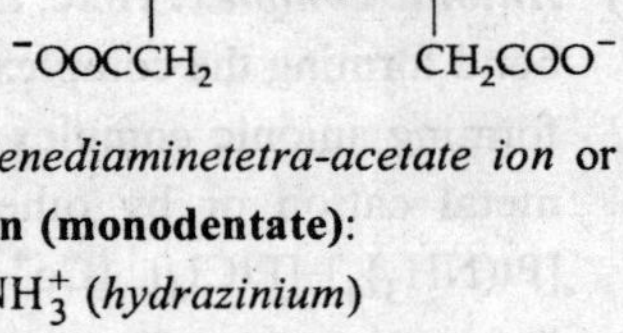 (*ethylenediaminetetra-acetate ion* or *edta*$^{4-}$) **Cation (monodentate):** $H_2NNH_3^+$ (*hydrazinium*) NO^+ (*nitrosonium*) NO_2^+ (*nitronium*)

WERNER'S COORDINATION THEORY

Werner in 1892 prepared a large number of complexes by reaction of NH_4OH and $CoCl_3$ and studied them. The results and observations of these complexes are listed as follows:

1. Luteocobaltic chloride, $CoCl_3$, $6NH_3$. It is an orange-yellow compound.
   (a) The compound on treatment with H_2SO_4 liberates all chlorine as HCl and solution (water) on treatment with $AgNO_3$ removes all chlorine indicating loose bonding between the cobalt ion and chloride ion.
   (b) The solid compound on heating with conc. HCl at 100ºC leaves no NH_3 indicating strong bond between cobalt ion and NH_3.
   (c) Conductance measurement indicates the presence of four ions in solution.
2. Roseocobaltic chloride, $CoCl_3$, $5NH_3$, H_2O (pink colour). Here
   (a) H_2O is removed on heating above 100°C,
   (b) all the chlorine atoms are precipitated with $AgNO_3$ as before,
   (c) No NH_3 can be removed by heating,
   (d) Conductivity measurement shows the presence of four ions. Thus, chlorides are loosely bound. NH_3 is held up strongly, H_2O molecule here has the same behaviour as NH_3.
3. Purpureocobaltic chloride, $CoCl_3$, $5NH_3$ (violet colour):
   (a) The solid compound on treatment with H_2SO_4 or aqueous solution of the component with $AgNO_3$ solution removes only two chlorine atoms. So, one chlorine atom is firmly bounded to cobalt ion like NH_3 and the other is loosely bounded.
   (b) No NH_3 can be removed from the compound.
   (c) Conductance measurement indicates the presence of three ions.
4. Two compounds violeocobaltic chloride (violet colour) and praseocobaltic chloride (green colour), both having the formula—$CoCl_3$, $4NH_3$, were prepared.
   (a) only one chlorine can be precipitated with $AgNO_3$.
   (b) only two ions are present as is evident from conductance measurement.
5. From these observations, Alfred Werner in 1892 proposed a theory, called Wener's theory of coordination compounds, regarding the nature of bonding in complex compounds for which he was awarded with Nobel Prize in 1910. This theory states that
   (a) Every metal possesses two types of valency—the primary or principal or ionizable valency and secondary or non-ionizable valency.
   (b) Every metal has a fixed number of secondary valencies or coordination number, e.g., cobalt has coordination number 6, iron has 6, and copper has 4 and so on.
   (c) Primary valency is satisfied by negative ions. Primary valency may be variable. Neutral molecules or negative groups satisfy secondary valencies. In some cases negative group may satisfy both, but in each case fulfillment of secondary valency is essential.

   Recently it has been found that some positive groups can also satisfy the secondary valency.
   (d) Secondary valencies are directed in space in certain definite directions around the central metal ion and as a result, the compound assumes some geometrical shape. Thus, coordination number 6 has the secondary valencies directed along the corners of a regular octahedron, coordination number 4 may have either along the corners of a regular tetrahedron or along the corner of a square planar. This special arrangement of secondary valencies results in different types of isomerism that will be described later on.

Application of this theory to the ammonia derivatives of $CoCl_3$ will clearly make us understand the nature of linkages among the ammonia and Co (III) ion. Here, dotted lines indicate the secondary valencies and the primary valencies are indicated by continuous lines.

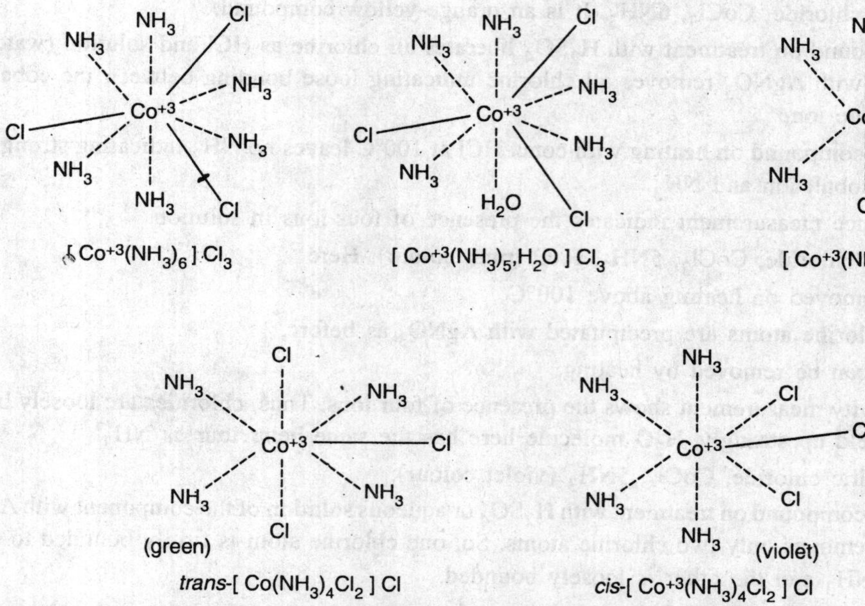

Fig. 2.1 : Diagrammatic representation of cobalt-ammine complexes.

In writing the formula of complex compounds, the metal ion and molecule that are held by secondary valencies, also called *coordination number*, are written within third bracket called *coordination sphere* or *complex moiety*. The metal ion or molecules or anions held in this third bracket are non-ionizable and cannot be detected by simple methods. The anions indicated by solid line or continuous line outside the coordination sphere are held by electrostatic attraction and are ionizable and hence, detectable.

COORDINATION NUMBER

The coordination number of a metal ion in a complex moiety is the number of ligands that surround the metal ion and held by coordinate valencies or secondary valencies. Here, the ligands are donor, i.e., Lewis base and metal ion is the acceptor, i.e., Lewis acids. The number of the secondary valencies of a metal ion can be determined by Effective Atomic Number Rule, i.e., E. A. N. Rule. The coordination number,

$$\text{C.N.} = \frac{\text{number of extra nuclear electron in next inert gas} - \text{number of extra nuclear electron in metal ion}}{2}$$

As for example, Fe^{+2} has 24 extra nuclear electrons in the orbits. The next higher inert gas Krypton has 36 extra nuclear electrons, so C.N. of $Fe^{+2} = \frac{36-24}{2} = 6$. Similarly, Cu^{+2} has 27 extra nuclear electrons, so, C.N. of Cu^{+2} will be $\frac{36-27}{2} = \frac{9}{2} = 4.5$. As coordination number cannot be a fraction, it will be 4. It is evident that through coordination, the metal ion will assume the electronic configuration of next higher inert gas.

But there are some compounds which do not obey the E. A. N. rule, e.g., $[Cr^{+3}(NH_3)_6]^{+3}$, $[Ni^{+2}(NH_3)_6]^{+2}$, $[Cu^{+1}Cl_2]^-$, etc.

In $[Cr^{+3}(NH_3)_6]^{+3}$ the electron on the Cr^{+3} is $(24-3+2\times6)=33$. In $[Ni^{+2}(NH_3)_6]^{+2}$, it is $(28-2+2\times6)=38$; in $[Cu^{+1}Cl_2]^-$, it is $(29-1+2\times2)=32$ and so on. Thus during complex formation, the trend of the metal ion is to attain the higher inert gas configuration, but it is not the necessary condition.

BONDING IN COORDINATION COMPOUNDS

The nature of bond formed by the metal ion and the ligand was given first by Pauling based on valence bond approach, later it was developed further.

According to this approach,

(a) The metal atom or ion should have some vacant orbitals in the valence shell equal in number to the coordination number of the metal ion. These orbitals will have equal energy. If the metal ion has no such orbitals available, the coordinating ligand will forcefully make available of such orbitals from the $(n-1)d$-shell and ns and p-shell. The orbitals then undergo hybridization to form hybrid orbitals stronger and of equal energy. These orbitals will have directional character and so the compound will have some geometrical configuration.

TABLE 2.05: *Electron configurations of some transition metal ions and their complexes*

Free ion/complex	3d	4s	4p	4d	Hybridization
	Electron configuration				
$Cr^{3+}(d^3)$	[•][•][•][ ][ ]	[ ]	[ ][ ][ ]	[ ][ ][ ][ ][ ]	
$Cr(NH_3)_6^{3+}$, $Cr(CN)_6^{3-}$	[•][•][•][••][••]	[••]	[••][••][••]	[ ][ ][ ][ ][ ]	d^2sp^3 i. o. c.
Mn^{2+}, $Fe^{3+}(d^5)$	[•][•][•][•][•]	[ ]	[ ][ ][ ]	[ ][ ][ ][ ][ ]	
$Mn(H_2O)_6^{2+}$, $Fe(H_2O)_6^{3+}$	[•][•][•][•][•]	[••]	[••][••][••]	[••][••][ ][ ][ ]	sp^3d^2 o. o. c.
$Fe(CN)_6^{3-}$	[••][••][•][••][••]	[••]	[••][••][••]	[ ][ ][ ][ ][ ]	d^2sp^3 i. o. c.
$FeCl_4^-$	[•][•][•][•][•]	[••]	[••][••][••]	[ ][ ][ ][ ][ ]	sp^3
Fe^{2+}, $Co^{3+}(d^6)$	[••][•][•][•][•]	[ ]	[ ][ ][ ]	[ ][ ][ ][ ][ ]	
$Fe(H_2O)_6^{2+}$, CoF_6^{3-}	[••][•][•][•][•]	[••]	[••][••][••]	[••][••][ ][ ][ ]	sp^3d^2 o. o. c.
$Fe(CN)_6^{4-}$, $Co(NH_3)_6^{3+}$	[••][••][••][••][••]	[••]	[••][••][••]	[ ][ ][ ][ ][ ]	d^2sp^3 i. o. c.
$Co^{2+}(d^7)$	[••][••][•][•][•]	[ ]	[ ][ ][ ]	[ ][ ][ ][ ][ ]	
$Co(H_2O)_6^{2+}$	[••][•][•][•][•]	[••]	[••][••][••]	[••][••][ ][ ][ ]	sp^3d^2 o. o. c.
$CoCl_4^{2-}$	[••][••][•][•][•]	[••]	[••][••][••]	[ ][ ][ ][ ][ ]	sp^3
$Ni^{2+}(d^8)$	[••][••][••][•][•]	[ ]	[ ][ ][ ]	[ ][ ][ ][ ][ ]	
$Ni(NH_3)_6^{2+}$	[••][••][••][•][•]	[••]	[••][••][••]	[••][••][ ][ ][ ]	sp^3d^2 o. o. c.
$Ni(CN)_4^{2-}$	[••][••][••][••][••]	[••]	[••][••][ ]	[ ][ ][ ][ ][ ]	dsp^2
$NiCl_4^{2-}$	[••][••][••][•][•]	[••]	[••][••][••]	[ ][ ][ ][ ][ ]	sp^3
$Zn^{2+}(d^{10})$	[••][••][••][••][••]	[ ]	[ ][ ][ ]	[ ][ ][ ][ ][ ]	
$Zn(NH_3)_6^{2+}$	[••][••][••][••][••]	[••]	[••][••][••]	[••][••][ ][ ][ ]	sp^3d^2 o. o. c.
$ZnCl_4^{2-}$	[••][••][••][••][••]	[••]	[••][••][••]	[ ][ ][ ][ ][ ]	sp^3

i. o. c. → inner orbital complex

o. o. c. → outer orbital complex

(b) The ligand should have an atom with lone pair of electron. During reaction as these ligands approach the metal ion, the overlap of metal hybrid orbital and the ligand orbital takes place through coordination. So the metal-ligand bond is semi-polar bond, the metal ion is the acceptor (acid) and the ligand is the donor (base). These electrons will occupy the empty metal orbitals and the bond is written as M ← L. Formation of such bond is ascertained by the measurement of the magnetic behaviour of the complex.

(c) If there be only donation of lone pair of electron of the ligand, there will be accumulation of unfavourable electron density over the metal ion and the metal ion should be unstable. Moreover, in such cases the M-L bond should be electrostatic in nature. If overlap of the metal ion orbital and the ligand orbitals are considered then excess electron density over the metal ion is delocalized to some extent to the empty ligand orbital and the bond becomes covalent in nature. Thus, electro-neutrality is maintained. The formation of certain complexes is described above as in the Table 2.05.

Bonding in $[NiCl_4]^{2-}$ and $[Ni(CN)_4]^{2-}$

In these compounds nickel is in +2 oxidation state and $3d^8$ electron arrangement. The electron distribution in free ion in weak ligand Cl^- and strong ligand CN^- are as follows:

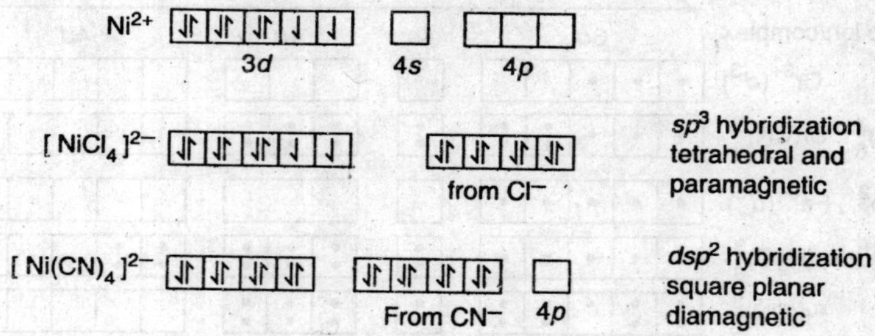

Fig. 2.2 : The diagrammatic representation of $[NiCl_4]^{2-}$ and $[Ni(CN)_4]^{2-}$ complexes.

Bonding in $[FeF_6]^{3-}$ and $Fe(CN)_6]^{3-}$

In these compounds Fe^{3+} has $3d^5$ configuration. The electron distributions in free ion and in the complexes are as follows:

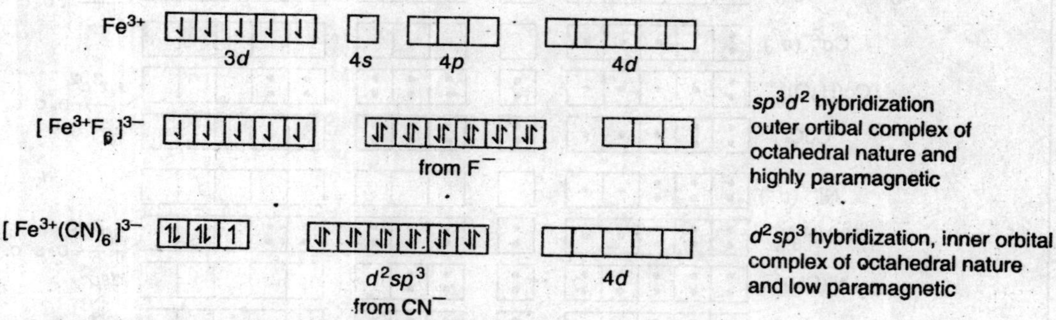

Fig. 2.3 : The diagrammatic representation of $[FeF_6]^{3-}$ and $[Fe(CN)_6]^{3-}$ complexes.

Besides these, complexes of coordination number five, three, two are also known but they are uncommon. Complexes of coordination number four may have sp^3 and dsp^2 hybridization having tetrahedral and square planar configuration. The Ma_4 or Ma_2b_2 tetrahedral complexes have four attachment sites identical and so, no geometrical isomers are possible. But for square planar arrangement of Ma_2b_2 only, two arrangements are possible, one having like ligands in one side or close to each other –*cis*-isomer and the other having like ligands far away – *trans*-isomers.

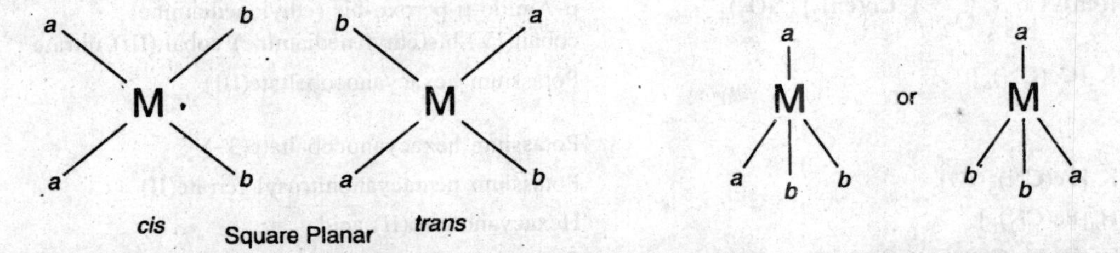

Fig. 2.4 : Trans- and cis-isomer in square planar complexes; Ma_4 or Ma_2b_2 in titrahedral complexes.

This *cis*- and *trans*-isomerism will be discussed in geometrical isomerism of complexes.

NOMENCLATURE OF COMPLEX COMPOUNDS

Complex compounds were named after their discoverers and some compounds were named on the basis of their colours. Comprehensive system of nomenclature proposed by Werner and later modified by IUPAC in 1990 are given as under:

(a) Like simple compounds, in complexes, the cationic part is named first and then the anionic part. In the cationic part, the ligands with number (bis, tris, etc.) are mentioned before the metal ion, the ligands are listed in alphabetical order of the initial letter irrespective of charge.

(b) Anionic ligands end with the letter "o" and neutral molecules bear the name of the molecule. Special name for H_2O is "aqua" and for NH_3 is "ammine".

(c) If the complex part is anionic, the suffix –ate is at the end. Oxidation number of the metal is written in the parentheses in Roman Number.

(d) When the complex is an acid, -ic is used at the end.

(e) For polynuclear complex, the bridging ligand is preceded by Greek letter μ.

(f) For geometrical isomers *cis-* or *trans-* is used.

(g) For optical isomers, the term dextro is designated by *d-* or (+) and for leavo form by *l-* or (–) is used. Some examples are given below in the Table 2.06.

TABLE 2.06: *Nomenclature of complexes*

Formula	*Name*
$[Cr(NH_3)_6]Cl_3$	Hexaaminechromium (III) chloride or Hexaaminechromium (3+) chloride
$[Co(en)_3]\ (NO_3)_3$	Tris (ethylenediamine) cobalt(III) nitrate
$[CoCl(en)_2(NH_3)]\ Cl_2$	Amminechlorobis (ethylenediamine) cobalt(III) chloride
$[CoCl(NO_2)\ (NH_3)_4]\ NO_3$	Tetramminechloronitrocobalt(III) nitrate
$[Co(en)_2\ (ox)]\ Cl$	Bis(ethylenediamine) oxalatocobalt(III) chloride
$[Co(NO_2)_3\ (NH_3)_3]$	Triamminetrinitrocobalt(III)
$[Cr(NH_3)_6]\ [Co(CN)_6]$	Hexaamminechromium(III) hexacyanocobaltate(III)
$[(NH_3)_5\ Co\text{-}O\text{-}Co(NH_3)_5]\ (NO_3)_4$	μ-Peroxobis(pentaamminecobalt(III) nitrate

contd. table...

Formula	*Name*
$[(en)_2 Co \left\langle \begin{matrix} NH_2 \\ O_2 \end{matrix} \right\rangle Co(en)_2] (NO_3)_4$	μ-Amido-μ-peroxo-bis (ethylenediamine)-cobalt(IV)-bis(ethylenediamine) cobalt(III) nitrate
$K_3[Co(CN)_6]$	Potassium hexacyanocobaltate(III) or Potassium hexacyanocobaltate(3–)
$K_3[Fe(CN)_5NO]$	Potassium pentacyanonitrosyl ferrate(II)
$H_4[Fe(CN)_6]$.	Hexacyanoferro(II) acid
$[(CO)_3Fe(CO)_3Fe(CO)_3]$	Tri-μ-carbonyl-bis(tricarbonyliron) (Fe–Fe)[a]
$[(CH_3Hg)_4S]^{2+}$	μ_4-Thiotetrakismethylmercury(II) ion or μ_4-Thiotetrakismethylmercury ion(2+)

(a) This compound has a Fe–Fe bond in addition to three bridging CO, the Fe–Fe bond is indicate by (Fe–Fe), this is followed in other cases also wherever there is evidence for a metal bond. Thus, $Mn_2(CO)_{10}$ = $(CO)_5$ Mn–Mn$(CO)_5$ is bis(pentacarbonylmanganese) (Mn–Mn).

CHELATE COMPOUNDS

It has been established from experiments that a complex formed by a metal ion with an unidentate ligand is less stable than that formed with a bidentate ligand. If the number of points of attachment of the ligand to the metal ion is more than two, the more stable is the complex. In this way, a bidentate or a polydentate ligand forms a ring with the metal ion. If the ring is five or six membered with the metal ion as one point, the strain in ring is less and more stable is the complex. These ring-forming ligands are called chelating group and the compound is called the chelate complex. e.g., $C_2H_5NH_2$ formed complexes with Co^{3+} and Cr^{3+} but these are unstable whereas ethylenediammine NH_2–CH_2–CH_2–NH_2 formed five membered ring and the compounds are much stable. Similarly, neither phenol nor benzaldehyde can form complexes but combination of the two salicylaldehydes form stable complexes forming six membered ring.

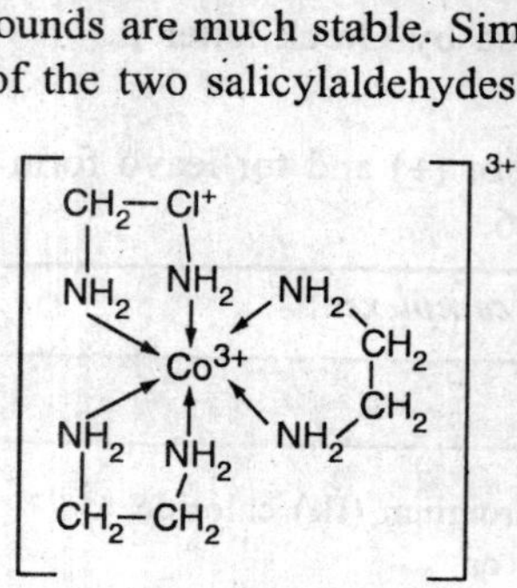

Five membered ring

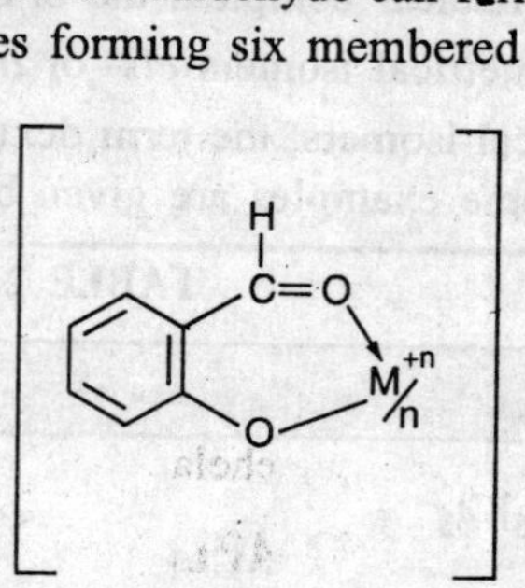

Six membered ring

Fig. 2.5 : 5-membered ring en complex of Co^{3+} and six membered ring salicylaldehyde complex of metal ion.

Similarly, acetone cannot form complex but acetylacetone can form very stable complexes.

Chelate complexes are of two types:

(a) Inner metallic complex of first order and

(b) Inner metallic complex of second order.

A polydentate ligand may possess two types of coordination position, one may simply attach to the metal ion through coordination and the other type simultaneously fulfils the primary and the secondary valencies of the metal ion. e.g., glycolic acid (NH_2CH_2COOH). Here, COO^- satisfies both primary and secondary valencies and NH_2 only secondary valency. In doing so if the complex formed is neutral or with no charge overall, then the complex is called inner complex of first order e.g. $[Cu^{+2}(NH_2CH_2COO)_2]^0$, $[Cr^{+3}$ (acetylacetonate)$_3]^0$ whose structures are as follows:

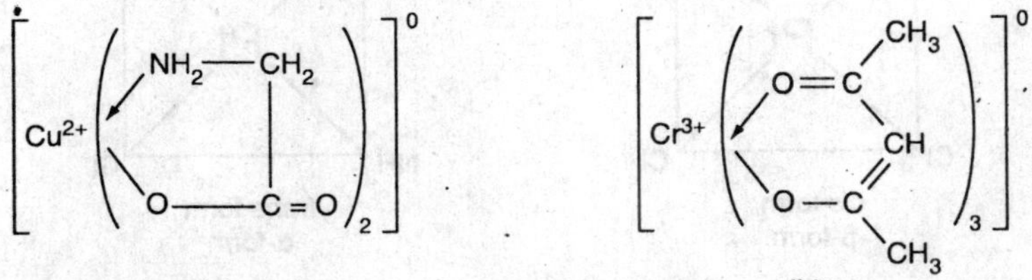

(a) (b)

Fig. 2.6 : (a) Structure of $[Cu^{2+}(NH_2CH_2COO^-)_2]^0$ and (b) structure of $[Cr^{3+}(acac)_3]^0$.

(b) If the complex has overall charge, i.e., the primary valency is not neutralized, the complexes are called the inner metallic complex of second order.
e.g., $Na[Cu^{2+}(acac)_3]$, $[Ti^{4+}(acac)_3]Cl$, etc.

Applications of chelate complexes

The chelate complexes have innumerable applications in analytical chemistry, biology, industry, agriculture, medicinal science, etc.

(a) Analytical chemistry

Ag^+ ion can be separated from the mixture of Fe^{3+}, Cu^{2+}, Pb^{2+}, Bi^{3+}. With EDTA, Ag^+ forms soluble chelate but others are not soluble. Gravimetrically, nickel can be estimated by formation of insoluble chelate with dimethyl glyoxime in slightly alkaline medium; colorimetric identification and estimation of copper can be done with cupferron. Hardness of water can also be determined with EDTA.

(b) In biology and in medical treatment

Chlorophyll is a magnesium chelate that helps in photosynthesis using atmospheric CO_2 and liberating O_2 in air, thus controlling the oxygen balance in air. From human body, lead can be removed by using calcium salt of EDTA, i.e., $Ca[Na_2EDTA]$ when lead forms stable $Pb[Na_2EDTA]$ and then goes out of human body through urine. Lead poisoning is treated by dilute aqueous EDTA solution (used as drink). Haemoglobin being an iron chelate carries oxygen to different parts of the body

ISOMERISM OF COMPLEXES

Like organic compounds, depending on types of linkages, position of the ligands, etc. inorganic complexes also exhibit different types of isomerism. These are discussed as follows:

Stereo isomerism

This isomerism is of two types; (a) geometrical or cis-trans isomerism and (b) optical isomerism.

(a) Geometrical or *cis-trans* isomerism

This type of isomerism occurs for square planar of type Ma_2b_2 or octahedral of type Ma_4b_2 only. As all other points of attachment being identical, so, no isomerism occurs for tetrahedral. For Square planar, two types of arrangements are possible. In (i) like ligands are close to each other and thus, is called *cis*-isomer and in (ii) the like ligands are far away to each other and called *trans*-isomer (Fig. 2.4). The compounds

$[Pt(NH_3)_2Cl_2]$ prepared from K_2PtCl_4 with NH_3, and from $[Pt(NH_3)_4]Cl_2$ with HCl, give two varieties of compounds—the former is β-form and the latter is α-form, both have different properties. Replacement

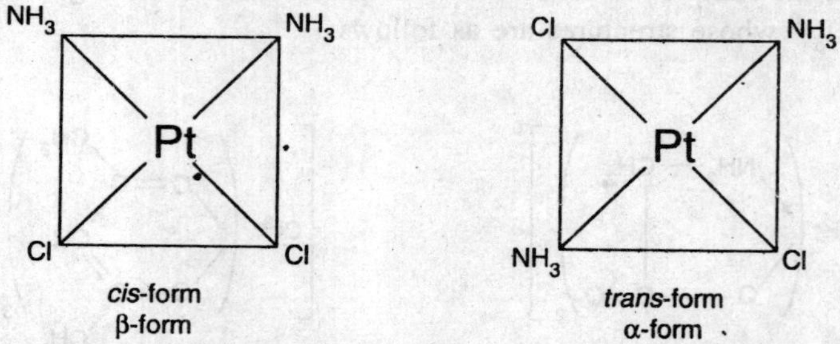

Fig. 2.7 : *Cis-* and *trans*-form of $[Pt(NH_3)_2Cl_2]$ complex.

of groups in α- and β-form by pyridine, indicates α-form of dichloro diammine platinum (II) is the *trans*-form and β-form is the *cis*-form.

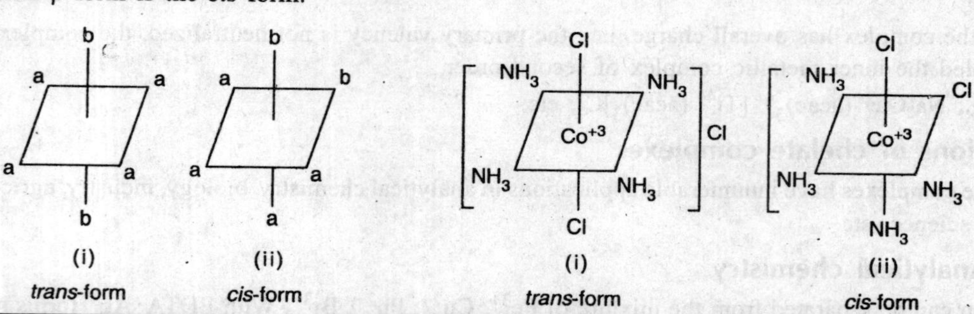

Fig. 2.8 : *cis-* and *trans*-form of Ma_4b_2, e.g., $[Co^{+3}(NH_3)_4Cl_2]Cl$.

For six coordinated octahedral complexes, Ma_4b_2 may have two arrangements, (i) is the trans form and (ii) is the cis form, e.g., Tetramine dichlorocobalt (III) chloride has the two forms, (i) is the *trans*-form and (ii) is the *cis*-form.

If one of the ligand is bidentate, due to steric factor the ligand occupies *cis*-position and one structure is possible. In case of two such ligands, they occupy the *cis*-position only, e.g., dichlorobis-ethylenediamine cobalt (III) chloride may have the following configuration:

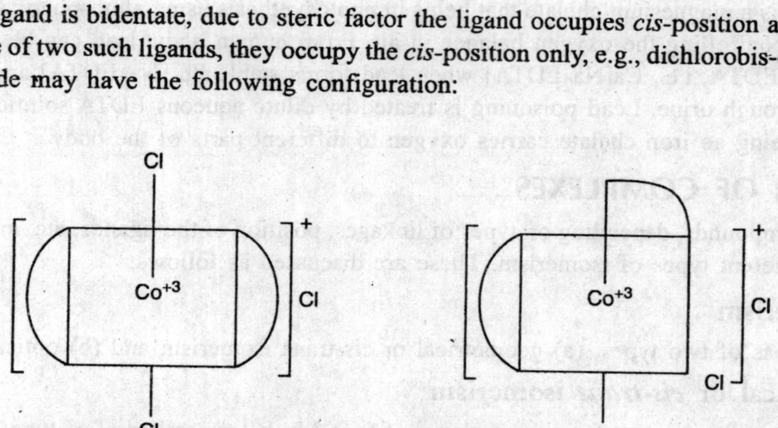

Fig. 2.9 : *cis*-and *trans*-form of $[Co^{3+}en_2Cl_2]Cl$.

(b) Optical isomerism

Like optically active organic compounds, if the inorganic complex have no symmetry (axis of, centre of-plane of-) in the molecule, then it can rotate the plane of polarized light in either direction–left (laevo) or right (dextro). If the compound has two bidentate groups, then the compound will have *trans-* and *cis-*form. The *trans-*form has the plane of symmetry and not optically active, but the *cis-*form can be represented in two ways according to the arrangement of the bidentate ligand. These two forms bear the mirror image relationship and are not superimposable. There are two optical isomers—one is dextro and the other is laevo. in Fig. 2.10.

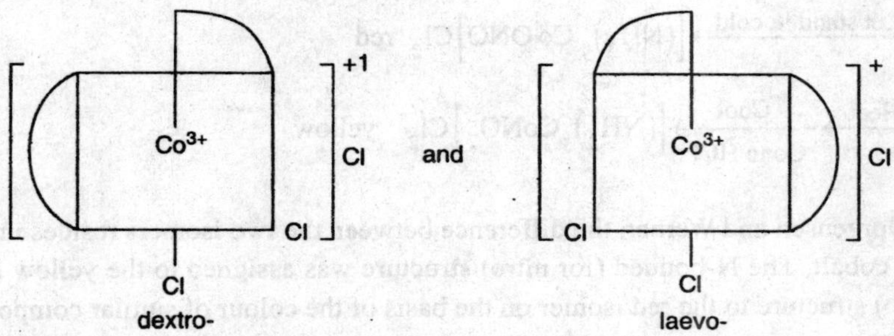

Fig. 2.10 : Optical isomerism in $[Coen_2Cl_2]Cl$.

Similarly, tris ethylenediaminecobalt (III) chloride can be represented in two forms and the both forms are optically active.

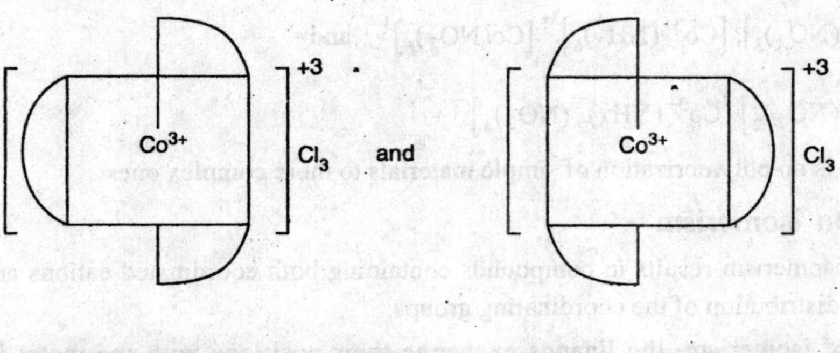

Fig. 2.11 : Two forms of tris-ethylenediaminecobalt (III) chloride.

Similarly, purely inorganic compounds of cobalt and rhodium have the optical activity, e.g.,

$$\left[Co^{3+}\left\{\left[\begin{matrix}HO\\HO\end{matrix}Co^{3+}(NH_3)_4\right]\right\}_3\right]^{6+} \quad \text{and} \quad \left[Rh\{(NH_2)_2SO_2\}_2(H_2O)_2\right]^-$$

A four coordinated square planar platinum compound has been known that has also optical activity.

LINKAGE STRUCTURAL ISOMERISM

It deals with ligands that are capable of bonding through one type of donor atom in one situation but a different atom in another complex. Some authors refer to this type of isomerism as "structural isomerism" but inasmuch as all isomerism is basically 'structural', the term linkage isomerism is preferable. The first example of this type of isomerism was provided by Jorgensen, Werner's contemporary. His method of preparation was as follows:

$$[Co(NH_3)_5Cl]Cl_2 \xrightarrow{NH_3} \xrightarrow{HCl} \xrightarrow{NaNO_2} \text{Solution A}$$

$$\text{Solution A} \xrightarrow{\text{Let stand in cold}} [(NH_3)_5 Co\,ONO]Cl_2 \ \text{red}$$

$$\text{Solution A} \xrightarrow{\text{Heat}} \xrightarrow[\text{Conc HCl}]{\text{Cool}} [(NH_3)_5 CoNO_2]Cl_2 \ \text{yellow}$$

According to Jorgensen and Werner, the difference between the two isomers resides in the linkage of the NO_2 group to the cobalt. The N-bonded (for nitro) structure was assigned to the yellow isomer and the O-bonded (for nitrito) structure to the red isomer on the basis of the colour of similar compounds.

In general the other types of isomerism for coordination compounds are less interesting than those discussed previously, but will be listed briefly to show the variety of possibilities.

a) Polymerization isomerism

Compounds are said to be polymerization isomers when stoichiometric composition of metal and ligand is the same but actual molecular compositions are multiple of simple stoichiometric ratio e.g.,

$$[Co^{3+}(NH_3)_3(NO_2)_3],\ [Co^{3+}(NH_3)_6]^{3+}[Co(NO_2)_6]^{3-} \text{ and}$$

$$[Co^{3+}(NH_3)_4(NO_2)_2][Co^{3+}(NH_3)_2(NO_2)_4]^{-}$$

Actually there is no polymerization of simple materials to more complex ones

b) Coordination isomerism

Coordination isomerism results in compounds containing both coordinated cations and anions having differences in the distribution of the coordinating groups.

In this type of isomerism, the ligands exchange their positions with the metal ions so they are different. e.g.,

$$[Co^{3+}(en)_3]^{3+}[Cr^{3+}(CN)_6]^{3-} \text{ and } [Cr^{3+}(en)_3]^{3+}[Co^{3+}(CN)_6]^{3-}$$

$$[Cu^{2+}(NH_3)_4]^{2+}[Pt^{2+}Cl_4]^{2-} \text{ and } [Pt^{2+}(NH_3)_4]^{2+}[Cu^{2+}Cl_4]^{2-}$$

Also when the metal ions are same,

$$[Cr^{3+}(NH_3)_6]^{3+}[Cr^{3+}(SCN)_6]^{3-} \text{ and } [Cr^{3+}(NH_3)_4(SCN)_2]^{1+}[Cr^{3+}(NH_3)_2(SCN)_4]^{1-}$$

$$[Pt(NH_3)_4][PtCl_4] \text{ and } [Pt(NH_3)_3Cl][Pt(NH_3)Cl_3]$$

c) Hydrate isomerism

In this type of isomerism water appears as ligand like NH_3 in the coordination position or it may appear in the lattice positions without being closely attached to the metal ion.

$$[Cr(H_2O)_5Cl]\ Cl_2 . H_2O \text{ and } [Cr(H_2O)_4Cl_2]Cl.2H_2O$$

$$[Co(NH_3)_4(H_2O)Cl]Cl_2 \text{ and } [Co(NH_3)_4Cl_2]Cl.H_2O$$

They have different color and physical properties.

d) Ionization isomerism

Here the compounds have same stoichiometric composition but yield different ions in solution. e.g.,

$$[Co(NH_3)_5Br]SO_4 \text{ and } [Co(NH_3)_5SO_4]Br$$

$$[Co(NH_3)_5NO_3]SO_4 \text{ and } [Co(NH_3)_5SO_4]NO_3$$

e) Coordination position isomerism

In poly nuclear complexes, coordination groups may be present in the same numbers but may arrange themselves differently with respect to the different metal ion nuclei present. This gives coordination position isomerism, e.g.,

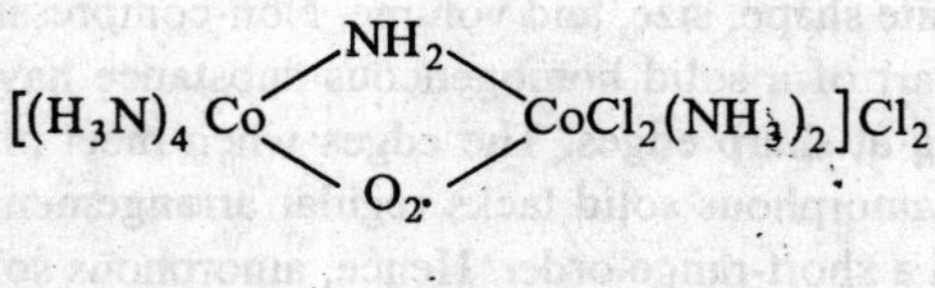

and

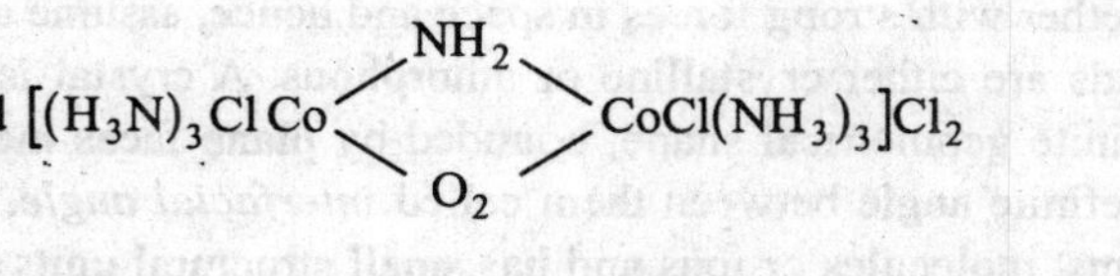

3

SOLID STATE CHEMISTRY

INTRODUCTION

LAWS OF CRYSTALLOGRAPHY

Solids are non-compressible rigid substances where atoms, molecules or ions are closely packed and held together with strong forces in space and hence, assume definite shape, size, and volume. Non-compressible solids are either crystalline or amorphous. A crystal is a part of a solid homogeneous substance having definite geometrical shape, bounded by plane faces meeting at sharp edges. The edges when meet make a definite angle between them called *interfacial angle*. An amorphous solid lacks regular arrangement of atoms, molecules or ions and has small structural units with a short-range order. Hence, amorphous solids are called super cooled liquid. The branch of science that deals with geometrical properties and structure of crystals is called crystallography. Crystallography is based on three fundamental laws.

1. The law of constancy of interfacial angle.
2. The law of rational indices.
3. The law of symmetry.

1. The law of constancy of interfacial angles

Interfacial angle is the angle between the two faces where they intersect. Crystals of same substance may vary in shape and size depending upon the condition at which crystallization occurs, but the interfacial angle will be always constant.

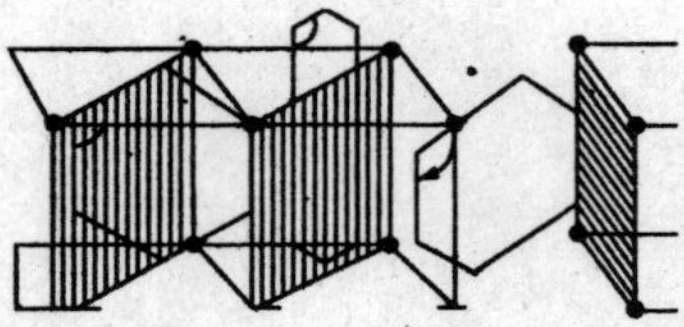

Fig. 3.1 : Interfacial angles in a crystal.

2. The law of rational indices

The crystals are three dimensional solids and so three mutually perpendicular axis X, Y, Z are chosen such that all faces of the crystal will either intercept these axes at a definite distance from the origin or will be parallel to it, i.e., the intercept is at infinity. According to this law, these intercepts (a, b, c) may not be the same, but bear a simple ratio among them. Suppose, OX, OY and OZ are the three mutually perpendicular axes in the crystal. The plane *ABC* cuts the intercepts *a*, *b* and *c* from the axes. Another plane *A'B'C'* cuts the intercepts at 2*a*, 2*b* and 3*c* respectively. Then, the ratio of the intercepts is 2:2:3, coefficients of *a*, *b* and *c* is called *Weiss indices* of the given plane. The reciprocal of the Weiss indices and multiplying them with the lowest common multiple gives the whole number, which are called *Miller indices* of the plane. So, Miller indices of the above *A'B'C'* plane are $\frac{1}{2}:\frac{1}{2}:\frac{1}{3}$ or 3:3:2 [Fig. 3.2(B)].

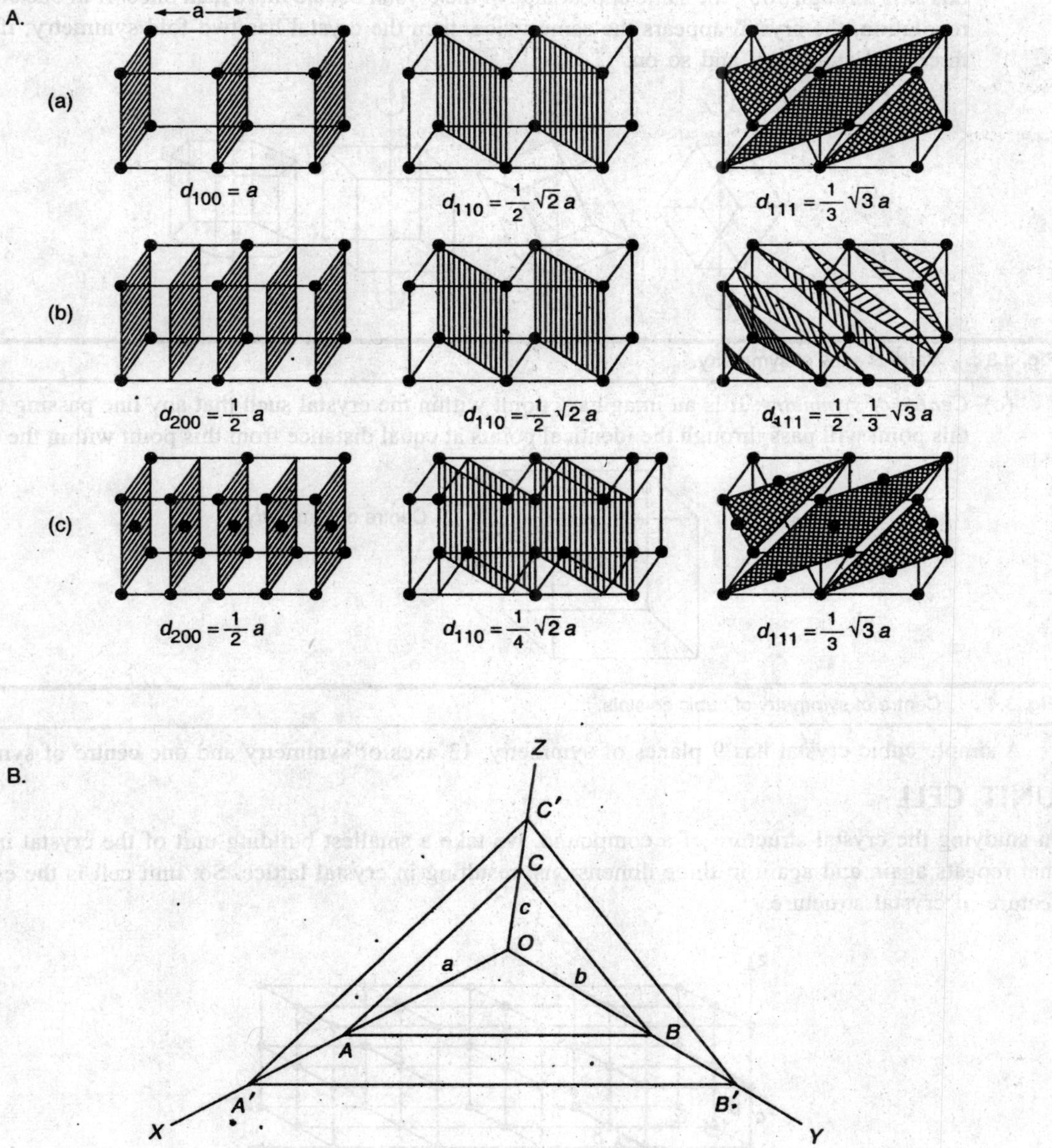

Fig. 3.2 : (A) Illustration of Miller indices of : (a) sc, (b) bcc, (c) fcc lattice.
(B) Crystallographic axes.

3. The law of symmetry

Crystals of the same substance possess same elements of symmetry. These are

(a) *Plane of symmetry*: It is an imaginary plane in the crystal that divides the crystal into two parts, one part will be the mirror image of the other.

(b) *Axis of symmetry*: It is an imaginary line through the crystal such that if the crystal is rotated about this axis through 360°, the same appearance of the crystal occurs more than once. If in one complete revolution, the crystal appears the same twice, then the crystal has two-fold symmetry; if thrice, three-fold symmetry and so on.

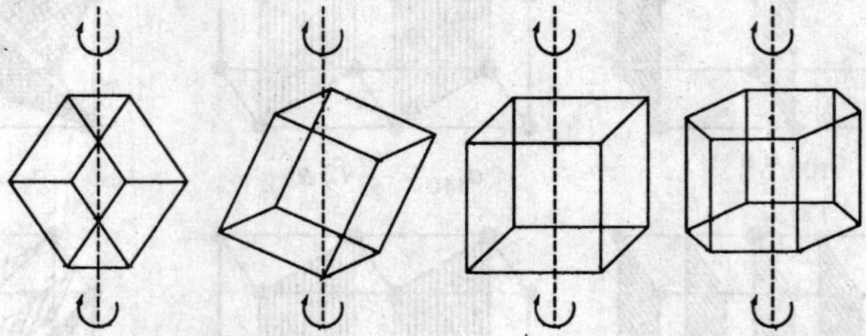

Fig. 3.3 : Various axes of symmetry.

(c) *Centre of symmetry*: It is an imaginary point within the crystal such that any line passing through this point will pass through the identical points at equal distance from this point within the crystal.

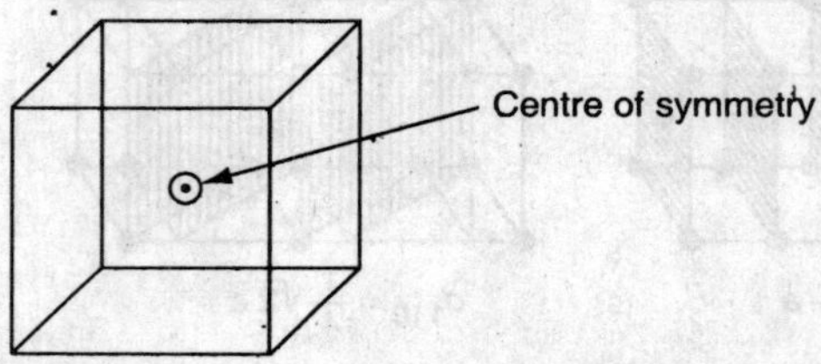

Fig. 3.4 : Centre of symmetry of cubic crystals.

A simple cubic crystal has 9 planes of symmetry, 13 axes of symmetry and one centre of symmetry.

UNIT CELL

In studying the crystal structure of a compound, we take a smallest building unit of the crystal in space that repeats again and again in three dimensions, resulting in crystal lattice. So, unit cell is the essential feature of crystal structure.

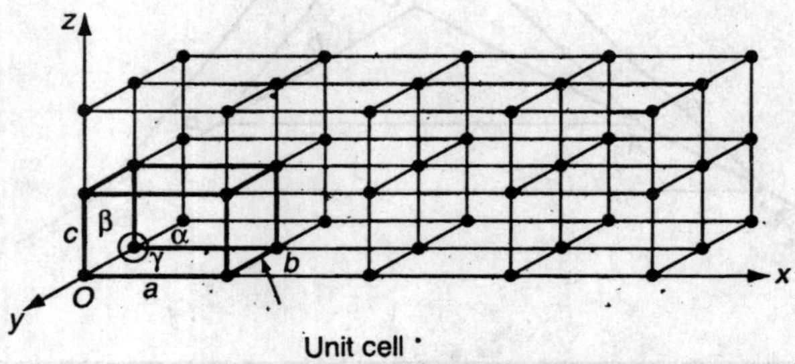

Fig. 3.5 : Crystal lattice and unit cell.

The unit cell is characterized by the lengths a, b and c and the angles α, β and γ, are collectively called the unit cell parameters. There are different types of crystals having different parameters which are listed below:

TABLE 3.01: *Crystal Systems*

System	*Axial characteristics*	*Angles*	*No. of space lattices*	*Examples*
1. Cubic	$a = b = c$	$\alpha = \beta = \gamma = 90°$	3	NaCl, KCl, CaF_2, Cu, ZnS, CsCl, CuO.
2. Tetragonal	$a = b \neq c$	$\alpha = \beta = \gamma = 90°$	2	SnO_2, TiO_2, $ZrSiO_4$.
3. Orthorhombic	$a \neq b \neq c$	$\alpha = \beta = \gamma = 90°$	4	$BaSO_4$, KNO_3, $MgSiO_3$, K_2SO_4, $CdSO_4$, AgBr.
4. Monoclinic	$a \neq b \neq c$	$\alpha = \gamma \neq 90°$ $\beta \neq 90°$	2	$CaSO_4$, $2H_2O$; $NaHCO_3Na_2SO_4$, $10H_2O$; monoclinic sulphur.
5. Triclinic	$a \neq b \neq c$	$\alpha \neq \beta \neq \gamma = 90°$	1	$CuSO_4$, $5H_2O$; $NaHSO_4$; $K_2Cr_2O_7$; H_3PO_3.
6. Hexagonal	$a = b \neq c$	$\alpha = \beta = 90°$ $\gamma = 120°$	1	PbI_2, Mg, Cd, Zn, ZnO, BN, SiO_2, HgS, CdS
7. Rhombohedral (or trigonal)	$a = b = c$	$\alpha = \gamma = 90°$ $\beta \neq 90°$	1	Graphite, ICl, Al_2O_3, calcite ($CaCO_3$), As, Sb, Bi.

The cube crystals where $a = b = c$ and $\alpha = \beta = \gamma = 90°$, are discussed here. The cubic crystal may be of three types:

1. *Simple cubic cell*: Here, 8 atoms or ions occupy the 8 lattice points. 2 atoms or ions along each axis and so total 6 surround every corner atom or ion. Hence, coordination number is 6 : 6.
2. *Body centered cubic cell*: In this type of crystal one atom or ion is at the centre and other atoms or ions occupy eight corners. Hence coordination number is 8 : 8.
3. *Face centred cubic cell*: Each surface of the cubic cell has one atom or ion at the centre. It is surrounded by 4 other atoms or ions very close in the same plane and one above and one below close to it. Hence, coordination number is 6 : 6.

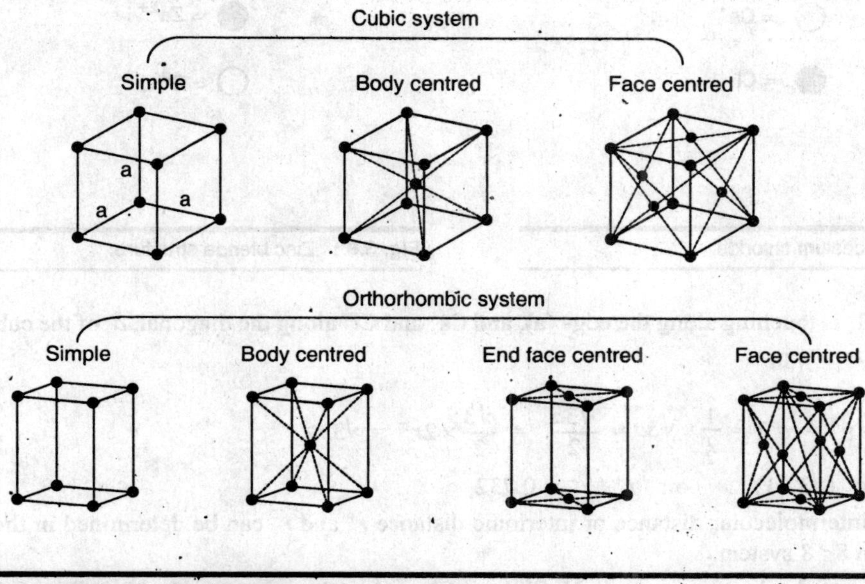

Fig. 3.6 : Seven Bravais lattices of the seven crystal systems.

In cubic system another crystal system is obtained as in ZnS. It is called *Wurtzite structure*. Here one atom or ion is at the centre of the cell and tetrahedrally surrounded by other atom or ion. Coordination number is 4 : 4. The types of these ionic crystals depend upon the radius ratio of the cation and anion $\frac{r^+}{r^-}$. Calculation of the radius ratio and the types of unit cell are given below.

The compound or the crystal is stable when the attractive force between the oppositely charged ions increases and that increases with increasing coordination number of the ion, and the repulsive force between the like ions when squeezed together. The optimum condition is when the ions in the crystals just touch each other, that depends on the relative size of the ions. Let us consider the case of CsCl, a body centered cubic lattice with coordination number 8 : 8 (where O are Cs^+, and Cl^- are •). (Fig. 3.7).

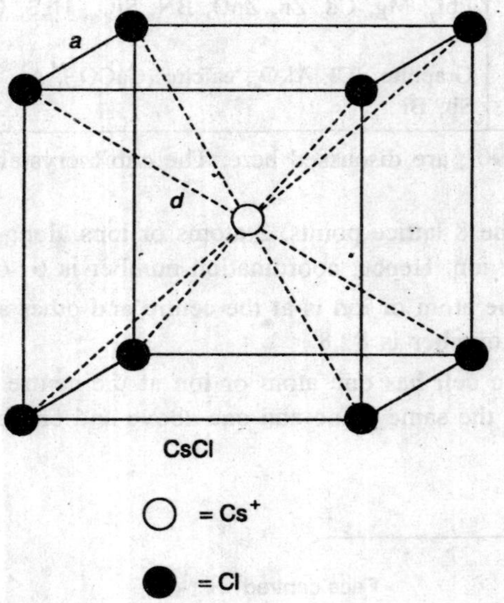

Fig. 3.7 : The cesium chloride.

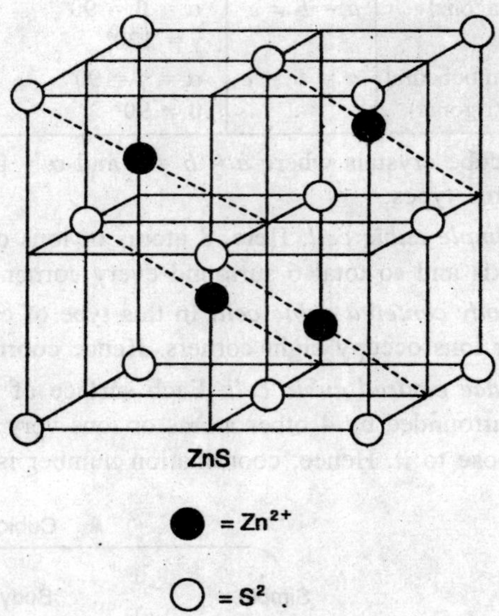

Fig. 3.8 : Zinc blende structure.

The two Cl^- is touching along the edge (a), and Cs^+ and Cl^- along the diagonal, *d*, of the cube. Diagonal, $d = \sqrt{3a^2} = \sqrt{3}a$

So $r^+ + r^- = d/2 = \frac{1}{2} \times \sqrt{3}a = \frac{\sqrt{3}a}{2} = \frac{\sqrt{3}}{2} \times 2r^- = \sqrt{3}r^-$

or, $r^+ = r^- (\sqrt{3} - 1)$ or, $r^+ / r^- = 0.732$.

Knowing intermolecular distance or interionic distance r^+ and r^- can be determined in the crystal, i.e., crystal radii in 8 : 8 system.

For face centred cubic system as in NaCl, here O are the Cl^- and ● are the Na^+. (Fig. 3.9).

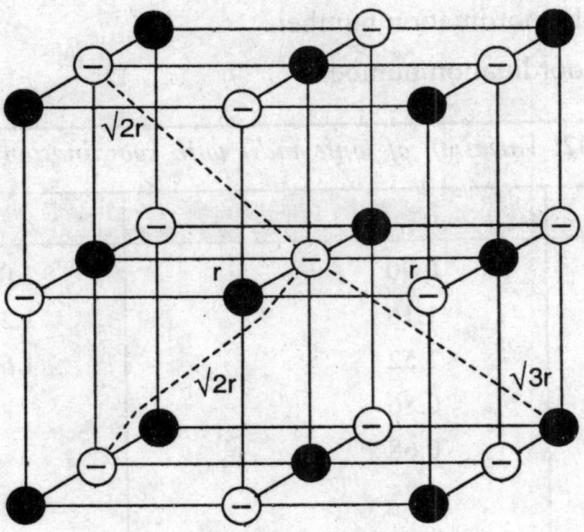

Fig. 3.9 : The sodium chloride structure.

Here, $d^2 = 2a^2$ or, $d = \sqrt{2}a$, $\therefore\ r^+ + r^- = \dfrac{d}{2} = \dfrac{\sqrt{2}a}{2} = \dfrac{a}{\sqrt{2}} = \dfrac{2r^-}{\sqrt{2}} = \sqrt{2}r^-$.

i.e., $r^+ + r^- = \sqrt{2}r^-$ or, $r^+ = \sqrt{2}r^- - r^- = (\sqrt{2} - 1)\ r^-$ or, $\dfrac{r^+}{r^-} = 0.414$.

For Wurtzite structure, i.e., ZnS where the crystal arrangement is tetrahedral, (Fig 3.8) the radius ratio can be calculated as shown below. The open circles are S^{-2} and the solid circle is Zn^{+2}. Here

$$r^+ + r^- = \frac{d}{2} = \sqrt{3}a/2 \quad \text{and} \quad 2r^- = \sqrt{2}a \quad \text{or,} \quad a = \frac{2r^-}{\sqrt{2}} = \sqrt{2}r^-.$$

$$r^+ + r^- = \frac{\sqrt{3}a}{2} = \frac{\sqrt{3}}{2} \times \sqrt{2}r^- = \frac{\sqrt{6}}{2}r^-$$

$$\therefore\ r^+ = \frac{\sqrt{6}}{2}r^- - r^- = r^-\left(\frac{\sqrt{6}}{2} - 1\right)$$

$$= (1.225 - 1)\ r^-.$$

$$\therefore\ \frac{r^+}{r^-} = 0.225.$$

For square planar crystal, Fig. 3.10.

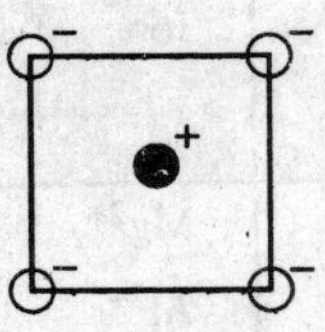

Fig. 3.10 : Square planar crystal.

$$r^+ + r^- = \frac{\sqrt{2}a}{2} = \frac{a}{\sqrt{2}} \quad \text{and} \quad 2r^- = a \qquad \therefore\ r^+ + r^- = \frac{a}{\sqrt{2}} = \frac{2r^-}{\sqrt{2}} = \sqrt{2}r^-.$$

$$r^+ = \sqrt{2}r^- - r^- = (\sqrt{2} - 1)\ r^- = 0.414r^- \qquad \therefore\ \frac{r^+}{r^-} = 0.414.$$

The radius of Cl^- changes with coordination number.

Variation of ionic radii with coordination number:

TABLE 3.02: *Variation of ionic radii with coordination number*

Ion	4	6	8	12
Li^+	0.73	0.90	1.06	
Na^+	1.13	1.16	1.32	1.53
K^+	1.51	1.52	1.65	1.78
Mg^{+2}	0.71	0.86		
Al^{+3}	0.53	0.68		
Ti^{+4}	0.56	0.75	0.88	
Fe^{+2}	0.77	0.92(HS), 0.75(LS)		
Fe^{+3}	0.63	0.79(HS), 0.69(LS)		

Thus ionic radius depends on the type of crystal arrangement or radius ratio. The table below shows the relation of radius ratio and crystal arrangement.

TABLE 3.03: *Relation of radius ratio and crystal arrangement*

Coordination Number	Crystal arrangements	radius ratio
2	Linear	Up to 0.15
3	Triangular	0.15-0.22
4	Tetrahedral	0.22-0.41
	Square Planar	0.41-0.73
6	Octahedral	0.41-0.73
8	Cube	> 0.73

Table of ionic radii of a few ions are given below:

TABLE 3.04: *Ionic radii of a few ions*

Ion	Goldschmidt empirical ionic radii	Pauling's calculated ionic radii	Ion	Goldschmidt empirical ionic radii	Pauling's calculated ionic radii
H^-	1.54	2.08	Mg^{2+}	0.78	0.65
Li^+	0.78	0.60	Al^{3+}	0.57	0.50
Be^{2+}	0.34	0.31	Si^{4+}	0.39	0.41
B^{3+}		0.20	P^{5+}	0.3-0.4	0.34
N^{3-}		1.71	P^{5-}		2.12
O^{2-}		1.40	S^{6+}	0.34	0.29
F^-		1.36	S^{2-}	1.74	1.84
Na^+	0.98	0.95	Cl^-	1.81	1.81

X-RAY STUDY OF CRYSTALS

The knowledge of a crystal structure can be best obtained from the X-ray diffraction study by Bragg. When light of short wavelength like X-ray (100 pm) falls on an object of same size range as the wavelength of the light, it is diffracted. Crystals have a large number of parallel and equidistant atomic planes, called *crystal gratings*. When X-ray falls on a crystal face, it penetrates into it and is scattered by atoms or ions. If the path difference between the two rays is an integral multiple of the wavelength of the X-ray, the two rays will be in phase, and the diffraction pattern will have higher intensity.

Thus, the path difference = $n\lambda$, where $n = 1, 2. 3$, etc.

Let two parallel beams *PA* and *OC* incident at an angle θ in the parallel planes at *A* and *C*. The perpendicular distance *AC* between the planes is *d*. The beams reflected along *AR* and *CS*. The path difference between the beams is $BC + CD$.

From the diagram, $BC + CD = AC \sin\theta + AC \sin\theta = 2d \sin\theta$

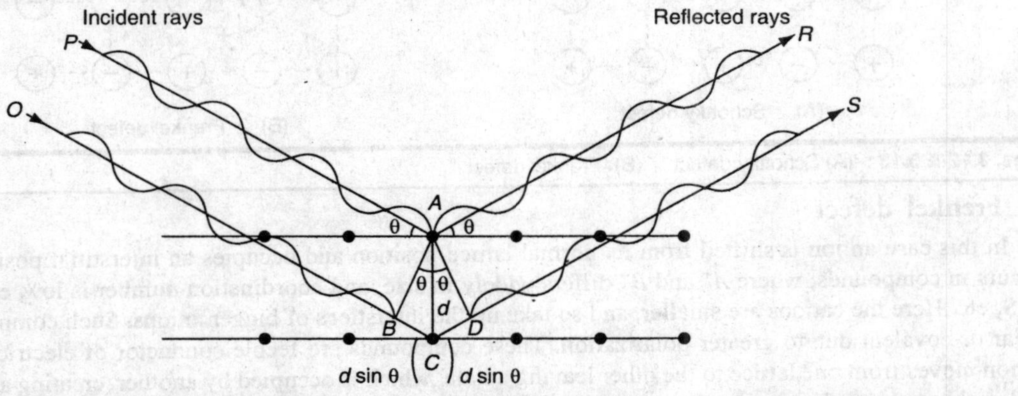

Fig. 3.11 : The incident and the reflected X-ray beams and the two parallel planes of atoms in a crystal

Bragg's law states $2d \sin\theta = n\lambda$, *n* is called the order of reflections; $n = 1$ is the first order, for $n = 2$ it is second order and so on.

In performing this experiment, the following assumptions are to be made:

1. The incident X-rays are reflected by parallel planes of atoms or ions and will follow the laws of reflection.
2. Part of incident X-ray is reflected by the planes.
3. When the reflected X-ray waves interfere, then diffraction pattern is obtained.
4. The reflections will be elastic reflection, i.e., the wavelength will not undergo any change.

CRYSTAL IMPERFECTIONS

Lattice Defects

In a real or ideal crystal of a compound, at absolute zero, every atom or ion is placed on an appropriate lattice point and each lattice point is occupied by right kind of ions or atoms. At temperature above this, the thermal vibration of ions or atoms causes the lattice defects that may be due to

(i) interchanging of ions or atoms from their lattice position,

(ii) the location of ions or atoms of any kind in interstitial positions or

(iii) the existence of some vacant sites from which any ion or atom is missing.

The first kind is uncommon for energetic reason, and observed only in metallic phases. Therefore, the defects in crystals are either stoichiometric defects or non-stoichiometric defects.

Stoichiometric defects

Here the ratio of cations and anions represents their chemical formula. Two types of defects may occur:

A. Schottky defect

Here the ratio of $\frac{A^+}{B^-}$ is same but same number of cations and anions are missing from the crystal lattice creating equal number of vacancies or holes. This type of defect is generally found in highly ionic compounds of high co-ordination number, where size of cation and anion are equal, e.g., NaCl, CsCl, etc.

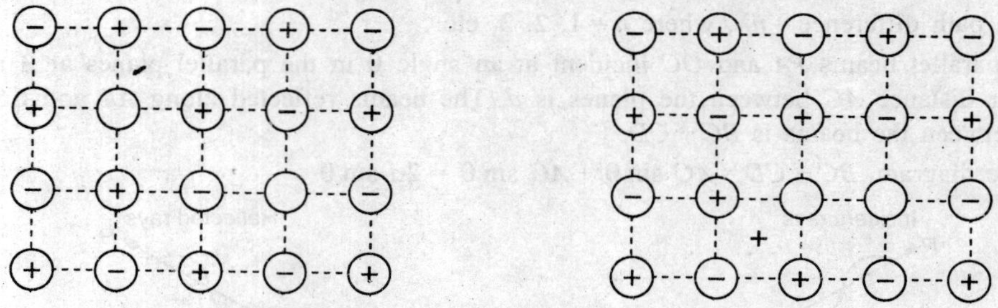

(A) Schottky defect (B) Frenkel defect

Figs. 3.12 & 3.13 : (A) Schottky defect (B) Frenkel defect.

B. Frenkel defect

In this case an ion is shifted from its normal lattice position and occupies an interstitial position. This occurs in compounds, where A^+ and B^- differs widely in size, and coordination number is low, e.g., AgBr, ZnS, etc. Here the cations are smaller, and so take up the interstices of bigger anions. Such compounds are polar or covalent due to greater polarization. These compounds are feeble conductor of electricity as the cation moves from one lattice to the other leaving a hole, which is occupied by another, creating a new hole and in this way conduction occurs.

Non-stoichiometric defects

Here the ratio of cation and anion does not follow the law of constant proportion, i.e., they are non-stoichiometric or Berthollide compounds. These can be represented by the formula $M_{1\pm x}X$, where M and X is generally bivalent or monovalent and M has variable valency, e.g., FeS, FeO, NiO, CuO, CuI, etc. and x is the fraction of vacant cation sites per formula weight that has been oxidized to higher valence state to maintain the electrical neutrality. Thus, FeS may be represented by $Fe^{+2}_{1-3x}\ Fe^{+3}_{2x}S^{-2}$. This metal deficiency defect can be diagrammatically represented as in Fig. 3.14.

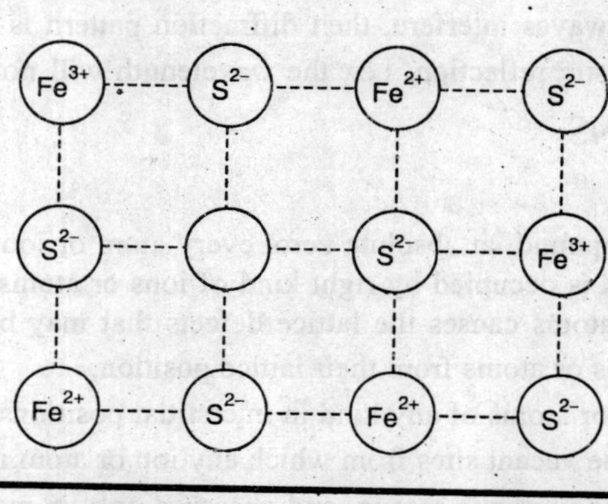

Fig. 3.14 : Defects in FeS.

Sometimes metal deficiency crystal defect also occurs when an extra anion occupies the interstitial position, and charge is balanced by oxidation of cation to higher valence state. Non-stoichiometric defect may also occur with metal excess. Anion is missing from lattice site and an electron occupies that position, e.g., NaCl with sodium vapour forming a yellow $Na_{1+x}Cl$ compound or KCl forms $K_{1+x}Cl$ (lilac colour) and electron occupies the vacant site, as in Fig. 3.15.

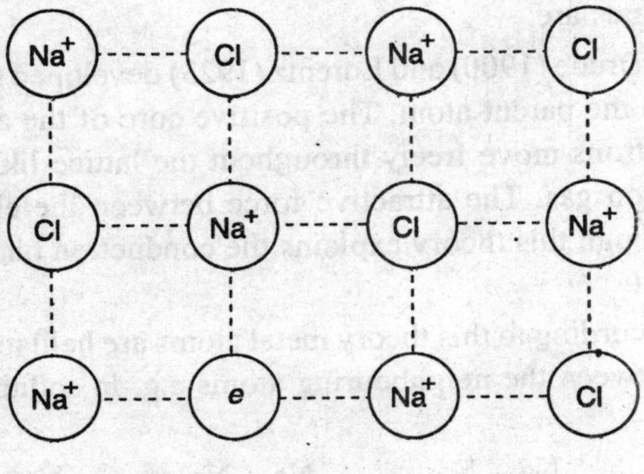

Fig. 3.15 : Defects in NaCl lattice showing "tapped ***e***"

ZnO lattice contains Zn^{+2} excess having formula $Zn_{1+x}O$. When ZnO is heated, it looses some oxygen, leaving vacant anion sites where electron occupies that sites. Now electrons are excited to higher energy level with absorption of energy and assume yellow colour. In small crystals when other metal atom replaces one metal atom, crystal defects are observed. Presence of these metal impurities imparts some special properties to the crystal. Thus Germanium metal containing some gallium (having less electron) leaves some electron holes and this gallium-doped germanium behaves as *p*-type semiconductor.

Similarly, arsenic (having more electron) doped germanium gives *n*-type semiconductors as negative charge flows through the crystal lattice. So $Fe_{1-x}O$ may behave as *p*-type semiconductor as electron transfer between Fe^{+2} and Fe^{+3} takes place and $Zn_{1+x}O$ behaves as *n*-type semiconductor as trapped electron flows through the crystal.

Thus, lattice defective compounds are very useful as semiconductors in transistors, as catalyst, in colour T.V., photography, etc.

METALLIC BOND

The metals have fewer electrons in their valence shell and show electropositive character. They have some characteristic properties as given below:

(a) The metals have high electrical conductivity–highest for univalent metals and decreases with higher valence metals, i.e., periodic function of atomic numbers. The conductivity decreases with temperature. The metals also show thermal conductivity.

(b) The metals can emit electrons when heated to certain temperature (thermoionic emission) or when radiation of short wavelength falls on it (photoelectric effect). The photoelectric emission is temperature independent up to some point, then thermoionic emission occurs. For this emission threshold energy is needed.

(c) Metals have metallic lustre, malleability and ductility. The metals can absorb light of all frequencies and immediately radiate it and hence the lustre. Malleability and ductility are due to the gliding of adjacent atomic planes over each other.

(d) All metals have high melting and boiling points and constant atomic heat. They are also hard indicating strong cohesive forces among atoms. Strong non-directional and non-rigid bonds hold atoms. Thus, the metals have close-pack structures with high coordination numbers.

There are some elements, which have borderline properties, i.e., they have some metallic character and some non-metallic characters, e.g., germanium, antimony, tellurium, etc. The metallic bond will be of such nature as to permit the transfer of free electron from one atom to other metal atoms and will thus hold 8-12 neighbours.

The theories of metallic bonds are:

1. *The electron gas model*: Drude (1900) and Lorentz (1923) developed the idea that the valence electrons of the metal atom leave the parent atom. The positive core of the atom enters into the joint crystal lattice and the free electrons move freely throughout the lattice like gas, thus the positive core are embedded in this electron gas. The attractive force between the electron gas and the positive ion stabilized the metal. Though this theory explains the conduction phenomena, but cannot explain the constancy in atomic heat.
2. *Valence bond theory*: According to this theory metal atoms are held together by electron pair bond and these bonds resonate between the neighbouring atoms e.g. in sodium metal the bonds are as in the figure:

Na – Na   Na – Na   ⟷   Na – Na   Na Na   ⟷ Na^+ Na – Na   Na

Na – Na   Na – Na   Na – Na | | Na Na^- – Na |

Na – Na   Na – Na   Na – Na   Na Na   Na   Na – Na   Na

This concept cannot explain the metallic properties clearly.

3. *Band theory*: Developed by Sommerfeld, Block and others. This approach is analogous to the molecular orbital theory where it is supposed that the *N* number of atomic orbitals of metal atoms combine to form *N* numbers of bonding and anti-bonding M.O.s. Bonding orbitals are called *energy bands*. The magnitude of the splitting depends on the internuclear distance and the nature of atomic orbitals. Greater the overlap of the A.O.s of neighbouring atoms, wider will be the energy bands and then the valence electrons of the atoms will occupy these bands in accordance with Pauli's Exclusion Principle. These bands are known as *Brillouin Zones*. These successive bands may have some forbidden zone or band gaps or may overlap to some extent. In some cases before the first Brillouin zone is filled, the electron flows to the second zone if the two bands overlap. This can be shown as follows:

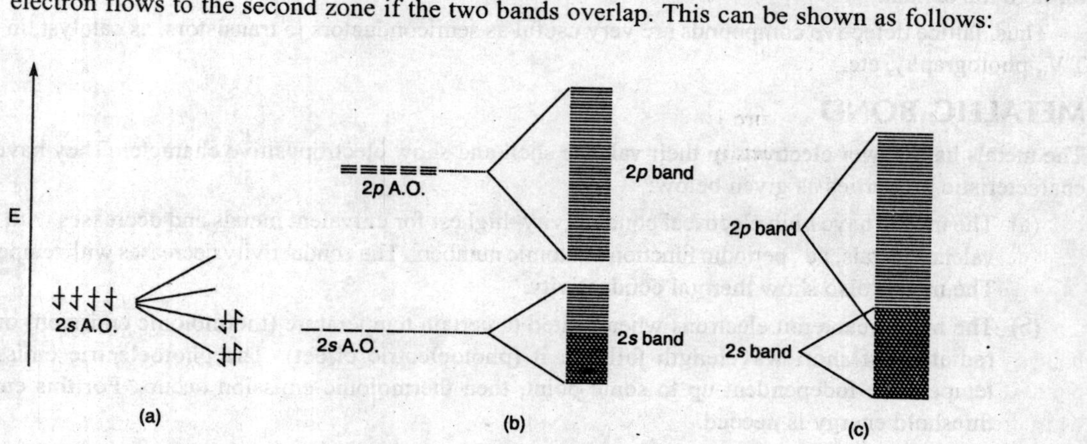

Fig. 3.16 : (a) Combination of four 2*s* A.O.s. (b) Combination of *n* atoms of Li. (c) Merger of 2*s* and 2*p* bands of Be.

Thus in case of Lithium, n atoms of Li will form energy bands; only $\frac{n}{2}$ are occupied by n electrons as regard to Pauli's Principle and others are vacant. In electric field, or when heated, these electrons will move to unfilled bands and indicate high conductivity. In beryllium with $2s^2$, $2p^0$, $2s$ bands are completely filled here, $2p$ band merges with $2s$ filled bands and under excitation electron from $2s$ will move to $2p$ bands and hence the metal shows conductivity. Here $2p$ bands are called conductor bands. In all cases of metallic conductors, the conductivity decreases with temperature.

SEMICONDUCTORS

In case of insulators, the energy gaps between the energy band / valence band and the conduction band is high and so transition of electron from the first to the second does not take place.But there are some elements where the energy gap between the valence band and the conduction band is low and some electronic movement from the first to the second may take place on heating, and shows conductivity to some extent. This property increases with temperature. These are called *semiconductors.* So, at sufficiently low temperature semiconductor may behave as an insulator.

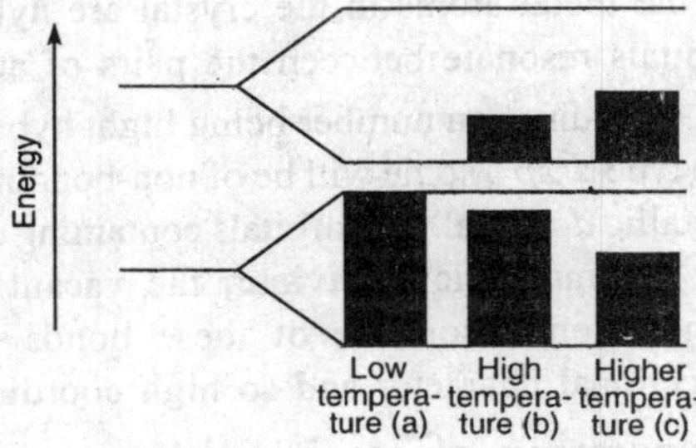

Fig. 3.17 : (a) At very low temp. (b) At high temp. (c) At very high temp.

If the energy gap between filled valence band and the empty conduction band is of the magnitude of the optical spectrum, then by absorption of light energy, some electron may move from the valence band to the conduction band causing conductivity on illumination, called *photo-conductivity*. This type of semiconductors is called *intrinsic semiconductors*. Pure germanium, silicon and grey tin are intrinsic semiconductors. For insulators, this energy gap ranges from 3.0 to 5.5eV, for semiconductors it ranges from 0.4 to 1.4eV. The specific conductivity of semiconductors is about 10^{10} to 10^{16} times higher than the insulators.

The conductivity of semiconductors can be increased by adding impurities (i.e., doping), like atoms of neighbouring group elements of the periodic group. This type of semi-conductivity is known as *extrinsic semiconductivity*. These types of semiconductors are of two types: (1) n-type semiconductors and (2) p-type semiconductors.

1. *n*-type Semiconductors

Si, Ge, etc. are semiconductors. If some As (arsenic) (having 5 valence electrons) is added ($1:10^9$ atoms ratio to Si) the As replacing the Si as host material has one electron in excess and is localized in a band called the donor band. This donor band is localized above the filled energy band and near the empty conduction band. If temperature is increased, some electron from the donor band will move to the empty conduction band and consequently some electron from the filled energy band will move to the donor band to fill the vacancy. As a result conduction continues and electron flow occurs throughout the crystal lattice of the host element. Since negative electron causes the conductivity it is called n-type semiconduction and the material as n-type semiconductor.

2. *p*-type Semiconductors

Here the doped material has one electron less (Ga, In) than the host semiconductor Si or Ge. The empty band of Ga, called the acceptor band (as it accepts electron from the host) is localized near the filled energy band of host Si or Ge crystal lattice. On heating, some electrons from the energy band move to acceptor band creating electron holes in the energy band, then electron from the neighbouring energy band moves to occupy the hole making another positive hole. So, as positive holes move through the crystal lattice of the host Si or Ge lattice causing conduction, this is called the p-type semiconductor.

Pauling's theory: In a metal lattice having high coordination number the neighbouring like atoms at equal distances linked to the central atom are not bonded by one electron or two electron bonds as the number of valence electrons are insufficient in number. Hence Pauling introduced Heitler-London's valence bond concept and exchange forces. According to this approach the bonding orbitals of the metal atoms in the crystal are hybridized and these hybrid orbitals resonate between the pairs of atoms.

Coordination number being high, hybridization that occurs is of the type d^2sp^3, some will be of non-bonding and some will be vacant metallic *d*-orbital. The orbitals containing one electron is responsible for paramagnetic behaviour, the vacant orbitals account for the conduction, resonance of these bonds gives stability and non-directional character and so high coordination number.

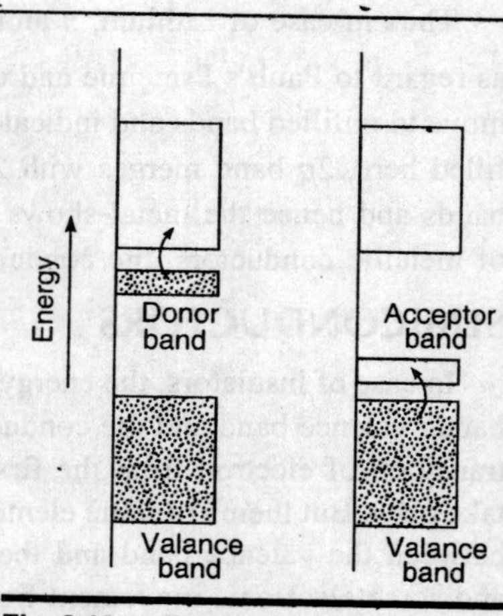

Fig. 3.18 : Band structure for doped extrinsic semiconductors (a) *n*-type, (b) *p*-type.

Preparation of semiconductors

As discussed in the preceding section, basic materials for the preparation of semiconductors of any type are extremely pure 99.9999% germanium or silicon. So, extreme precautions are taken to prepare such materials.

The following methods are available:

(a) ***Distillation***: Here, the starting materials are extremely pure $GeCl_4$ and $SiHCl_3$ for germanium and silicon. For germanium, $GeCl_4$ is taken (boiling point: 83.1°C) in a series of distillation stills, and a layer of HCl is taken over the charge. On heating, first impurity arsenic is removed. Then, Cl_2 gas is passed through the charge heated electrically when $GeCl_4$ distills off and is collected in a receiver kept in ice. Pure $GeCl_4$ thus obtained is then treated with extreme pure water when GeO_2 is obtained. This GeO_2 is reduced in an atmosphere of H_2 when element germanium [Ge] is obtained. This element Ge is subjected to further purification to get 99.9999% pure element.

(b) ***Zone refining***: This is the most successful and useful method. The basic principle of the method is that the impurities are more soluble in the melt than in the solid material. Hence, when impure molten metal is crystallized, the crystals are more pure. A systematic diagram of a vertical zone refiner is shown.

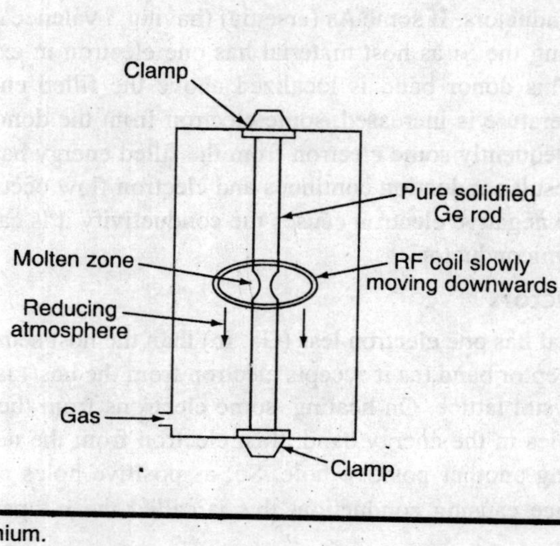

Fig. 3.19 : Zone refining of germanium.

A germanium rod already purified is clamped vertically and heated by R.F. coil at 1000°C in a reducing atmosphere. The R. F. heating coil is slowly moved from top to bottom when impurities are collected towards the molten bottom leaving highly pure germanium at the top. When purification is complete, the lower molten portion, where impurities are concentrated is cut off. The upper pure portion is again purified in the similar fashion. About 1 impure atoms to 10^{12} atoms of germanium can be obtained.

These two methods give the crystals of semi-conducting material of different sizes (polycrystalline). In practice, we require single crystal of the material. This is done in this way.

(c) *Czochralski crystal pulling technique*: Here, the polycrystalline semi-conducting material is taken in a Quartz chamber heated by R.F. coil when it melts. Then a pure single crystal attached to a puller rod is lowered just to touch the melt. Temperature is lowered to just above the melting point. The crystal seed is rotated slowly and pulled up slowly when melt in contact with the crystal seed solidifies producing the orientation identical with the seed crystal. The rotation speed is 100 rpm and pulled rate is 1.5-3 cm per minute. The rod is then cut to pieces of 0.25mm thick.

Doping

In this technique, calculated amount of dopant [about 1 atom of boron or phosphorous to 10^8 atoms of germanium or silicon] is added to the melt before crystal pulling stage. This is done (1) either by bombarding semiconductor material with electrically controlled beam of high-energy ions of dopant. This causes the implantation of dopant atoms in the crystal lattice of semiconductor or (2) by diffusion of dopant impurity in solid or gaseous state in the crystal lattice of the semiconductor without melting.

SEMICONDUCTOR DEVICES

Based on the electrical properties of semiconductors, numerous devices have been developed that have replaced the old type vacuum tubes. Unlike vacuum tubes, the semiconductor devices–called solid state devices –are small in size, low power consumption and no warm up times, e.g., transistors, rectifiers, photo voltaic cells etc.

1. *p-n* junction diode as rectifier

A rectifier or diode is an electrical device that allows the current to flow in one direction only, i.e., alternating current is transformed into direct current. Previously, this operation was carried out using vacuum tube diode.

Let us suppose, one thin *p*-type semiconductor is in contact with a thin *n*-type one. At this stage a poor rectifier results, no electrical potential is there, as holes will be the dominant in the *n*-region. When a battery is used with positive terminal to the *p*-side and negative terminal to the *n*-side, the holes on *p*-side and electrons in *n*-side are attracted to the junction, encounter one another, recombine and annihilate one another as stated below:

$$\text{Electron} + \text{Hole} \rightarrow \text{Energy.}$$

The electron crosses the junction to fill a hole and creates another new hole. In this way, current flows in one direction. If the battery terminals are connected in reverse way (reverse bias), the electrons and the positive holes are drawn away from the junction, recombination will not occur to an appreciable extent and the junction will be now highly insulating.

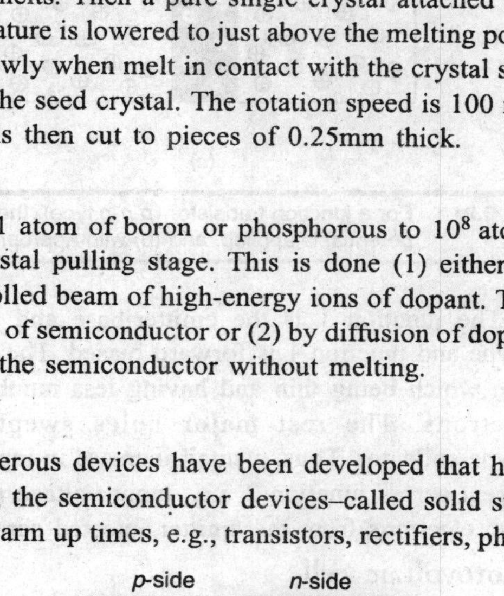

Fig. 3.20 : For a *p-n* rectifying junction, representations of electron and hole distributions for: (a) no electrical potential, (b) forward bias, and (c) reverse bias.

2. Junction transistors

The junction transistors are composed of either *n-p-n* or *p-n-p* type. We shall discuss the principle of *p-n-p* type here. They perform two types of functions (i) they can amplify the electrical signal like the vacuum tube (triode). They serve as switching device in computers for processing and storage of information. Here, as in Fig. 3.21, a very thin *n*-type semiconductor is sandwiched in between two *p*-type semiconductors; one is emitter and the other a collector region.

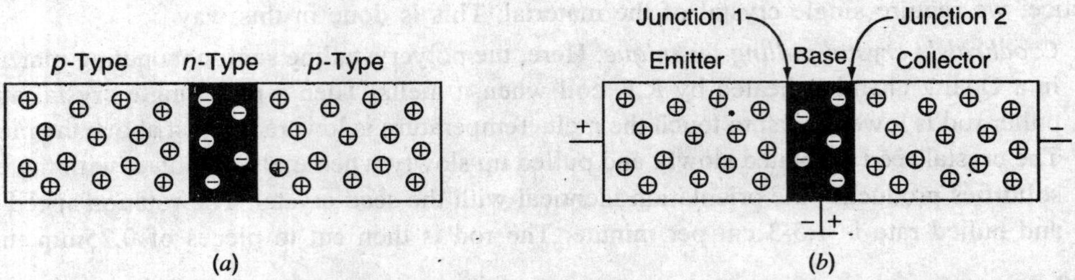

Fig. 3.21 : For a junction transistor (*p-n-p* type), the distributions and directions of electron and hole motion, (a) when no potential is applied, and (b) with appropriate bias for voltage amplification.

The junction 1 is the emitter-base and junction 2 is the collector-base junctions. The emitter is *p*-type and junction 1 is forward biased. The large number of holes through junction 1 enters the *n*-type base which being thin and having less number of electrons, some holes combine with the majority of electrons. The rest major holes swept the base and cross the junction 2 and into the *p*-type collector. Thus, a small increase in input voltage in emitter base circuit produces large increase in current across junction 2, i.e., input voltage is amplified. Similarly, if the junction transistor is of *n-p-n* type, electrons from the emitter injected across the base and holes are injected into the collector.

Photovoltaic cell

Semiconductors of *n-p* junction are used to make cells, i.e., photo voltaic cell to convert radiant solar energy to electrical energy. In a solar cell, a thin layer (3×10^{-6} m) *n*-type silicon doped with arsenic / phosphorous is in contact with a thick layer (4×10^{-4} m) *p*-type silicon doped with boron. The two are connected by an external circuit. When photons of sunlight strike the *n*-type silicon, the excited electrons get promoted to the conduction band, i.e., the layer now acts as an intrinsic semiconductor. These conduction electrons can easily cross the *n-p* junction; the potential difference between the two layers is created. As more and more solar light falls on the semiconductor, more and more electrons are set free. When the terminals of the semiconductors are connected to external circuit electric current flows. This type of device is called photo voltaic cell and an array of a number of such interconnected photo voltaic cell is called *solar battery*. Small solar units of 50-100 watts are used in telephone, television, relay power beacons, light in buoys in the harbour.

Micro miniaturized semiconductor devices

Computer chips, which can store several mega bytes of data on a single chip, are now a day's easily available in the market. These chips are equivalent to thousands of single crystal transistors junction. The steps of preparation of such micro-chips are as follows:

1. Initially an *n*-type semiconductor is prepared from single crystal of silicon, and then it is cut into species as thin slice.
2. Then surface layer of SiO_2 is made by heating the slice in air and the surface is coated with photosensitive film, called *photo resist*.

3. The photo resist is covered with a mask and the slice is exposed to UV light. The parts of the photo resist, which are exposed to light, are changed and removed by treatment with acid, the unexposed parts remain protected by photo resist.
4. To remove the exposed areas of SiO_2, the slice is treated with HF.
5. Then, surface is exposed to the vapour of Indium or any other elements of Group III. The areas covered by SiO_2 film are unaffected, but in the portions where the silicon itself is exposed, some silicon atoms are replaced forming a layer of *p*-type semiconductor.
6. Steps 2 and 4 are repeated using a different musk. The exposed area of silicon is then exposed to the vapour of a Group V element when a layer of *n*-type semiconductor is produced.
7. Steps 2 and 4 are repeated using musk to prepare the opening in which the metal can be deposited to "wire together" the various semiconductors so produced into an integrated circuit.
8. At last, the connecting pins are soldered on to chips so that it may plug into a socket circuit board. After testing, the faulty chips are rejected.

4

FUEL

INTRODUCTION

In recent years man's dependence on energy has increased tremendously mainly because of the increase in the standard of living and rapid technological advances. Today energy comes chiefly from the fossil fuels such as petroleum, coal and natural gas. Since these sources of energy will never last forever hence their proper utilization is the main concern of these days. In this chapter we shall discuss the basic understanding of the various fuels available to the mankind. This understanding is necessary also for their proper utilization and conservation for future use.

Fuel is a combustible substance, usually organic in nature, containing mainly carbon and hydrogen that on combustion in air liberates a large amount of heat together with the production of mainly carbon dioxide and water. This heat can be used commercially for domestic and industrial purpose. The heat energy liberated results from the chemical combination, i.e., rearrangement of valence electrons of the atoms of the combining elements or compounds. The fuel has higher heat content in it than that of the products, the difference in energy in them is liberated as heat. The primary or main sources of fuels in nature are coal (solid) and petroleum oil (liquid). These occur in the earth's crust and hence, called fossil fuels.

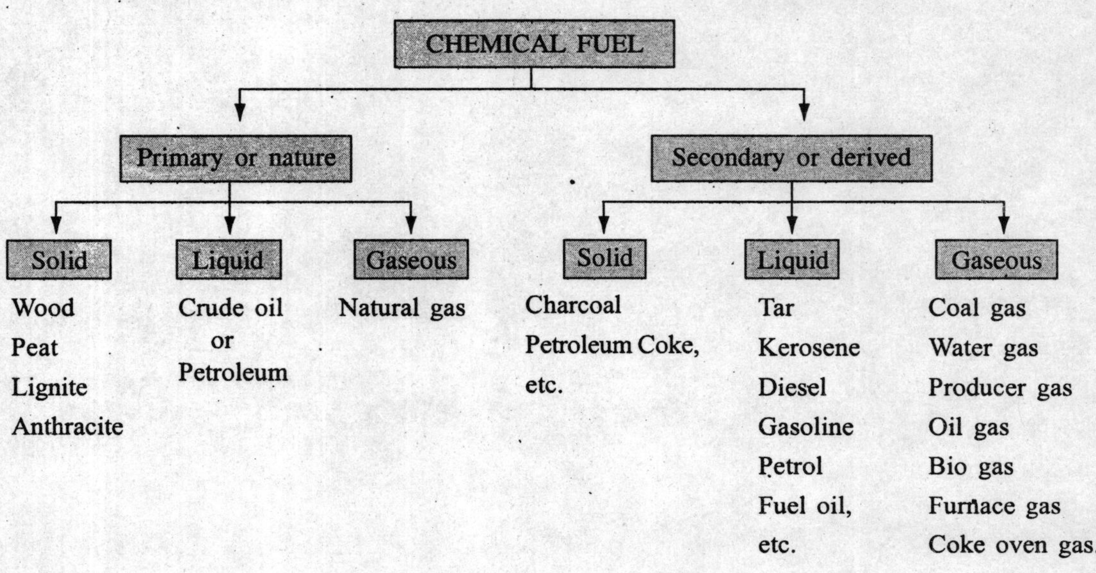

On the basis of availability and physical state, fuels are broadly classified into two types:

1. Primary or natural fuels. Fuels which are found in nature as such are called natural fuels, e.g., wood, coal, peat, petroleum and natural gas.
2. Secondary or derived fuels. Fuels, which are prepared artificially generally from primary fuels, are called secondary or derived fuels. Examples of this type of fuels are coke, charcoal, gasoline, petrol, kerosene, diesel oil, coal gas, producer gas, etc.

On the basis of physical state of aggregation, there are three types of fuels, viz., solid, liquid, and gaseous. Besides these, there is colloidal fuel which is the suspension of finely powdered coal in some liquid fuel generally in the ratio 3:2 by weight. To avoid separation of coal from the liquid fuel some stabilizing agent is added. Such fuels have high calorific value and easy to handle. This type of fuel is used in the furnace in the manufacture of cement, in ceramic firing, brick firing, etc.

3. In recent years nuclear fuels are known. Atomic nuclei undergo atomic fission and fusion liberating tremendous energy due to fragment of its mass loss obeying the formula, $E = mc^2$.

1 a.m.u. = 980 Mev.

CALORIFIC VALUE

Calorific value of a fuel is the total quantity of heat liberated from the combustion of a unit mass (in case of solid or liquid fuel) or volume (in case of gaseous fuel) of the fuel when burnt completely in air or oxygen. The different units of calorific values are:

1. British Thermal Unit (B.Th.U.)

It is the amount of heat required to raise the temperature of 1 pound of water through 1°F (60-61°F). It is the British system unit.

1 B.Th.U = 1 lb × 1°F = 453.6 g × 5/9°C = 252 cal = 0.252 kcal.

or, 1 kcal = 3.968 B.Th.U.

2. Calorie and kilo calorie

Calorie is the amount of heat required to raise the temperature of 1 g of water through 1°C (15°–16°C) and kilo calorie is the amount of heat required to raise the temperature of 1 kg of water through 1°C (15°–16°C). It is a bigger unit of calorie. 1000 cal = 1 kcal. 1000 g = 1 kg.

Centigrade heat unit (C.H.U.) is the quantity of heat required to raise the temperature of 1 pound of water through one degree centigrade. Thus 1 kcal = 3.968 B.Th.U. = 2.2 C.H.U.

3. Higher or gross calorific value(HCV)

Generally fuels contain hydrogen. During combustion hydrogen is converted into steam. If the net heat liberated is measured at room temperature, this steam is condensed to water giving out latent heat of condensation. This heat is included in the measurement and gives higher value. This is called higher or gross calorie value (HCV). Thus, HCV is the total heat produced on complete combustion of unit mass/volume of fuel and the products are cooled to room temperature 15°C or 59°F. But, if the products of combustion are not cooled to room temperature the gaseous products escape with hot combustion gases. So, less amount of heat is available. This is called lower or net calorific value (LCV). Thus, LCV is the amount of heat liberated on complete combustion of unit mass/volume of fuel and the products are permitted to escape.

LCV = HCV – latent heat of water vapour formed. Since 1 part by mass of hydrogen produces 9 parts by mass of water, hence LCV = HCV – mass of H_2 × 9 × latent heat of vaporization.

The latent heat of steam is 587 kcal/kg or 1060 B.Th.U./lb of water vapour formed at room temperature.

The units of calorific values are Calorie/gm or kcal/kg or B.Th.U/lb in case of solid or liquid fuel and in case of gaseous fuel it is kcal/m³ or B. Th.U./ft³

1 B.Th.U./ft³ = 9.3 kcal/m³   1kcal/m³ = 0.1077 B.Th.U./ft³   1 kcal/kg = 1.8 B.Th.U./lb

MEASUREMENT OF CALORIFIC VALUE

Bomb Calorimeter

It is used to measure the calorific value of solid and liquid fuels. It is a cylindrical stainless steel vessel called bomb, provided with a gas tight lid. Combustion of fuel takes place here. The lid is fitted with two stainless steel electrodes and an oxygen inlet. To one of the electrodes a stainless steel crucible fitted with a ring is attached. The bomb is placed in a copper calorimeter, which is surrounded by an air jacket and then a water jacket to prevent heat loss due to radiation. The calorimeter is provided with an electrically operated stirrer and a Beckmann thermometer capable of recording temperature up to 1/100th of a degree as shown in Fig.4.1(a).

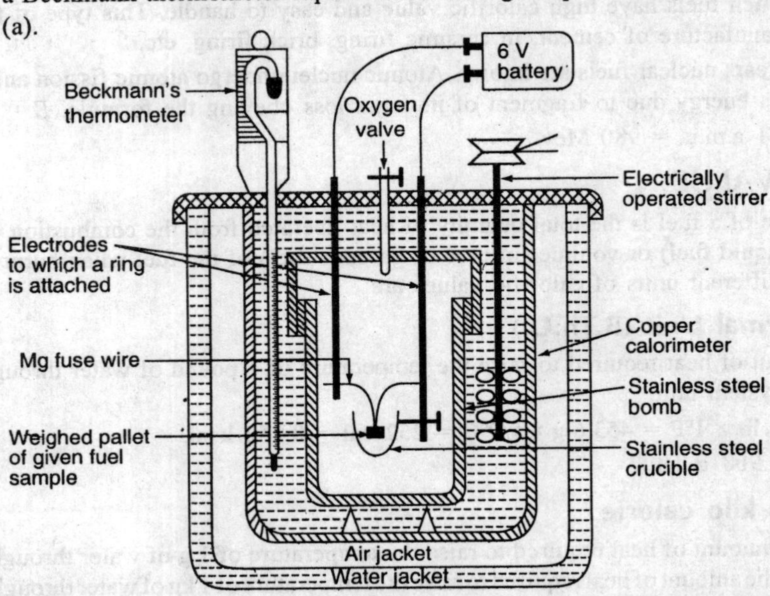

Fig. 4.1 : (a) Bomb calorimeter.

About 0.5 to 1.0 g fuel is taken in the crucible and then supported over the ring. A fine magnesium wire touches the fuel and then connects the electrodes. The bomb is closed with the lid and made air tight. Then, it is filled with oxygen under 25 atmospheric pressure. The bomb is then lowered into the calorimeter containing known amount of water. Initial temperature is noted. The electrodes are connected to the 6-volt battery. The sample burns, heat is liberated, the calorimeter is stirred continuously and final maximum temperature is noted.

Calculation: Let X g of fuel is used for combustion and L is its higher calorific value in cal/g. So that total heat = $X L$ cal.

Heat absorbed by water in the calorimeter, bomb, and thermometer will be equal to the heat given out. Let W g water was taken and the water equivalent of calorimeter, stirrer, bomb, etc. = w g, the initial and final temperatures are t_1 and t_2.

So, we can write, $XL = (W + w)\ (t_2 - t_1)$.

i.e., HCV of fuel, $L = (W + w)\ (t_2 - t_1) / X$ cal/g.

For accurate result, heat liberated by the fuse wire and heat liberated by the combustion of S, N to form H_2SO_4 and HNO_3 are included in this HCV. By measuring the amount of H_2SO_4 and HNO_3 formed, the correction factor for 1 mg of S, 2.25 cal and 1ml of HNO_3 (N/10), 1.43 cal are introduced. All are exothermic reactions.

Boy's Gas Calorimeter

The calorific value of gaseous fuel or easily vaporized liquid fuels is determined by **Boy's gas calorimeter.** It consists of a suitable burner that can burn a gas of known volume at a pressure flowing with uniform rate of 3-4 litres per minute. The chimney or combustion chamber surrounds the burner. The chimney is surrounded by copper tubes both inside and outside through which water at a constant rate flows. In the outer tube water enters from the top to the bottom and then in the inner one from the bottom to the top.

(b)

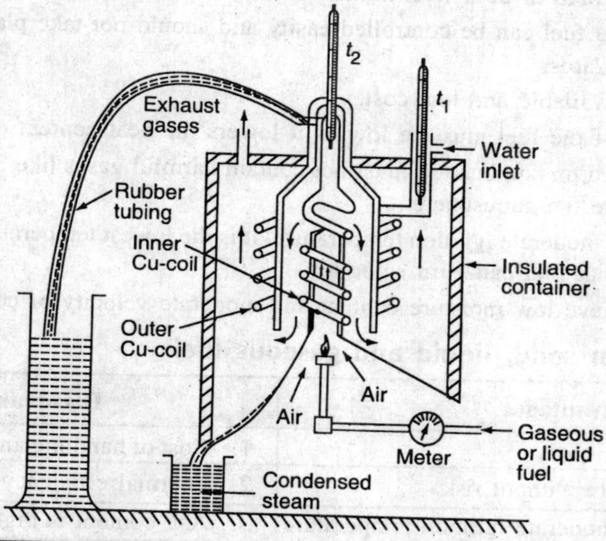

Fig. 4.1 : (b) Boy's gas calorimeter

In this type of arrangement all heat liberated by combustion is absorbed by water flowing through copper tube and the water formed during combustion is cooled and collected as shown in Fig. 4.1(b). The whole system is enclosed in an insulated container. The temperature of inlet and outlet water flowing through the Cu-tube is noted. The apparatus is adjusted such that after initial warming up of the apparatus, the rate of fuel burning, water circulation and the exit water leaves the apparatus nearly at atmospheric pressure.

Let the gas of V volume burns at STP in time t, and W mass of water flowing through Cu-coil, the temperatures of inlet and outlet water are T_1 and T_2, m is the mass of steam condensed after burning of gas during the same time t. L is the H.C.V. of fuel.

So, heat absorbed by water = $W(T_2 - T_1)$ and heat produced is VL.

Since $VL = W(T_2 - T_1)$,

or, $L = \frac{W}{V}(T_2 - T_1) = \text{H.C.V.}$

The water condensed per unit volume of gas is $\frac{m}{V}$ and latent heat per unit volume of gas is $\frac{m \times 578}{V}$ kcal/m^3

Now, L.C.V. = H.C.V – latent heat

$$= \left[\frac{W}{V}(T_2 - T_1) - \frac{m \times 578}{V}\right]$$

$$= \frac{1}{V}\left[W(T_2 - T_1) - m \times 578\right] \text{ kcal/m}^3$$

CHARACTERISTICS OF A GOOD FUEL

A good fuel should have the following characteristics:

1. It should have high calorific value so as to attain the desired temperature by the heat liberated at the time of combustion.
2. Easy to transport.
3. It should not produce smoke.
4. The fuel can be stored in bulk with minimum cost.
5. The burning of the fuel can be controlled easily and should not take place spontaneously, which may cause fire hazards.
6. It will be easily available and low cost.
7. The ash content of the fuel must be low as it lowers the heat content of the fuel.
8. The gases produced on combustion must not contain harmful gases like SO_2, CO, H_2S, PH_3, NO_2, etc. as they pollute the atmosphere.
9. The fuel will have moderate ignition temperature. It is the lowest temperature at which the fuel must be preheated so that fuel can burn smoothly.
10. The fuel should have low moisture content and moderate velocity of combustion.

Comparison between solid, liquid and gaseous fuels

Solid	Advantages	Disadvantages
1.	Easy to transport	1. Cost of handling and cost of labour are high
2.	Easy to store without risk	2. Thermal efficiency is low
3.	Possesses moderate ignition temperature	3. Ash content is high
4.	Low cost of production	4. Combustion cannot be controlled
		5. Require large excess of air for complete combustion
		6. Their applications are limited

Liquid	Advantages	Disadvantages
1.	High calorific value per unit mass or volume	1. Cost of the fuel is more high
2.	Burns without leaving any ash	2. Special costly storage tank required
3.	Combustion is easier and firing can be controlled	3. Greater risk of fire hazard
4.	labour cost and transport cost are low	4. It has bad odour
5.	Can be stored easily	5. Burner for combustion and spraying equipment are specially constructed
6.	Loss of heat is low	6. Sprayer may choke easily
7.	It requires less excess of air for combustion	
8.	Field of application is wider	
9.	Space for combustion is less	

Gaseous	Advantages	Disadvantages
1.	They can be conveyed easily by pipelines and cost of handling will be low	1. Large storage tanks are required
2.	The calorific value is highest as heat content is high and hence high temperature can be attained easily	2. Cost of production is high relative to others
3.	They can be kindled and extinguished quickly	3. Highly inflammable, so chances of fire hazards are high
4.	Preheating can be easily done by hot waste gases	
5.	They burn without soot	
6.	No special burner is needed for combustion	
7.	Impurity content is very low	
8.	They have wide applicability	
9.	Air supply for combustion is low	
10.	No choking of pipe line is occurred	
11.	No pollution effect	

Problem 1: A coal sample contains C = 60%, O = 33%, H = 6%, S = 0.5%, N = 0.3%, ash = 0.2%. Calculate the H.C.V and L.C.V. of coal. Latent heat of water vapour = 587 kcal/kg.

$$\text{H.C.V.} = \frac{1}{100}\left[8080\,C + 34500\left(H - \frac{O}{8}\right) + 2240\,S\right]$$

$$= \frac{1}{100}\left[8080 \times 60 + 34500\left(6 - \frac{33}{8}\right) + 2240 \times 0.5\right] = 5506 \text{ kcal/kg.}$$

$$\text{L. C. V.} = \text{H. C. V.} - \frac{9H}{100} \times 587 \text{ kcal/kg.}$$

$$= 5506 - \frac{9 \times 6}{100} \times 587 \text{ kcal/kg} = (5506 - 317) = 5189 \text{ kcal/kg.}$$

Problem-2: In Boy's gas calorimeter experiment, following data were obtained:

Vol. of gas used = 0.1 m^3 at S.T.P., mass of water heated from 20°C to 33°C is = 25 kg, mass of steam condensed = 0.025 kg. calculate H.C.V. and L.C.V. per m^3 at STP. Heat libarated in condensing water vapour and cooling the condensate to 580 kcal/kg.

$$\text{H. C. V.} = \frac{w(t_2 - t_1)}{V} = \frac{25(33-20)}{0.1} = 3250 \text{ kcal / m}^3$$

$$\text{L.C.V.} = \text{H.C.V.} - \frac{m}{V} \times 580 = 3250 - \frac{0.025 \times 580}{0.1} = 3250 - 145 = 3105 \text{ kcal/m}^3.$$

SOLID FUEL

1. WOOD: It is the easily available natural fuel obtained everywhere and used as fuel from ancient time. It has about 25-50 % moisture; when dried contains 15% moisture. Dry wood has C ≈ 52-55%, H_2 ≈ 6%,

$O_2 \approx$ 40-43% ash 1%. Its ignition temperature is low; calorific value is 3500-4500 kcal/kg or 8500-9200 B.Th.U./lb. After combustion the ash left has different salts which are used in various purposes.

2. COAL: Coal is the main solid fuel. It is a highly carbonaceous matter, formed as a result of change of vegetable matter, mainly plant under favourable conditions like high temperature and pressure under the earth when buried for crores and crores of years. During this period wood changes to solid coal together with some liquid and gaseous matters. The liquid goes deeper to the earth and is trapped there. This is the petroleum oil. The gaseous matters try to go up and are trapped under the earth. Hence, in coal mines and petroleum mines natural gas is found. Chemically coal is a complex chemical substance containing different percentage of carbon, hydrogen, oxygen, nitrogen, sulphur, etc., depending on the variety as given in Tables 4.1 and 4.2. It contains about more than 300 types of organic matters. According to the geologist two theories are proposed for the formation of coal:

(a) *In situ theory*: Dense forests of ancient times were buried under the earth due to earthquake or similar natural calamities and were changed to coal at higher temperature and pressure. Great purity of coal seams explains this theory.

(b) *Drifted or transported theory*: According to this theory, big trees, etc. are carried by rivers to big lakes or to some deep low lands. These huge piles of wood are then buried under ground. In absence of oxygen, bacterial decomposition, temperature and pressure change, the cellulose matter changes to coal liberating gases like carbon dioxide and methane. Great thickness of coal supports this theory.

The different stages of this transformation of wood to coal, called coalification, are

Wood ⟶ Peat ⟶ Lignite ⟶ Bituminous ⟶ Anthracite

The carbon content increases from left to right whereas moisture, volatile matter content decreases. This has been shown in Table 4.1 and Table 4.2 for air-dried fuels.

TABLE 4.1

Fuel	% of carbon	% of volatile matters	Calorific value kcal/kg	Main applications
Anthracite	92-98	2-8	8650-8700	Used in metallurgical purposes where no smoke and high local heat is desired. It is also used in household purposes and steam raising.
Bituminous	78-86 and 70-75	14-22 and 22-31	8000-8500 and 7000-7500	Used for steam generation in thermal plants and for domestic heating. It is also used for making coal gas and metallurgical coke.
Lignite	—	—	6500-7100	Used for steam generation in thermal power plants and for the production of producer gas.
Peat	—	55	4120-5400	Used when there is deficiency of high rank coals.
Wood	—	50	4000-4500	Domestic fuel.

The percentage of carbon, hydrogen, oxygen, sulphur, nitrogen in different grade of coal is given in Table 4.2.

TABLE 4.2

Grade	% Carbon	% Oxygen	% Hydrogen	% Nitrogen	% Sulphur
Anthracite	90-98	3	2-10	0.7	0.6
Bituminous	83-93	1-9	4-5	2-4	1-1.5
Lignite	70-73	22-25	4.5-5.5	0.5-1	0.5-0.8
Peat	55-60	30-35	6-6.5	1.5-2	0.8-1

Depending on the purpose of use of coal as fuel, the following points are to be considered.

1. Coal should have high calorific value so that sufficient heat can be obtained from it to attain the required temperature. It should have low moisture and ash contents. Composition of ash and its fusion temperature is most important in metallurgical and boiler purposes. Knowing the composition of ash, flux is selected in metallurgical purpose. If the fusion temperature of the ash is the working temperature, then ash fuses, forms lumps together with fuel, called *clinker* and hinders the air supply. This fused mass may stick to the tube of the boiler. Low melting ash may penetrate to the refractory lining of the boiler, induces non-uniform distribution of heat and hence reduces the life of the boiler.
2. Calorific intensity: It is the maximum temperature reached when coal is completely burnt. This depends on the quantity, nature of coal and the specific heat of the gaseous products formed.
3. If the coal on heating in absence of air, becomes soft, porous, hard and strong, suitable for metallurgical purpose, it is called coking coal. If it fuses to form large coherent mass, then it is called caking coal. Thus, all coking coals are caking coals, but not all caking coals are coking coals.
4. The sulphur and the phosphorous content of coal must be as low as possible as these elements affect badly the properties of the metals produced.

COAL ANALYSIS

Quality of coal is assessed depending on the analysis of coal that is done in two ways:

(A) Proximate analysis: Here percentage of moisture, volatile matter, ash and carbon is determined.

(a) **Moisture:** Moisture in coal lowers the effective calorific value of coal as it takes latent heat of evaporation from the fuel. Coal must not contain more than 10% moisture.

About 1g of air dried finely powdered coal is taken in a constant weighed crucible and placed inside an electrically hot air oven and heated at 105-110°C for about one hour and then cooled in a desiccator and weighed. Loss in weight is due to the amount of moisture.

$$\% \text{ of moisture} = \frac{\text{loss in weight}}{\text{weight of coal taken}} \times 100.$$

(b) **Volatile matter:** During burning of coal, volatile matter escapes away unburned, lowering the calorific value. The volatile matter burns with long flame and produces smoke. Volatile matter may be combustible (such as methane, hydrogen, carbon monoxide and other hydrocarbon) and non-combustible (such as carbon dioxide nitrogen, etc.) and these are not dissociable. Coal having high volatile matter, i.e., low rank coal (50% volatile matter) is suitable in coal gas manufacturing but do not cake well. Coal having medium quantity of volatile matter (26-30%) produces hard and strong coke on carbonization. Low volatile matter content coal do not cake at all and unsuitable for coke making.

The dried sample of carbon left after removing of moisture is covered with the lid and heated to 925 ± 20°C for seven minutes in a muffle furnace, cooled first in air, then in desiccator and then weighed. The second loss in weight of the sample is due to the volatile matter.

$$\% \text{ of volatile matter} = \frac{\text{second loss in weight}}{\text{weight of coal taken}} \times 100.$$

(c) **Ash:** It is the undesirable non-combustible constituent of coal. It causes lowering of calorific value, hindrance to the flow of air and heat. Fused ash, called clinkers, blocks the interspaces of grates on which coal is burnt causing obstruction of air supply and so, burning of coal becomes irregular. It causes wear of furnace wall, burning apparatus and feeding mechanism. Ash contains mainly oxides of inorganic compounds. The approximate compositions are

SiO_2	40-60%	MgO	0.5-4.0
Al_2O_3	20-35%	CaO	1-1.5%
Fe_2O_3	2-2.5%	$Na_2O + K_2O$	1-4%
SO_3	0.2-15%		

The residual coal after removing volatile matters is then heated without lid at (700 ± 20)°C for 30 min in the muffle furnace. The crucible is then cooled in air first and then in desiccator, and weighed finally after repeated heating and cooling until constant weight is obtained.

$$\% \text{ of ash} = \frac{\text{weight of ash left}}{\text{weight of coal taken}} \times 100.$$

(d) **Fixed carbon:**

% of carbon = 100 – % of (moisture + volatile matter + ash).

(B) Ultimate analysis of coal: It involves the determination of carbon, hydrogen, oxygen, nitrogen, sulphur and ash.

(a) **Carbon and hydrogen:** Percentage of carbon increases from lignite to anthracite. Higher the percentage of carbon and hydrogen better is the quality of coal.

About 1-2 g accurately weighed coal is burnt in a combustion chamber in a current of oxygen when CO_2 and H_2O produced are carried out and absorbed first in weighed fused $CaCl_2$ tube for H_2O and then into weighed KOH bulb for CO_2.

$$\underset{12}{C} + O_2 = \underset{44}{CO_2} \quad \text{and} \quad \underset{2}{H_2} + \tfrac{1}{2}O_2 = \underset{18}{H_2O}$$

$$2KOH + CO_2 = K_2CO_3 + H_2O \qquad CaCl_2 + 2H_2O = CaCl_2,\ 2H_2O$$

From the increase in weight of potash bulb and $CaCl_2$ tube, percentage of carbon and hydrogen is calculated.

$$\% \text{ of C} = \frac{\text{increase in weight of potash bulb} \times 12 \times 100}{\text{weight of coal taken} \times 44}$$

$$\% \text{ of } H_2 = \frac{\text{increased weight of } CaCl_2 \text{ tube} \times 2 \times 100}{\text{weight of coal taken} \times 18}.$$

(b) **Nitrogen:** It has no calorific value and hence undesirable. Hence, good coal should contain nitrogen as low as possible.

Nitrogen is determined in Kjeldehl's flask taking about 1g of coal using a chemical balance together with concentrated H_2SO_4 and K_2SO_4 as catalyst and then heating until the solution becomes clear. At this stage, nitrogen is converted to ammonium sulphate. The solution is then boiled with excess KOH solution and the liberated NH_3 is absorbed in known volume of standard acid solution. The unconsumed acid is then back titrated against standard NaOH solution. From the volume of acid consumed by NH_3 and the strength of acid, nitrogen is estimated.

$$\% \text{ of N} = \frac{\text{volume of acid consumed} \times \text{normality} \times 1{\cdot}4 \times 100}{\text{weight of coal taken}}$$

(c) **Sulphur:** Its presence in coal is undesirable. During combustion it produces acid fumes of SO_2 and SO_3 that corrodes the equipment and also pollutes the atmosphere. Again, if this coal is used in the production of coke used for the metallurgy of iron, this sulphur is transferred to iron and induces the bad effects in the quality of steel.

(d) It is determined from the washings obtained from the known weight of coal used in bomb calorimeter for the determination of calorific value of coal. In this operation all sulphur is converted to sulphuric acid. So the washings are treated with acidified barium chloride solution when H_2SO_4 is precipitated as $BaSO_4$. It is filtered, washed and heated to a constant weight.

From the weight of $BaSO_4$, sulphur is determined.

$$\% \text{ of S} = \frac{\text{weight of BaSO}_4 \times 32 \times 100}{\text{weight of coal taken in bomb calorimeter} \times 233}$$

The sulphur content is usually 0·5-1·5%.

Ash: Ash is determined as in proximate analysis.

Oxygen: It is determined by the defference.

% of O_2 = 100 – percentage of (C + H + S + N + ash).

BANDED COAL

Microscopic examination of coal shows the presence of organic matter called Macerals of different nature. Sometimes the markings or bands are viewed in naked eye. Depending on the types of these markings coal is classified in three categories. This is called classification on Petrographic composition.

(a) **Vitrain:** The coal contains the markings that are shiny and lustrous.

(b) **Fusain:** It contains the markings that are non-shiny black layers. This coal is brittle.

(c) **Attrital:** It contains both the markings of the first two types.

If the coal does not contain the above markings or bands then coal is called *non-banded* or *granular coal*. Vitrain is the purest form and has resulted from almost complete conversion of plant tissue into a jelly. It yields a very little amount of ash approximating 10%. Fusain resuls from the aerobic decomposition of vegetable matters, which emerges above the surface of water. Ash content is 30%.

PULVERIZED COAL

Solid coal has less surface area, so contact with oxygen is less and hence combustion rate is slow. If coal is used in finely powdered form (pulverized), its surface area increases and contact between coal and air is easier and hence combustion rate increases. The coal is pulverized either in ring roll mill or high-speed impact mill or ball mill. Coal containing volatile matter is used here as the volatile substance liberated quickly and burns and helps the burning of fixed carbon.

Pulverized coal has the following advantages:

1. Transportation is easy.
2. Rate of combustion can be controlled easily.
3. Heating can be started or stopped, at will and hence wastage can be controlled.
4. Low-grade coal with high ash content can be used.
5. Temperature rise is uniform.
6. Thermal efficiency is high.
7. Cost of handling is low.
8. Clinker is not formed.

But the disadvantages are powdered ash is carried out (75%) with the flue gas and deposited on the surrounding area causing nuisance. But using dust catcher this can be controlled.

If pulverized coal mixed with fuel oil is used, the potential heat content per unit volume exceeds that of oil alone. The specific gravity of the mixture is above 1.0 and hence, can be stored under water, and fire hazard is minimized. The coal oil mixture will be in suspension form either grinding the coal in fine form or by adding some stabilizer like petroleum soap, lime resin soap or alkali salt of fatty acids.

CARBONIZATION OF COAL

Carbonization is the process by which undesirable and volatile matters are removed from coal by heating in absence of air; as a result percentage of carbon increases. This is destructive distillation.

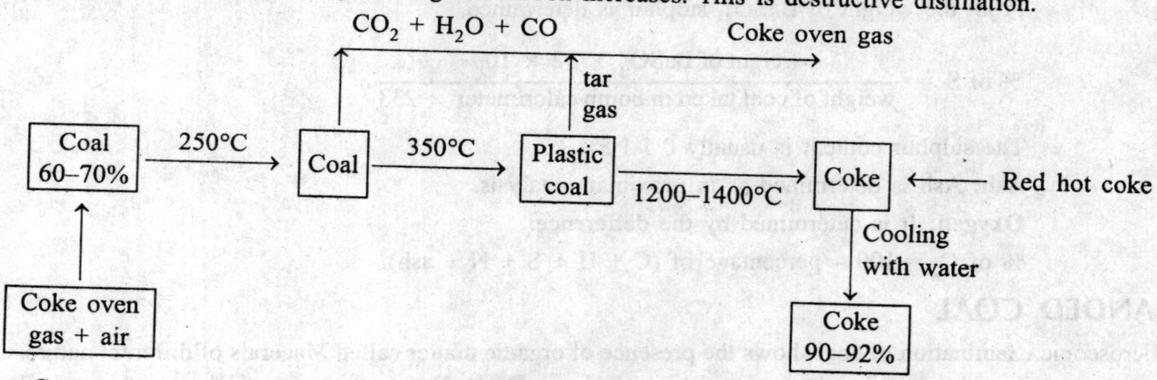

Conversion of coal to coke is done by two methods

1. High temperature carbonization using 900-1200°C and
2. Low temperature carbonization using 500-750°C.

In high temperature process carbonization yields coke having properties like

I) high porosity, strength, purity, etc
II) suitable for metallurgy process
III) carbon content is 85-90% and
IV) volatile matter is 1-3%.

The low temperature process yields coke having properties like

i) not mechanically strong, so not used in metallurgical purpose
ii) burns with smokeless hot and radiant fire and hence used for domestic purpose
iii) the gaseous fuel has high calorific value.

A comparative chart of high temperature and low temperature carbonization is given below:

TABLE 4.3

Low temperature carbonization	High temperature carbonization
1. Temperature used 500–750°C.	1. Temperature used 900–1200°C.
2. 75–80% coke is produced.	2. 65–70% coke is produced.
3. Good for domestic purpose.	3. Good for metallurgical purpose.
4. Not mechanically strong.	4. Hard and good mechanical strength.
5. Volatile matter is 5–10%.	5. Volatile matter is 1–3%.
6. Coal gas produced is 130–150 m^3/ton.	6. Coal gas is produced is 300–390 m^3/ton.
7. Calorific value of gas is 6500–9500 kcal/m^3.	7. Calorific value gas is 5400–6000 kcal/m^3.
8. Produces soft coke.	8. Produces hard coke.
9. 20–40 gallon/ton light oil and tar is produced.	9. 10–17 gallons/ton light oil and tar is produced.
10. 12-15 lbs/ton ammonium sulphate is produced.	10. 24–26 lbs/ton ammonium sulphate is produced.

High temperature carbonization

High temperature carbonization is done in a battery of ovens each of 25 ft high, 40-45 ft long and 10-18 inches wide rectangular chamber made of silica bricks and each having 3-4 openings at the lids through which pulverized coal is introduced. In between two ovens, annular space is there where hot oven gas ignites to heat the coal. The volatile matters escape through exit pipes, the coke produced is cooled by water and then collected. A sectional view of the coke oven is given below.

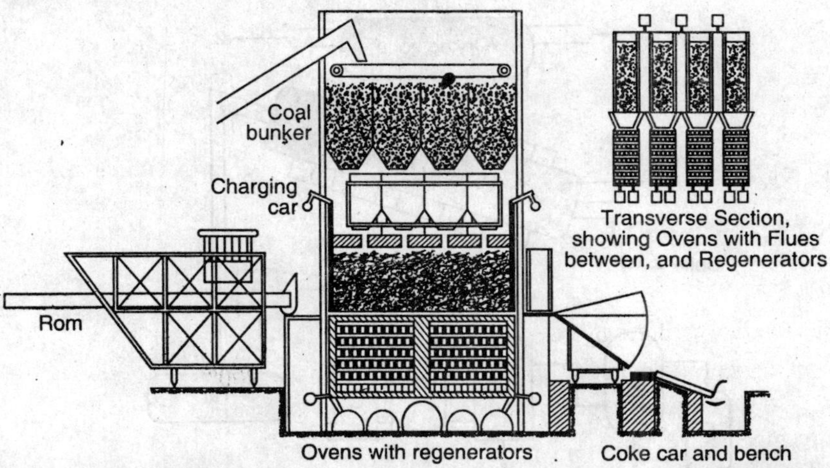

Fig. 4.2 : Modern Coke Oven.

High temperature process is used in the production of coal gas from low grade bituminous coal. Details will be given in coal gas preparation.

Low temperature carbonization

The object of the process is to produce smokeless free-burning soft coke and to produce tar suitable for the conversion into liquid fuel to use as motor spirit, etc. 30-40% volatile matter containg bituminous coal on low temperature carbonization yields the following:

(a) **Semi coke:** Yield is about 15lbs per ton of coal. It contains 10-15% volatile matter, hence burns easily without smoke.

(b) **Tar and oils:** Yield is about 15-20 gallons per ton. The constituents of tar are (i) 40-50% neutral oil, i.e., unsaturated hydrocarbon, naphthenes and paraffins but no benzene compounds. (ii) 20-30% tar acids, i.e. phenols and (iii) 1-2% bases. Tar oil on heating gives soft pitch having more free carbon.

(c) **Light oils:** Yield is about 3 gallons per ton of coal. It contains 50% saturated, 30% unsaturated hydrocarbons and 20% aromatic hydrocarbons.

(d) **Ammonia liquor:** Yield is low and can produce 8-10 lbs of ammonium sulphate.

(e) **Gas:** Yield is 3500-7000 cu ft per ton of coal. Calorific value is 600-800 B.Th.U./cu ft. The gas has the composition: $CO_2 = 1.3\%$, $CO = 1.9\%$, $CH_4 = 22.8\%$, $C_2H_4 = 2.5\%$, $H_2 = 56\%$, $N_2 = 6\%$ and $O_2 = 0.05\%$.

(f) The difficulty in carrying out the L.T.C. is that coal is a weak conductor of heat. Hence, external heating causes the outer layer carbonization at higher temperature. Duration of cooking time decreases with decrease in the width of the oven; e.g., cooking time required by 13″ wide oven is 13-15 hours whereas 20″ wide oven requires 28 hours.

The flow diagram of low temperature carbonization is given in Fig. 4.3.

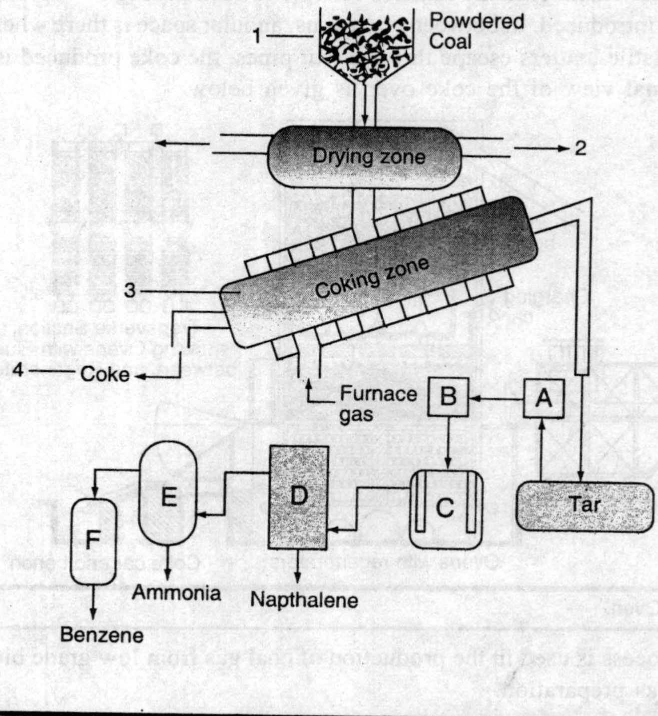

Fig. 4.3 : Flow diagram of low temperature carbonization.

1. Hooper, 2. Drying zone, and 3. Coking zone, 4. Coke cooling zone.

A, B – gas coolers, C – electrostatic precipitator, D – naphthalene recovery tower, E – $(NH_4)_2SO_4$ collecting zone, F – benzene collecting zone.

METALLURGICAL COKE

Coke used for metallurgical process should have the following qualities:

(a) **Purity:** The following impurities must be as low as possible:

(i) Moisture: Low moisture keeps the heating expenses less.

(ii) Ash will be low. 2% excess gangue is required for removal of 1% ash for slag formation.

(iii) P_4 and S should be low as they produce undesirable acid fumes that pollute the atmosphere and also affect the quality of metal.

(b) **Porosity:** Oxygen can come in contact with coal easily and burns completely

(c) **Strength:** Coke should be compact, hard and strong to withstand the presence of the ore, fuel and flux in the furnace

(d) **Size:** Big size coke causes non-uniform heating and small size will result in choking, hence moderate size is needed.

(e) **Combustibility:** Coke should burn easily. Generally, if cooking is carried out at 800-900°C, good quality coke is obtained. Calorific value should be high.

MANUFACTURE OF METALLURGICAL COKE

In modern type oven (Otto Hoffman by-product oven) is used to increase the thermal efficiency of process and to collect the valuable by-products like coal gas, tar, ammonia, benzol, oil, etc. The oven made of silica is 30-35″ long, 10-12″ high and 1-1.5″ wide erected side by side with vertical flues in between them to form a sort of battery. Each oven has a charging hole at the top and a refractory lined cast iron door for gas offtake and discharging coke. Finely crushed coal is introduced into the oven and heated to 1200°C by burning producer gas preheated by regenerative principle. Heating continued for about 11-18 hours. When evolution of volatile gases ceases, red hot coke is taken out and quenched by water spray (wet quenching) or by dry quenching where red hot coke is taken in a closed chamber and cooled by nitrogen. In dry quenching the coke becomes stronger, dense graphitized and non-reactive, cheaper drier, and contains less dust.

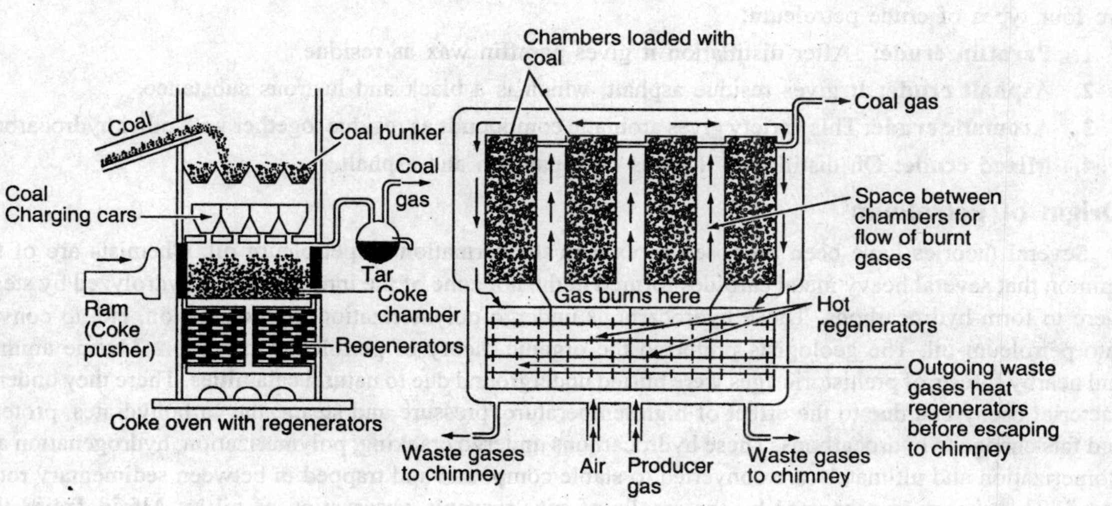

Fig. 4.4 : Otto Hoffman's by-product coke oven with regenerators

The volatile gas coming out as coke oven gas contains tar, ammonia, naphthalene, benzene and H_2S, and they are recovered.

JET FUEL

Jet fuel is a type of aviation fuel designed for use in jet-engine aircraft. It is a clear to straw colored fuel, based on either an unleaded paraffin oil (Jet A-1), or a naphtha-kerosene blend (Jet B). It is similar to diesel fuel, and can be used in either compression ignition engines or turbine engines. Jet fuel is a mixture of a large number of different hydrocarbons, possibly as many as a thousand or more. The range of their sizes (molecular weights or carbon numbers) is restricted by the requirements for the product, for example, freezing point or smoke point. Kerosene-type jet fuel (including Jet A and Jet A-1) has a carbon number distribution between about 8 and 16 carbon numbers; wide-cut or naphtha-type jet fuel (including Jet B), between about 5 and 15 carbon numbers

LIQUID FUELS

Liquid fuel is chiefly petroleum oil (*Petra* means rock and *oleum* means oil, so it is called *rock oil*). Small amount of liquid fuel is also obtained as shale oil, coal tar oil, and synthetic oil.

Shale oil: Shale oil occurs in a large variety of sedimentary rocks containing organic matters which when heated yields oil or is extracted with solvents to get petroleum. There are another variety of rocks, known as kerogen rocks, tarhanites, log heads, etc., which are known to have shale oil. The third variety of rock includes coals of the peat-anthracite series. The second class and the third class yield 50% and 15% of the weight respectively. Extraction of oil from kerogen and others is a laborious process.

Petroleum Oil

It is dark greenish-brown viscous oil found deep in earth's crust floating upon a layer of brine and a layer of gas on top of it. The main constituents of petroleum oil are hydrocarbon, straight chain paraffin, naphthalene, and olefins and aromatics and small amount of hydrocarbon containing oxygen, nitrogen and sulphur. Although chemical nature of petroleum varies with place to place where it is found, but ultimate composition is C = 79-87%, H = 11-14%, S = 0.1-4% and O + N = 0.1-0.5%. The reason for this consistency is that petroleum contains only eight or nine homologous series of compounds. Proportion of the individual compounds may vary slightly from place to place but ultimate result remains more or less constant. There are four types of crude petroleum:

1. **Paraffin crude:** After distillation it gives paraffin wax as residue.
2. **Asphalt crude:** It gives residue asphalt, which is a black and lustrous substance.
3. **Aromatic crude:** This variety gives aromatic compounds as residue together with other hydrocarbon.
4. **Mixed crude:** On distillation it gives both paraffin and asphalt.

Origin of Petroleum

Several theories have been proposed to explain the formation of petroleum oil. Chemists are of the opinion that several heavy metal carbides formed in the hot zone of the inner earth are hydrolyzed by steam there to form hydrocarbons. Those hydrocarbons undergo polymerization, hydrogenation, etc, to convert into petroleum oil. The geologists preferred the organic theory of petroleum formation. Marine animals and nearby forests of prehistoric ages were buried underground due to natural calamities. There they undergo bacterial decay and due to the effect of high temperature, pressure and steam, the carbohydrates, proteins and fats change to hydrocarbons. These hydrocarbons undergo cracking, polymerization, hydrogenation and isomerization and ultimately are converted to stable compound and trapped in between sedimentary rocks as oil. This theory is supported by the results of spectroscopic observation of oil by Alfred Treibs that petroleum oil contains chlorophyll and porphyrins (haemoglobin base).

Refining of crude oil: Crude oil cannot be marketed as such. Before marketing, it is treated under specific process to have desired product. The process is called refining of crude oil and the plant for it is called refineries. The crude oil, which is an emulsion of oil and salt water, is allowed to pass through highly charged electrodes (Cottrell's process) when water droplets coagulate to form large drops and separate out. Then oil is treated with CuO, when sulphur compounds in the oil become sulphur-free forming CuS black solid which is removed by filtration. The crude oil is then heated in a retort at 400°C, when volatile matter escapes leaving asphalt or coke. The gaseous product is then passed through a fractionating column. The column is a tall cylindrical tower containing a number of stainless steel shelves with small chimney as in Fig 4.5.

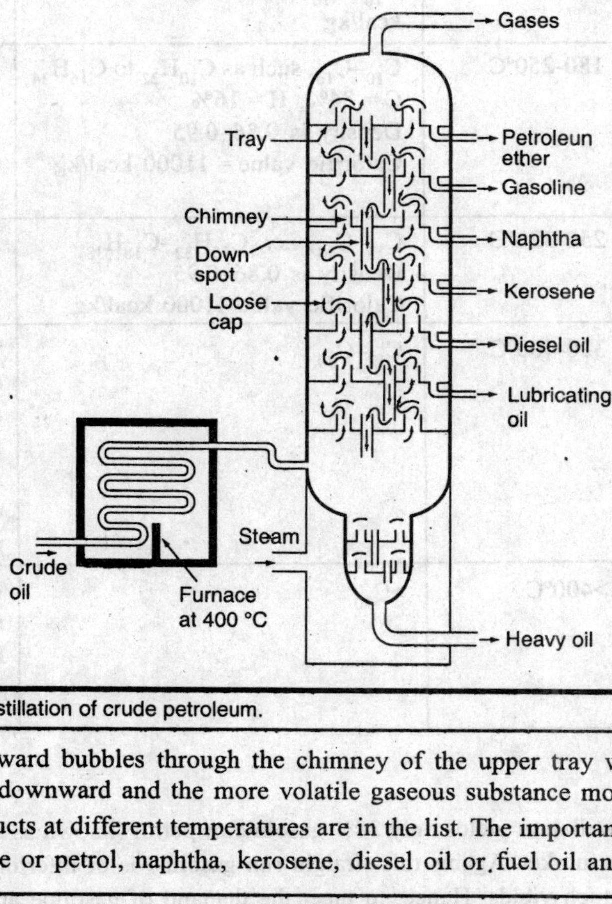

Fig. 4.5 : Fractional distillation of crude petroleum.

The gas going upward bubbles through the chimney of the upper tray where higher boiling liquid condenses and flows downward and the more volatile gaseous substance moves upward.

The different products at different temperatures are in the list. The important liquid fuels obtained from petroleum are gasoline or petrol, naphtha, kerosene, diesel oil or fuel oil and heavy oil.

TABLE 4.4

Name of fraction	Boiling range	Approx. composition in terms of number of carbon atoms	Uses
1. Uncondensed gas	Below 30 °C	C_1–C_4, C_2H_6, C_3H_6, iso-C_4H_{10}	L.P.G. as domestic and industrial fuel
2. Petroleum ether	30-70 °C	C_5–C_7	As a solvent

contd.

Name of fraction	Boiling range	Approx. composition in terms of number of carbon atoms	Uses
3. Gasoline or petrol or motor spirit	40-120°C	C_5–C_9, C_5H_{12}–C_8H_{18} C = 84%, H = 15%; N + S + O = 1%, Calorific value is 11250 kcal/kg	Used as fuel for internal combustion engines of automobiles and aeroplanes
4. Solvent Naphtha or solvent spirit	120-180°C	C_{10}–C_{16} calorific value = 11000 kcal/kg	As solvent and in dry cleaning
5. Kerosene oil	180-250°C	C_{10}–C_{16}, such as $C_{10}H_{22}$ to $C_{16}H_{34}$ C = 84%, H = 16% Density is 0.86–0.95 Calorific value = 11000 kcal/kg	Used as domestic fuel and jet engine fuel, as illuminant and for preparing laboratory gas.
6. Diesel oil or fuel oil	250-320°C	C_{15}–C_{18}, i.e., $C_{15}H_{32}$–$C_{18}H_{38}$ Density is 0.86–0.95 Calorific value 11000 kcal/kg	Used as diesel engine fuel
7. Heavy oil from which lubricating oil, petroleum jelly, grease, and paraffin wax are obtained	320-400°C	C_{17}–C_{30}	Used to get gasoline after cracking. These are used as lubricant, in cosmetics and in medicines, Wax is used as candle, boot polish, etc.
8. Solid residue is asphalt petroleum coke.	>400°C	>C_{30}	Used as waterproof for roofs, fuel, arc light rods, etc.

Cracking

Crude oil on fractional distillation yields only 20% gasoline or petrol (known as straight run gasoline) that has much demand in the market. Agains this straight run gasoline is of inferior quality. The middle and heavy fractions are of much yields. Hence, to meet the demand of gasoline and to have the better quality higher molecular weight hydrocarbons are decomposed to lower molecular weight or lighter ones of low boiling points. This process of decomposition is called cracking; e.g.,

$$\underset{\text{Boiling point: } 174^\circ C}{C_{10}H_{22}} \xrightarrow{\text{cracking}} \underset{\text{Boiling point: both } 36^\circ C}{C_5H_{12} + C_5H_{10}}$$

Two processes carry out cracking:

(a) Thermal cracking and (b) catalytic cracking. They are discussed below.

(a) **Thermal cracking:** Here high molecular weight hydrocarbons from petroleum, usually $C_{10}H_{22}$ fraction, are heated to high temperature when they break down to lower ones forming gasoline. There are two processes to carry out the thermal cracking.

(i) **Liquid phase cracking:** Heavy oil fraction is run into the hot pipe of the cracker at 470-500°C at a pressure of 100 kg/cm^2, contact time 0.1-0.5 seconds. Sometimes steam is mixed with heavy oil to maintain pressure. The product is then separated by fractionation. The yield is about 50-60% of octane number 65-70.

(ii) **Vapour phase cracking:** Here low boiling oil is vaporized at 600-650°C and passed through the cracker at a pressure of 10-20 kg/cm^2, contact time is still less. The gasoline produced is of better quality and better anti-knock property. Here cracking proceeds via free radical mechanism as proposed are given below.

1. Initial stage $C_{10}H_{22} \rightarrow C_8\dot{H}_{17} + C_2\dot{H}_5$

2. Propagation stage $C_2\dot{H}_5 + C_{10}H_{22} \rightarrow C_2H_6 + C_{10}\dot{H}_{21}$

Next, cleavage of bonds. $CH_3\,(CH_2)_6\,CH_2CH_2\,\dot{C}H_2 \rightarrow CH_3\,(CH_2)_6\dot{C}H_2 + C_2H_4$

$$CH_3\,(CH_2)_6\dot{C}H_2 \rightarrow CH_3\,(CH_2)_4\dot{C}H_2 + C_2H_4$$

$$CH_3\,(CH_2)_4\dot{C}H_2 \rightarrow CH_3\,(CH_2)_2\dot{C}H_2 + C_2H_4$$

$$CH_3\text{-}(CH_2)_2\dot{C}H_2 \rightarrow CH_3\,\dot{C}H_2 + C_2H_4$$

3. Termination stage $CH_3\dot{C}H_2 \rightarrow C_2H_4 + \dot{H}$

$$2\dot{H} \rightarrow H_2$$

$$2\dot{C}H_3 \rightarrow C_2H_6$$

$$\dot{C}H_3 + C_2\dot{H}_5 \rightarrow C_3H_8$$

(b) **Catalytic cracking:** Selective catalyst under suitable condition can decompose high molecular weight hydrocarbons to lighter one. Two types of catalytic methods are used.

(I) *Fixed bed catalytic cracking*:, Catalyst used in this process is either granulated aluminium silicate or artificial clay mixed with zirconium oxide heated to 420-450°C. Pressure used here is 1.5kg/cm^2. When oil vapour passes through the cracker, about 40-45% gasoline and 2-4% carbon are formed. Carbon gets absorbed in the catalyst and deactivates it. After 8-10 hr, regeneration is done by burning off the deposited carbon. The vapours produced are passed through fractionating column and pure gasoline is collected.

(II) *Moving bed catalyst*: Here the oil vapour is mixed with finely powdered clay and alumina-silica mixture (that behaves like a fluid) is forced into cracker and then into cyclone separator where the gaseous product passes out and leads to the fractionating column and the separator retains the catalyst powder.

Catalytic cracking has the following advantages:

(i) Here, naphthenic compounds crack preferentially, hence, the nature of the gasoline is mere paraffinic.

(ii) Gasoline obtained is better and yield is high.

(iii) Much lower pressure is used, hence, installation cost is low.

(iv) The process can be controlled easily to get the desired products.

(v) It contains aromatics, hence, better antiknock characteristics.

SYNTHETIC PETROL

The gaseous product obtained from cracking of heavy oil, after separation of gasoline, mainly contains ethylene, propylene, butylene and alkanes like methane, ethane propane, etc. These gases under high temperature and pressure undergo polymerization to form higher hydrocarbons resembling gasoline called polymer gasoline or synthetic petrol.

$$\underset{\text{Propene}}{CH_3-CH=CH_2} + \underset{\text{butane-1-ene}}{CH_3-CH_2-CH=CH_2} \longrightarrow \underset{\text{5-methyl hexane-1-ene}}{CH_2=CH-CH_2-CH_2-CH\ (CH_3)_2}$$

KNOCKING

Gasoline obtained from the fractional distillation of crude petroleum is not always suitable for the internal combustion engine. Gasoline for this purpose must be such that burning would be smooth and the flame must spread rapidly. It has been found that if the gasoline contains normal straight chain hydrocarbon, its combustion becomes irregular and ignition becomes very rapid. The rate of oxidation becomes so great that the last portion of fuel-air mixture ignited instantaneously producing an explosive violence to the engine known as KNOCKING. This causes the loss of efficiency of the engine. For example, if the gasoline contains *n*-heptane, it is the most inferior fuel and knocking is high or antiknock value is zero. If it contains 2:2:4-trimethyl pentane, i.e., isooctane, its knocking property is very little, i.e., antiknock value is very high, assumed to be 100. **Octane number** or octane rating grades quality of gasoline. Thus, octane number of fuel of an internal combustion engine is defined as the percentage by volume of isooctane in a mixture of isooctane and *n*-heptane, which matches with the knocking characteristics of the fuel under test. Say, a fuel has "Octane 80" means the fuel has the combustion characteristics that matches with the characteristics of the fuel containing mixture of 80 volume isooctane and 20 volume *n*-heptane. The knocking tendency of hydrocarbons increases of the order *n*-paraffin > branched chain paraffins > olefins > cyclo paraffins > aromatics. The reasons of knocking tendency of normal hydrocarbon and branched chain compound can be understood in this way. The normal chain compounds have greater surface area, and intramolecular atomic interactions are less, whereas branched compounds have less surface area, and intramolecular atomic interactions are higher. So, surface energy of the branched chain compounds are higher and hence, rupture of bonds are easier and regular, and knocking tendency is less.

The octane number of a fuel is determined by combusting the fuel under test in an internal combustion engine at 600 rpm at 52°C.The value thus obtained is known as Research Octane Number (RON). Then the fuel is burnt in the same engine at 900 rpm at 150°C. The value is known as Motor Octane Number (MON). The average of the two is a measure of Antiknock Index.

Improvement of antiknock property

The antiknock property of fuel can be improved by addition of some chemical compounds – mostly used lead tetraethyl, TEL. For motor fuel 0.5ml TEL and for aviation fuel 1-1.5 ml TEL per litre of petrol is added. For motor fuel, with TEL, some amount of 1, 2-dibromo ethane or 1, 2- dichloro ethane is added. Otherwise, deposition of finely divided PbO particle occurs that is harmful for engine. Addition of the chloro and bromo compounds forms volatile $PbBr_2$ or $PbCl_2$ and this escapes with exhaust gas. Diethyl telluride is also used for the purpose.

Alternate method is to add isopentene, isooctane, ethylbenzene, isopropylbenzene, methyl tertiary butyl ether (MTBE) to increase the octane rating. MTBE is more suitable as the oxygen in ether group helps better combustion of the fuel. Thus, unleaded petrol has high octane rating without addition of lead compound. For use of this unleaded fuel the engine is provided with a catalytic converter containing Rhodium metal that converts toxic gasses CO and NO to harmless CO_2 and N_2 and also converts unburnt hydrocarbon to CO_2 and H_2O.

LPG

The first fraction of the distillation product of petroleum is the LPG. It is also obtained from the cracking of heavy oil and from natural gas. Main constituents of Liquefied Petroleum Gas, LPG are n-butane, isobutane, butylene and propane. Under ordinary pressure, they are gas but can be liquefied and bottled easily under pressure. Trade name of LPG is Indane, HP, etc. The calorific value is around 27800 kcal/m^3, which are 3 times that of natural gas and 7 times that of coal gas. The use of this gas is safe, as it does not contain carbon monoxide. Some amount of organic sulphides (mercaptans) is mixed with the gas to recognize the gas leak. Fire hazards are as great as any other gaseous fuel. Its use as motor fuel is advantageous over gasoline as it is cheaper, highly knock resistant and also load sensitivity is high.

KEROSENE OIL

It is fractionation product of petroleum between 175-280°C, the fraction contains C_{11}–C_{14} hydrocarbons. Its flash point is above 24°C and hence safe to use for domestic purpose as stove oil and as an illuminant. In excess air it burns with smokeless blue flame, otherwise soot is formed. It has characteristic smell. The constituent compounds are undecane $C_{11}H_{24}$, dodecane $C_{12}H_{26}$, tridecane $C_{13}H_{28}$ and tetradecane $C_{14}H_{30}$. Generally, it also contains sulphur compound like thiophene that must be removed before using as jet engine fuel. This is done by washing it with alkali or by the action of H_2SO_4 and $(NH_4)_2S_x$.

DIESEL OIL OR GAS OIL

It is fractionation product of petroleum ranging 300-400°C, mainly containing C_{12}–C_{18} hydrocarbons. It must not contain side chain hydrocarbons and aromatics. It is more viscous than kerosene oil but less than lubricating oil. Its calorific value is 11000 kcal/m^3. Diesel oil is exploded in the diesel engine by application of heat and pressure. Ignition quality of the hydrocarbon in diesel oil is of the order *n*-alkanes > naphthalene > alkenes > branched alkenes > aromatics. The quality of diesel oil or fuel efficiency is expressed in terms of **"Cetane number"**. Cetane is the *n*-hexadecane, $C_{16}H_{34}$. It has good ignition value taken as 100 and 1-methyl naphthalene has least ignition value taken as zero. So, cetane number of diesel oil is the percentage by volume of cetane in a mixture of cetane and 1-methyl naphthalene that has the same ignition characteristics as the diesel oil in question. Thus, if the cetane number of diesel oil is 60, then the diesel oil in question has the same ignition characteristic as the mixture of 60% by volume of cetane and 40% 1-methyl naphthalene. Depending on the cetane number, diesel oil is high speed (cetane number is higher than 45 and maximum 60. It requires 1/500th second to heat the diesel drop to ignition point), low speed (cetane number is 25) and medium speed (cetane number is 35).

POWER ALCOHOL

Absolute ethyl alcohol 25% mixed with petrol is used as fuel in internal combustion engine. This mixture is called power alcohol. Ethyl alcohol has octane number 90, petrol has 60-70; so, octane number of the blend increases. It also absorbs some trace of moisture in petrol and increases the efficiency of the fuel but decreases the net calorific value. Alcohol is easily oxidized to acid and cause corrosion to the engine. Hence, its use is limited in internal combustion engine.

NATURAL GAS

It is the only gaseous fuel collected from nature and found on top of coal or petroleum oil mine. Other gaseous fuels are derived from solid or liquid fuels. The composition and calorific value varies widely.

The main constituents are CH_4 and C_2H_6, which is more than 90%. Sometimes natural gas contains some low boiling hydrocarbons like propane, propene, butene, and butane. This is called **wet gas". These higher hydrocarbons are removed and used as LPG. Sometimes, natural gas is admixed with crude oil, and then it is called **dry gas**. Before use, water, dust, H_2S, CO_2, N_2 and liquid hydrocarbons are removed from it, Its calorific value is 12000-14000 kcal/m^3. It is an excellent domestic fuel; it is applied for the manufacture of a number of chemicals, for producing carbon black (filler for rubber) and hydrogen.

It is liquefied under pressure and stored in bottle and sold under the name **Compressed Natural Gas**, (CNG). It has less pollution effect on ignition, as it does not liberate N_2, S, etc. and hence, better fuel relative to petrol and diesel oil.

COAL GAS

Coal gas has calorific value 4900 kcal/m^3. It is used as illuminant in cities, as domestic fuel, etc. Its average composition can be given as

(i) *Non-illuminating but heat producing substances*: $H_2 = 45\text{-}50\%$, $CH_4 = 30\text{-}50\%$, $CO = 5\text{-}10\%$.

(ii) *Illuminants*: $C_2H_4 = 3\%$, $C_2H_2 = 2\%$, and trace amount of C_6H_6.

(iii) *Diluents*: $N_2 = 2\text{-}10\%$, $CO_2 = 0\text{-}2\%$, $O_2 = 1\%$

(iv) *Impurities*: H_2S, HCN, CS_2, NH_3, etc.

The composition varies depending on the type of coal used and temperature of combustion of coal.

Coal gas is manufactured in higher yield by heating coal in retorts in absence of air (destructive distillation) at 1300°C (high temperature carbonization). At lower yield it can also be obtained by heating at 600-650°C. Besides coal gas, other products like tar, ammoniacal liquor, coke and gas carbon are obtained. High temperature process yields coal gas of higher yield and hard coke suitable for metallurgical purpose. The gas is manufactured in the way as given in Fig. 4.6.

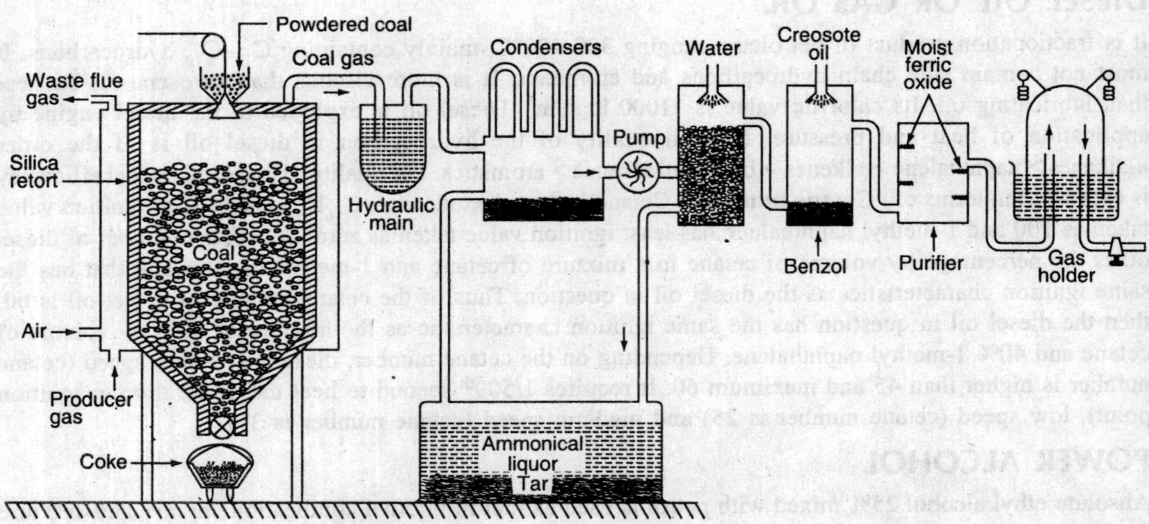

Fig. 4.6 : Manufacture of coal gas.

The retort used are either horizontal, vertical or inclined type, 18′-25′ long, 18″-25″ wide and 13″-16″ deep, made of fire clay or silica to withstand high temperature. The gas coming out is first passed through hydraulic main and then through condenser where most of tar and ammonia condenses and separates in two layers. The gas is then scrubbed with water in scrubber where any remaining tar and ammonia separates out. Then it is scrubbed with creosote oil to remove benzol, naphthalene, etc. The gas is then passed through a tower containing moist ferric oxide Fe_2O_3, xH_2O where H_2S is removed.

$Fe_2O_3, xH_2O + 3H_2S \rightarrow Fe_2S_3 + (x+3)\ H_2O$

Sulphur is separated from this spent oxide by passing air through it.

$Fe_2S_3 + (x + 3)\ H_2O + \frac{3}{2}O_2 \rightarrow Fe_2O_3, x\ H_2O + 3H_2O + 3S$. Thus, purified gas is stored in gasholder.

PRODUCER GAS OR AIR GAS

It is mainly a mixture of combustible gases CO and N_2 with little H_2. CO = 20-30%, N_2 = 50-55%, H_2 = 10-12%, and CO_2 = 3%. Its calorific value is 1300 kcal/m^3. The calorific value can be increased up to 4000kcal/m^3 by passing O_2 (not air) and some super heated steam.

The gas is prepared by passing air through white-hot carbon at 1100°C.The furnace called producer is made of steel lined with firebricks. The hot coke rests on a grating through which air passes. The bed is generally above 4" thick when CO is the main product; if it is less than 4", CO_2 production is higher. So, coke bed has three distinct zones as below:

(a) **Ash zone:** It is the lowest portion of bed containing mostly ash that protects the grate from heat and also increases the temperature of incoming air and steam when they pass through it.

(b) **Combustion zone:** It is above the ash zone and it is the oxidation zone where CO and CO_2 are produced.

$$C + O_2 \rightarrow CO_2 + 97\ \text{kcal}$$
$$2C + O_2 \rightarrow 2CO + 29.5\ \text{kcal}$$

Temperature of this zone is 1100°C.

(c) **Reduction zone:** At this zone temperature is maintained at 1000°C, at which carbon reduces CO_2 and steam to liberate CO and H_2.

$CO_2 + C = 2CO \quad -36$ kcal

$C + H_2O = CO + H_2 \quad -29$ kcal

$C + 2H_2O = CO_2 + 2H_2 \quad -19$ kcal

All these reactions are endothermic and so do not allow rising the temperature of the producer. As a result, the ash does not fuse and the lining of the producer is protected. This is the reason why some super-heated steam mixed with air is passed.

(d) **Distillation zone:** It is the uppermost zone; temperature is 400-800°C. Here, the outgoing air gas preheats the down coming coal. As a result the volatile matter of coal escapes with the producer gas.

The ratio CO/CO_2 increases with temperature as shown in Table 4.5.

TABLE 4.5

Temperature°C	850°C	900°C	950°C	1000°C	1050°C	1100°C
CO	93.7	97.8	98.7	99.4	99.6	99.8
CO_2	6.3	2.2	1.3	0.6	0.4	0.2

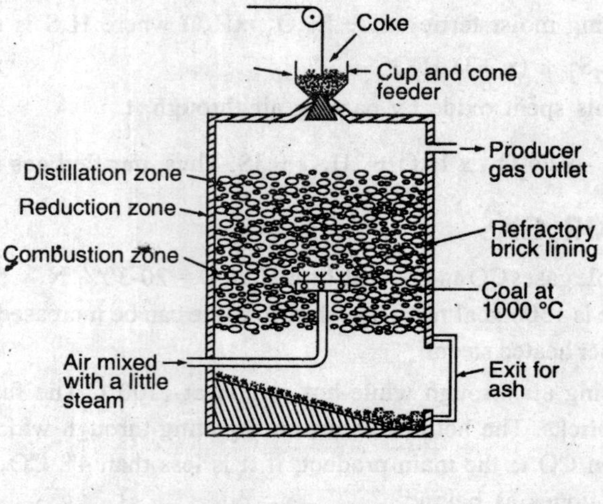

Fig. 4.7 : Gas producer.

WATER GAS OR BLUE GAS

It is a mixture of nearly equal volume of CO and H_2 produced by passing steam through white-hot carbon at 1300°C. If the temperature falls to 800-900°C, then CO_2 proportion increases. So, the condition is to be maintained such that temperature of bed does not fall below 1000°C. Its composition is CO = 43%, H_2 = 51%, CO_2 = 3% N_2 = 3%. As the gas burns with blue flame, it is called blue water gas or blue gas. But composition varies with temperature of the bed as given below:

Temperature	% Decomposition of steam	H_2	CO	CO_2
900°C	42	62	15	23
1010°C	94	49	50	1
400°C	99.4	51	48	1

Its calorific value is 2800 kcal/m³. It is used as a source of hydrogen, as illuminating gas, and as reducing agent in metallurgy. The reaction which is taking place is shown below.

(a) Supplied steam reacts with red-hot coke at 1100-1300°C.

$C + H_2O = CO + H_2 - 29$ kcal

This being an endothermic reaction, temperature begins to fall and CO_2 will be produced. Hence, to increase the temperature of the bed, steam line is put off and air is passed through when exothermic reaction, $C + O_2 = CO_2 + 97$ and $2C + O_2 = 2CO + 59$ Kcal takes place and temperature of the bed rises to desired value. Thus, initially steam is passed for 8 min and then air for 2 min alternately. The gases produced during second operation are vents out separately.

The reactor is a 4 m high 3 m wide steel vessel lined inside with fire bricks and refractory bricks. It is provided with a cup and cone feeder at the top, and outlet for the gas is near the top. Near the base, inlet pipes for air and steam and at the base an opening for taking out the ash are provided.

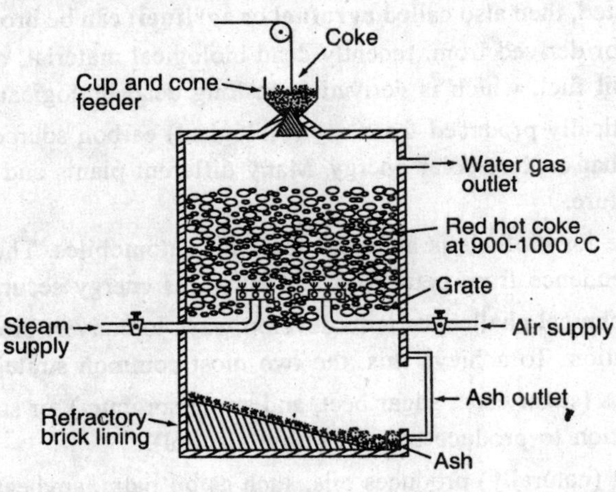

Fig. 4.8 : Water gas producer.

SEMI WATER GAS

It is prepared by passing the mixture of steam and air –the proportion adjusted in such a way that the heat change in exothermic and endothermic reaction is adjusted and the temperature of the bed remains constant. Its average composition is H_2 = 12%, CH_4= 3%, CO = 30%, CO_2 = 2%, N_2 = 53% and calorific value is 1700 kcal/m³. It is used as a source of N_2 and H_2 for the manufacture of NH_3, as fuel, etc.

CARBURETTED WATER GAS

To increase the calorific value of water gas from 2800 kcal/m³ to about 4500 kcal/m³, some saturated and unsaturated hydrocarbons obtained by cracking of crude oil are mixed with it in the carburetter. Its average composition is H_2 = 35%, CO = 25% saturated and unsaturated hydrocarbon = 35% and N_2 + CO = 5%.

The water gas coming out of the reactor is passed through two-tower carburetters, where crude oil vapour mixed through super heated steam and cracking of oil occurs, thus, percentage of hydrocarbon, mainly olefins, increases. The gas is freed from tar and H_2S and stored in gasholder.

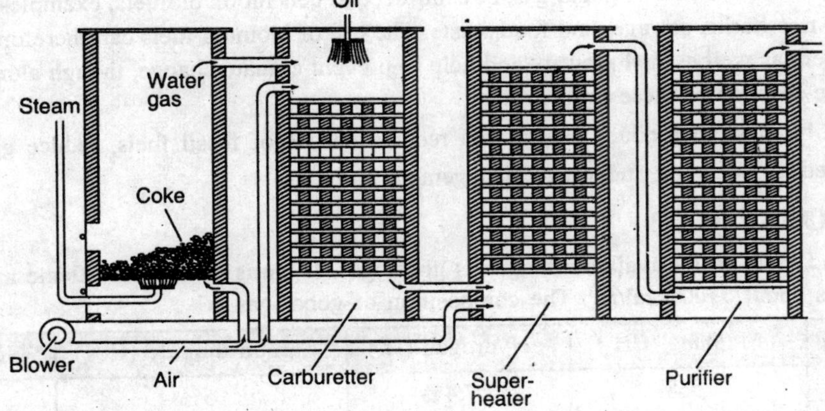

Fig. 4.9 : Carburetted water gas producer.

BIOFUEL: (if cultivated, then also called **agrofuel or agrifuel**) can be broadly defined as solid, liquid, or gas fuel consisting of/or derived from, recently dead biological material, most commonly plants. This distinguishes it from fossil fuel, which is derived from long dead biological material.

Biofuel can be theoretically produced from any (biological) carbon source. The most common by far is photosynthetic plants that capture solar energy. Many different plants and plant-derived materials are used for biofuel manufacture.

The most common use for biofuels is as liquid fuels for automobiles. The use of renewable biofuels provides increased independence from petroleum and enhances energy security.

One of the greatest technical challenges is to develop ways to convert biomass energy specifically to liquid fuels for transportation. To achieve this, the two most common strategies are:

1. To grow sugar crops (sugar cane, sugar beet, and sweet sorghum), or starch (corn/maize), and then use yeast fermentation to produce ethanol (ethyl alcohol).
2. To grow plants that (naturally) produces oils, such as oil palm, soybean, algae, or jatropha. When these oils are heated, their viscosity is reduced, and they can be burned directly in a diesel engine, or the oils can be chemically processed to produce fuels such as biodiesel.

Wood and its byproducts can also be converted into biofuels such as woodgas, methanol or ethanol fuel.

Biomass is material derived from recently living organisms. This includes plants, animals and their by-products. For example, manure, garden waste and crop residues are all sources of biomass. It is a renewable energy source based on the carbon cycle, unlike other natural resources such as petroleum, coal, and nuclear fuels.

Animal waste is a persistent and unavoidable pollutant produced primarily by the animals housed in industrial sized farms. It was found that microorganisms turn farm waste into alternative energy, providing farmers with a simple way to treat their waste and convert it into energy.

There are also agricultural products specifically grown for biofuel production which include corn, switchgrass, soybeans, rapeseed, wheat, sugar beet, sugar cane, palm oil, miscanthus, sorghum, cassava, and jatropha. Hemp has also been proven to work as a biofuel. Biodegradable outputs from industry, agriculture, forestry, and households can be used for biofuel production, either using anaerobic digestion to produce biogas as in the case of gobar gas, or using second generation biofuels; examples include straw, timber, manure, rice husks, sewage, and food waste. The use of biomass fuels can therefore contribute to waste management as well as fuel security and help to prevent climate change, though alone they are not a comprehensive solution to these problems.

Using waste biomass to produce energy can reduce the use of fossil fuels, reduce greenhouse gas emissions and reduce pollution and waste management problems.

BIOGAS (GOBAR GAS)

The cheapest and most easily available biogas is gobar gas. It burns with a blue flame and its average calorific value is about 5300 kcal/m^3. The composition of gobar gas is

Constituents	Methane (CH_4)	Hydrogen (H_2)	Carbon dioxide (CO_2)	Nitrogen (N_2)
Percentage	55%	7.4%	5%	2.6%

Production: Gobar gas is manufactured in gobar plant, which consists of (i) a 'well' constructed of masonry works (also known as 'digester').The digester is usually built below the ground level,(ii) a 'gas holder' which covers digester and made up of weld steel sheets and (iii) a 'pipeline',10cm diameter (shown in Fig. 4.10).

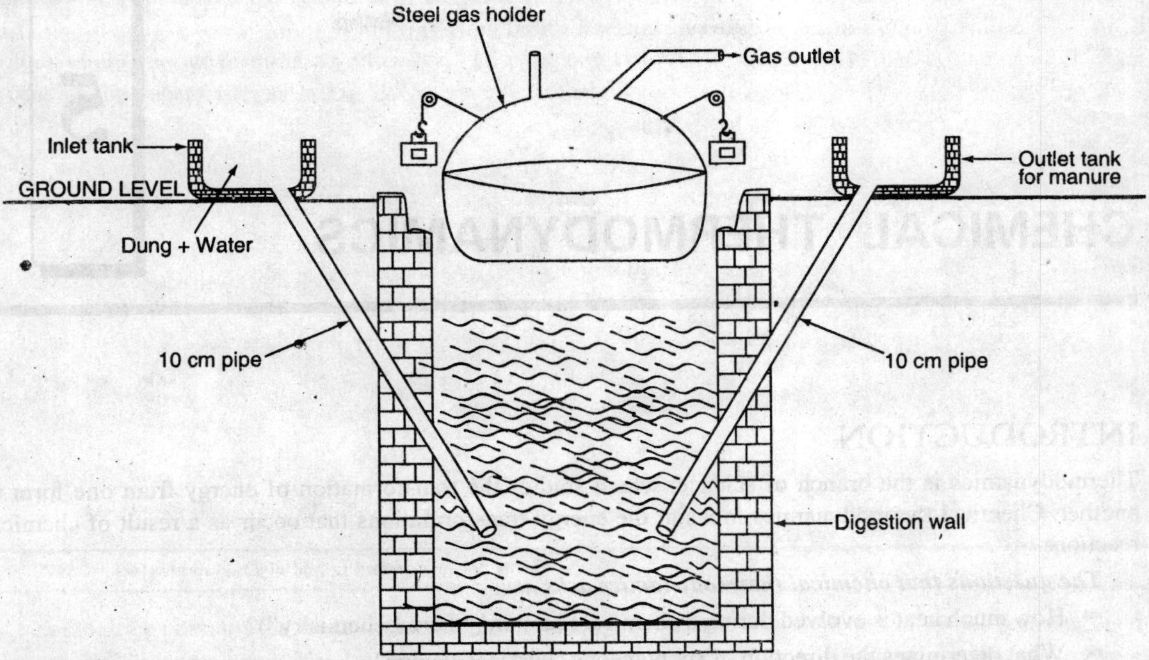

Fig. 4.10 : Gobar gas plant.

Cattle dung in the form of slurry (made by mixing equal parts of water) is poured in digester. Anaerobic bacteria present in the dung digest this slurry in a process called anaerobic fermentation. The optimum temperature for this fermentation is 34-48°C. The gas generated, due to continuous decay, is mostly methane and is collected in gas holder. It is interesting to note that in addition to cattle dung gobar gas plant can also digest human refuge, poultry, sweeps etc.

Advantages

1. Heat generated by direct burning of 1kg of dry cattle dung is 23.4 kcal but if the same amount of cattle dung is converted first into gobar gas, (160 l gas per kg dung), it can supply 188 kcal of heat.
2. Gobar gas does not contain poisonous gas (like CO) as an ingredient
3. Gobar gas is free from smoke, dust, dirt, etc; hence by its use environment and utensils remain comparatively clean
4. By producing gobar gas, we in fact are optimally utilizing waste.
5. It can provide flame temperature of 540°C, with proper burners.

Applications

1. It is used as domestic fuel in many villages
2. It is also used for lighting and power purposes
3. Gobar gas also gives simultaneously excellent yield of good manure which has 2% nitrogen content as against 0.75% in firm yard manure.

Limitation

Gobar gas should be used within a very short distance.

CHEMICAL THERMODYNAMICS

5

INTRODUCTION

Thermodynamics is the branch of science, which studies the transformation of energy from one form to another. Chemical thermodynamics looks at the energy transformations that occur as a result of chemical reactions.

The questions that chemical thermodynamics asks are:

- How much heat is evolved during a chemical reaction ("thermochemistry")?
- What determines the direction of spontaneous chemical change?
- What determines the extent of chemical change?

All reactions do not result in the complete conversion of reactants to products. Some reactions will proceed only so far before the state of chemical equilibrium is achieved. Chemical equilibrium is said to exist if no further tendency for overall chemical change is observed – the ratios of the concentrations of products to reactants remain fixed.

So, chemical thermodynamics attempts to answer the important question, "What determines the position of chemical equilibrium?"

SOME BASIC CONCEPTS

THE SYSTEM AND THE SURROUNDINGS

The system may be defined as the particular region of space that we are making measurements on. It may be defined by physical boundaries, like a test-tube or a bomb calorimeter, or the system may simply be defined by a set of Cartesian coordinates specifying a particular volume in space.

The definition of the surroundings follows naturally–the surroundings are everything that is not the system.

One way of summarizing the above is as under:

The universe = the system + the surroundings

All thermodynamic systems may be classified in the following manner—

System

Open | Closed | Isolated

Open System: In an open system, the transfer of both energy and mass between system and surroundings can take place. For example, a glass of water; water can be poured out of the glass or added to the glass, again the water in the glass may be heated or cooled thus enabling the transfer of both mass and energy.

Closed System: In a closed system, no transfer of mass is possible, only energy transfer may take place. For example, if we seal the mouth of the glass containing the water – no mass transfer can take place. However the water in the glass may still be heated or cooled.

Isolated System: In an isolated system, both energy and mass transfer are all impossible. For example, if the water is placed inside a thermos flask and the mouth of the flask is sealed then it becomes an isolated system.

The boundary walls of a system that separates it form its surroundings may be of two types –

Diathermal wall: allows the transmission of heat through them into and outside the system. All closed systems have diathermal walls.

Adiabatic wall: does not allow the transmission of heat through them. Isolated systems are bounded by adiabatic walls.

PROPERTIES OF A SYSTEM

The experimentally measurable physical characteristics of a system are called the thermodynamic properties or parameters. Various properties like pressure, volume, temperature, composition, surface tension, viscosity, refractive index, etc. may be associated with a system. However, the properties of fundamental interest which are sufficient to define a thermodynamic system are pressure, volume, temperature and composition. For a homogeneous system of given composition only three parameters (P, V, T) are sufficient and since these three are interrelated if any two are specified, the third automatically becomes fixed.

The properties of a system are of two types:

Extensive: Mass dependent properties

Intensive: Mass independent properties.

TABLE 5.1

Extensive	Intensive
Volume	Molar volume
Number of moles	Density
Mass	Refractive index
Energy	Molar enthalpy
Free energy	Molar entropy
Internal energy	Free energy per mole
Enthalpy	Specific heat
Heat capacity	Pressure
Entropy	Temperature

THERMODYNAMIC PROCESSES

When one or more parameters defining the thermodynamic system change, the system is said to have changed its state. The path of change of state is called a process. Let us examine a few very common processes–

Isothermal process – Temperature remains constant during the process; $dT = 0$.

Adiabatic process – Heat is neither absorbed nor released; $q = 0$.

Isochoric process – Volume remains constant during the process; $dV = 0$.

Isobaric process – Pressure remains constant during the process; $dP = 0$.

Cyclic process – After a series of transformations the system comes back to its initial state.

Reversible process – A reversible process may be defined as a process, which can be reversed without causing any permanent change in the system or the surrounding. The process occurs in such a way that the properties of the system remain uniform. It occurs at a very slow rate so that enough time is allowed for the system to maintain equilibrium. It occurs through a series of small infinitesimal steps. In actual practice it is not possible to perform experiments under such an ideal situation. Vaporization of water at 100°C and 1 atm pressure in a closed vessel may be approximated as a reversible process.

Irreversible process – The changes are produced rapidly such that the system does not get enough time to attain equilibrium. All natural processes are irreversible and hence spontaneous. Once undergone the process cannot be reverted without changing the properties of the surroundings.

INTERNAL ENERGY (*U*)

We are going to be talking about our chemical system losing and gaining energy, so it is logical to introduce a quantity, which represents the total energy of the system. This is known as the internal energy, *U*. As with all types of energy, the unit of U is the joule. Now, the internal energy is defined as the total energy of our system–chemical, nuclear, heat, gravitational, any other type of energy one can think of, the sum of all the system's energy. Although we cannot measure the absolute value of the internal energy of a system, we can measure changes in the internal energy. Internal energy is a state function, i.e., its change is not governed by the process of change and only depends on the initial and final state of the system.

$$\Delta U = U_f - U_i,$$

where U_f = internal energy of the final state

U_i = internal energy of the initial state.

THE FIRST LAW OF THERMODYNAMICS

The universality of this law is based on human experiences. In its broadest form it may be stated as follows: The total energy of the Universe is constant. Energy can neither be created nor be destroyed. It can only be changed from one form to another. Supposing that a system in its initial state is associated with energy U_i and heat q is given to it, the heat absorbed by the system may be used to increase its internal energy or perform work or both. If the energy of the system in the final state is U_f, then $(U_f - U_i) = \Delta U$ is the increase in internal energy. The law of conservation of energy demands that:

Heat absorbed = increase in energy + work done by the system, i.e.,

$$q = \Delta U + (-w) = \Delta U - w.$$

The minus sign appears because the work done by the system is taken as $(-w)$ since it will decrease the internal energy of the system.

Thus $\Delta U = q + w$ (5.01)

or, in the differential form $dU = đq + đw$. (5.02)

Both q and w are measured in joules. q and w are path dependent functions and are hence not perfect differentials. Therefore they are written in the form $đq$ and $đw$.

Now let us investigate the fate of the first law for various processes–

Isothermal Process

The temperatures of the initial and final states are the same. That is, the total kinetic energy of all the gas molecules does not change. Hence, the internal energy change of the isothermal expansion must be zero. From the First Law,

$$0 = q + w.$$

Adiabatic process

Heat is neither absorbed nor rejected.

$dU = đw$.

Isochoric process

$dV = 0$; therefore, $dU = đq$.

Work (w)

Scientists define work w to be the product of force F acting through a distance h:

$$w = \text{force} \times \text{distance moved in the direction of force}$$

$$\text{i.e., } w = \underset{(m)}{m}\underset{(g)}{g} \times \underset{(h)}{h} = \text{kg} \times \text{ms}^{-2} \times \text{m} = \text{kg m}^2\text{s}^{-2}$$

$$1 \text{ kg m}^2 \text{ s}^{-2} = 1 \text{ joule.}$$

Now let us consider work against an opposing force.

Consider a gas contained in a cylinder of cross section A and fitted with an ideal piston. The gas at pressure P_{in} expands against an external pressure P_{ext} as shown in Fig. 5.1.

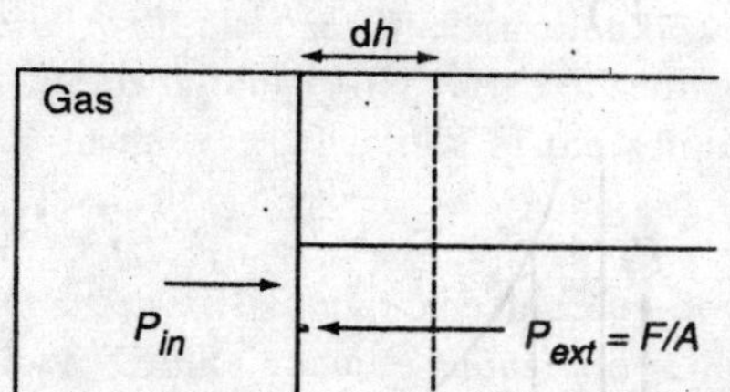

Fig. 5.1 : Work done against an external pressure P_{ext}.

The direction of movement depends on our choice of internal and external pressures.

(i) If $P_{in} > P_{ext}$: expansion work against P_{ext}.

(ii) If $P_{in} < P_{ext}$: compression work against P_{in}.

(iii) If $P_{in} \approx P_{ext}$ equilibrium, reversible work of expansion.

For an infinitesimal change in volume, dV (where $dV = Adh$)

Work done by the system $(-đw) = PdV$.

For **a reversible isothermal expansion** from volume V_1 to volume V_2 the work done by the system is

$$(-w) = \int_{V_1}^{V_2} PdV.$$

Now we know that

for an ideal gas, $PV = nRT$.

Therefore, $P = nRT/V$.

$$\text{Therefore,} \quad (-w) = \int_{V_1}^{V_2} (nRT/V)\,dV$$

$$= nRT \ln (V_2/V_1). \qquad (5.03)$$

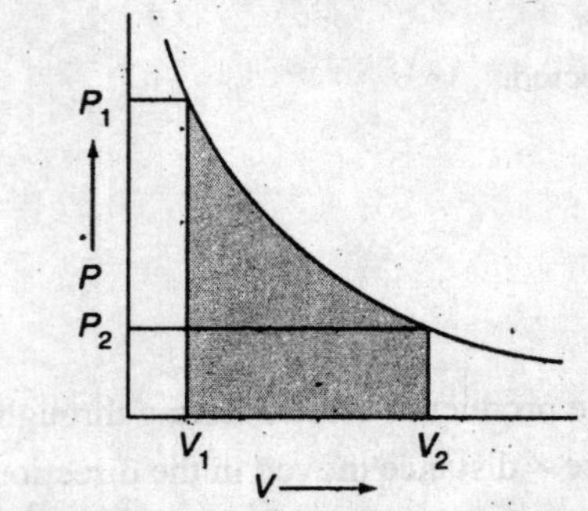

Fig. 5.2 : Reversible isothermal expansion–shaded area under the curve shows the work done.

For a gas obeying van der Waals equation (real gas),

$P = nRT/(V - nb) - an^2/V^2$

$(-w) = nRT \ln [(V_2 - nb)/(V_1 - nb)] + an^2 [1/V_2 - 1/V_1]$ (5.04)

For **irreversible** work done on the system,

work done on the system $= P_{ext}(V_2 - V_1)$

i.e., $w = -P_{ext}(V_2 - V_1)$ (5.05)

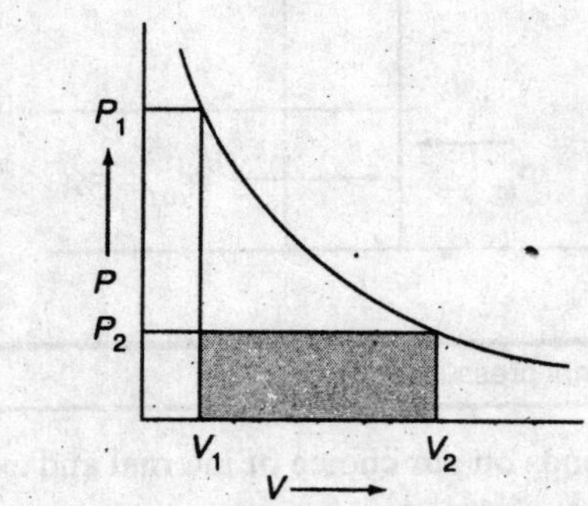

Fig. 5.3 : Irreversible isothermal expansion–shaded area under the curve shows the work done.

Problem: A gas expands against a variable external pressure given by $P = 10/V$ atm, where V is the volume of the gas at each stage of expansion. Further in expanding from 10 to 100 litres, the gas undergoes a change in internal energy $\Delta U = 100$ cal. What is the work done and how much heat has been absorbed?

Solution: Work done against a variable pressure is given by

$$(-w) = \int_{V_1}^{V_2} P\,dV$$

Here $P = 10/V$ atm.

So, $(-w) = \int_{V_1}^{V_2} (10/V)\,dV = 10 \ln (V_2/V_1) = 10 \ln (100/10) = 23.03$ litre-atmosphere

$\therefore (-w) = 557.3$ cal.

Since from first law, $\Delta U = q + w$,

$\therefore$ heat absorbed $(q) = 100 + 557.3 = 657.3$ cal.

Problem: Calculate the work done by 5 moles of an ideal gas during expansion from 5 atm at 25°C to 2 atm at 50°C against a constant pressure of 0.5 atm. If for the gas $C_P = 5$ cal deg^{-1} mole^{-1}, calculate ΔU and q.

Solution: $(-w) = P_{ext}(V_2 - V_1)$

Here $P_{ext} = 0.5$ atm, $V_2 = nRT_2/P_2 = (5 \times 0.082 \times 323)/2 = 66.22$ litres

and $V_1 = nRT_1/P_1 = (5 \times 0.082 \times 298)/5 = 24.44$ litres.

So $(-w) = 0.5\,(66.22 - 24.44) = 20.89$ litre-atm $= 505.54$ cal.

Now $C_P = 5$ cal. So, $C_V = 3$ cal (for ideal gas $C_P - C_V = R$)

$\Delta U = nC_V(T_2 - T_1) = 5 \times 3\,(323 - 298) = 375$ cal.

Since from first law of thermodynamics, $\Delta U = q + w$, so, $q = 880.54$ cal.

Adiabatic Reversible process

Relation between P, V and T

We know that for adiabatic process, $đq = 0$.

Therefore, $dU = đw$ (from first law);

Now $dU = nC_V dT$ and $đw = -PdV$

Therefore, $nC_V dT = -PdV$

$= -nRT\, dV/V$

or, $C_V\, dT/T = -RdV/V$.

Integrating between limits T_1 and T_2 and limits V_1 and V_2 respectively, we get

$C_V \ln(T_2/T_1) = -R\ln(V_2/V_1)$

or, $T_2/T_1 = (V_1/V_2)^{R/C_V}$

or, $T_2/T_1 = (V_1/V_2)^{\gamma-1}$

or, $T_2(V_2)^{\gamma-1} = T_1(V_1)^{\gamma-1}$

i.e., $T(V)^{\gamma-1} =$ constant. (5.06)

Replacing T in the above equation, we get

$PV(V)^{\gamma-1}/nR =$ constant

or, $PV^{\gamma} =$ constant (5.07)

Again, replacing V in equation,

$P(nRT/P)^{\gamma} =$ constant

or, $P^{1-\gamma}\,T^{\gamma} =$ constant. (5.08)

Adiabatic Reversible Work

Work done by n moles of ideal gas $= -đw = -nC_V dT$

$= nC_V(T_1 - T_2)$

$= C_V(P_1V_1 - P_2V_2)/R$

$= [P_1V_1 - P_2V_2]/\gamma - 1$ (5.09)

Problem: N_2 gas is expanded reversibly and adiabatically from a volume of 1 litre at 0°C and 1 atm to a volume of 2 litres. If $C_V = 20.8$ joule-mole^{-1}K^{-1} and $C_P = 29.1$ joule-mole^{-1}K^{-1}, then assuming ideal behaviour calculate the final temperature and pressure. What are q, w and ΔU?

Solution: $\gamma = C_P / C_V = 1.4$.

From adiabatic reversible relation, $P_1 V_1^{\gamma} = P_2 V_2^{\gamma}$. So, $P_2 = 0.38$ atm.

Again, $T_1 (V_1)^{\gamma-1} = T_2 (V_2)^{\gamma-1}$ So, $T_2 = 206.9$ K

In adiabatic process $q = 0$

$(-w) = C_V(T_1 - T_2)$ / mole = 1381.5 J / mole

$\Delta U = C_V(T_2 - T_1)$ / mole = –1381.5 J / mole.

Comparison between isothermal & adiabatic processes:

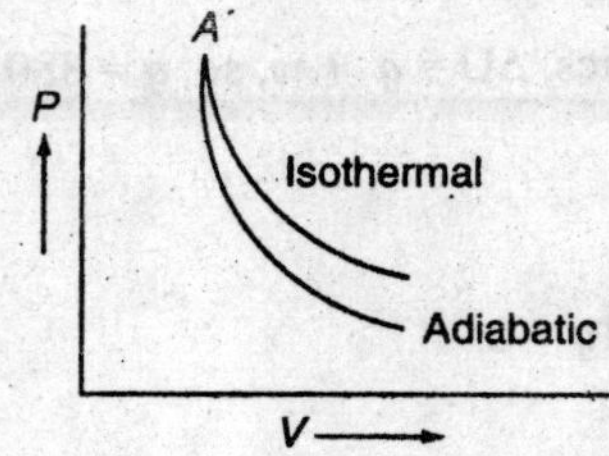

Fig. 5.4 : *P-V* curve for isothermal and adiabatic processes.

For adiabatic expansion of an ideal gas, PV^{γ} = constant (K, say)

For isothermal expansion of the same PV = constant (K', say)

The slope of P-V curve is obtained by dP/dV

For adiabatic change, $P = K/V^{\gamma}$ Therefore, $dP/dV = (K/V^{\gamma+1})(-\gamma) = -\gamma P/V$

For isothermal change, $P = K'/V$. Therefore, $dP/dV = -K'/V^2 = -P/V$

In both the changes the slope is negative. Since γ is greater than unity, the slope of adiabatic P-V curve will be steeper than that in isothermal one.

ENTHALPY

We know that the heat absorbed at constant volume is equal to the change in internal energy. Most chemical reactions, however, are carried out not at constant volume, but at constant pressure. That is, chemical reactions are normally carried out in flasks or test tubes under (relatively) constant pressure of the atmosphere. So, it would be useful if there exists some state function whose change equals the heat absorbed by a system at constant pressure. Enthalpy, H, is that state function.

Enthalpy is defined as follows,

$$H = U + PV, \tag{5.10}$$

where U is the internal energy, P is the pressure and V is the volume of the system.

The change in enthalpy is equal to the heat absorbed by the system at constant pressure.

Now let us look at the change in this expression as enthalpy changes by an infinitesimally small amount dH

$$dH = dU + PdV + VdP. \tag{5.11}$$

At constant pressure, dP, the change in P is zero, so the last equation becomes

$$dH = dU + PdV.$$

In words, this is the change in enthalpy at constant pressure which is equal to the change in the internal energy plus the product of the pressure and the change in volume of the system.

Relation between q_P and q_V

From first law of thermodynamics, $\Delta U = q + w$

or, $\Delta U = q + (-P\Delta V)$

or, $q = \Delta U + P\Delta V$.

At constant volume $\Delta V = 0$.

Therefore, q_V (heat absorbed at constant volume) $= \Delta U$.

Again, $H = U + PV$

or, $\Delta H = \Delta U + P\Delta V + V\Delta P$.

At constant pressure $\Delta P = 0$.

Therefore, $\Delta H = \Delta U + P\Delta V = q_P$ (heat absorbed at constant pressure).

Thus we can write,

$q_P = q_V + P\Delta V$.

In a gaseous reaction if n_2 and n_1 represent the numbers of resultant and reactant molecules, then

$P\Delta V = P(n_2 - n_1)RT/P = \Delta nRT$.

Therefore $q_P = q_V + \Delta nRT$. (5.12)

HEAT CAPACITY C_P AND C_V AND RELATIONSHIP BETWEEN THEM

Heat capacity at constant volume may be defined as the heat required to increase the temperature of the system by 1°C at constant volume.

Therefore, $C_V = (\partial q/\partial T)_V = (\partial U/\partial T)_V$.

Again, the heat required to increase the temperature of the system by 1°C at constant pressure is the heat capacity at constant pressure.

Therefore, $C_P = (\partial q/\partial T)_P = (\partial H/\partial T)_P$.

$C_P - C_V = (\partial H/\partial T)_P - (\partial U/\partial T)_V$

$= (\partial U/\partial T)_P + P(\partial V/\partial T)_P - (\partial U/\partial T)_V$. (5.13)

We know that $U = f(V, T)$.

Therefore, $dU = (\partial U/\partial T)_V dT + (\partial U/\partial V)_T dV$.

Dividing throughout by dT and imposing the condition of constant pressure, we get

$(\partial U/\partial T)_P = (\partial U/\partial T)_V + (\partial U/\partial V)_T (\partial V/\partial T)_P$

or, $(\partial U/\partial T)_P - (\partial U/\partial T)_V = (\partial U/\partial V)_T (\partial V/\partial T)_P$. (5.14)

Inserting the value of equation (5.14) in equation (5.13), we get

$C_P - C_V = (\partial U/\partial V)_T (\partial V/\partial T)_P + P(\partial V/\partial T)_P$.

$= [(\partial U/\partial V)_T + P] (\partial V/\partial T)_P$

For ideal gas, $(\partial U/\partial V)_T = 0$ and $(\partial V/\partial T)_P = nR$.

Therefore, for 1 mole ideal gas, $C_P - C_V = R$. (5.15)

HERMOCHEMISTRY

Thermochemistry is the study of the heat evolved or absorbed in chemical reactions. It is generally oncerned with the heat exchange accompanying transformations, such as mixing, phase transitions, chemical eactions, etc., which includes calculations of such quantities as the heat capacity, heat of combustion, heat of ormation, etc. The *laws of thermochemistry* rest on two statements:

1. Lavoisier and Laplace's law (1782): The heat exchange accompanying a transformation is equal and opposite to the heat exchange accompanying the reverse transformation.

2. Hess's law (1840): The heat exchange accompanying a transformation is the same whether the process occurs in one or several steps.

Example: The enthalpies of combustion of carbon, hydrogen and ethane are -394 kJ mol^{-1}, -286 kJ mol^{-1} and -1542 kJ mol^{-1} respectively

(1) $C(s) + O_2(g) \rightarrow CO_2(g)$ $\Delta H = -394$ kJ mol^{-1}

(2) $H_2(g) + ½O_2(g) \rightarrow H_2O(l)$ $\Delta H = -286$ kJ mol^{-1}

(3) $C_2H_6(g) + (7/2)O_2(g) \rightarrow 2CO_2(g) + 3H_2O(l)$ $\Delta H = -1542$ kJ mol^{-1}

The enthalpy of formation of ethane $2C(s) + 3H_2(g) \rightarrow C_2H_6(g)$ may be calculated as follows:

2 × (1)	$2C(s) + 2O_2(g) \rightarrow 2CO_2(g)$	$\Delta H = 2(-394$ kJ)
3 × (2)	$3H_2(g) + (3/2)O_2(g) \rightarrow 3H_2O(l)$	$\Delta H = 3(-286$ kJ)
Reverse (3)	$2CO_2(g) + 3H_2O(l) \rightarrow C_2H_6(g) + (7/2)O_2(g)$	$\Delta H = 1542$ kJ mol^{-1}
Add:	$2C(s) + 3H_2(g) \rightarrow C_2H_6(g)$	$\Delta H = -104$ kJ mol^{-1}

Therefore, $\Delta H_f = -104$ kJ mol^{-1}.

Every reaction can be considered as a series of bond breaking steps followed by bond forming steps. Bond breaking always requires energy. All bonds have different 'strengths' of energies. If bond forming releases more energy than was required in the bond breaking, then overall energy is released to the surroundings, i.e., an exothermic reaction. If bond forming releases less energy than the amount required in bond breaking, then overall energy will have been absorbed from the surroundings, i.e., an endothermic reaction.

An exothermic reaction has a negative ΔH, e.g., combustion, neutralization, dissolving NaOH pellets.

An endothermic reaction has a positive ΔH, e.g., mixing ammonium thiocyanate and barium hydroxide crystals.

Variation of enthalpy change and internal energy change with temperature: Kirchhoff's equation

If a system undergoes a change from a given state to another state both the internal energy and heat content would alter.

So, $U_f - U_i = \Delta U$ and $H_f - H_i = \Delta H$,

where suffices i and f represent initial and final state respectively.

Differentiating these with respect to temperature at constant volume and constant pressure respectively, we get

$$[\partial(\Delta U)/\partial T]_V = [\partial U_f / \partial T]_V - [\partial U_i / \partial T]_V$$
$$= C_{Vf} - C_{Vi} = \Delta C_V.$$

Again,
$$[\partial(\Delta H)/\partial T]_P = [\partial H_f / \partial T]_V - [\partial H_i / \partial T]_V$$
$$= C_{Pf} - C_{Pi} = \Delta C_P$$

or, $\int \partial(\Delta U) = \int \Delta C_V \, dT$

and $\int \partial(\Delta H) = \int \Delta C_P \, dT$.

Integrating between limits T and 0 K

$$\Delta U - \Delta U_0 = \int \Delta C_V \, dT \quad (5.16)$$

$$\Delta H - \Delta H_0 = \int \Delta C_P \, dT. \quad (5.17)$$

These relations are called *Kirchhoff's equations*.

Problem: The combination of nitrogen and oxygen at 27°C occurs according to $2N_2 + O_2 = 2N_2O$; $\Delta H = 39.4$ kcal. What will be the heat of formation at 227°C? Use the relation $C_P = a + bT$, where C_P is the molar heat capacity at constant pressure, and data are provided.

Gas	a	$b \times 10^3$
N_2	6.52	1.52
O_2	6.15	3.1
N_2O	7.0	7.1

Solution: For the reaction $2N_2 + O_2 = 2N_2O$

$$\Delta C_P = 2\,C_P(N_2O) - 2\,C_P(N_2) - C_P(O_2)$$

$$= 2[7 + 7.1\times 10^{-3}\,T] - 2\,[6.52 + 1.52 \times 10^{-3}T] - [6.15 + 3.1 \times 10^{-3}T]$$

$$= -5.19 + 8.06 \times 10^{-3}T.$$

$$\Delta H_{500} - \Delta H_{300} = \int_{300}^{500} \left(-5.19 + 8.06\times 10^{-3}\,T\right) dT$$

$$\Delta H_{500} - 39400 = -5.19\,(500 - 300) + 8.06 \times 10^{-3}\,[500^2/2 - 300^2/2]$$

or, $\Delta H_{500} = 39.01$ kcal.

This is for the reaction, $2N_2 + O_2 = 2N_2O$.

So heat of formation = 39.01 / 2 kcal = 19.51 kcal.

SECOND LAW OF THERMODYNAMICS

The first law tells us that energy of the universe is constant and that diferent forms of energy are interconvertible. However, this law does not tell us whether a change would occur at all or not, or, if it occurs, to what extent. The first law also does not tell us about the direction in which any change will take place. These insufficiencies of the first law formed the basis of the second law. Different scientist have expressed the law differently. A common enunciation of second law of thermodynamics by Kelvin-Planck is given below:

It is impossible to construct an engine that, operating in a cycle, will produce no effect other than the extraction of heat from a reservoir and the performance of an equivalent amount of work.

An equivalent statement by Clausius

It is impossible for a self-acting machine, unaided by any external agency, to convey heat from a body at a low temperature to one at a high temperature.

It is an experience that complete conversion of heat into work is impossible without leaving a permanent change elsewhere.

If the second law were not true, it would be possible to propel a ship across the ocean by extracting heat from the surrounding air. There is nothing in the first law to preclude the possibility of converting this heat completely into work. The second law is, therefore, not a deduction from the first law, but stands by itself as a separate law of nature, referring to an aspect of nature different from that described by the first law. The first law denies the possibility of creating or destroying energy; the second law denies the possibility of utilizing energy in a particular way. The continuous operation of a machine that creates its own energy and thus violates the first law is called a perpetual motion of the first kind. The operation of a machine that utilizes the internal energy of only one heat reservoir, thus violating the second law, is called a perpetual motion of the second kind.

CARNOT CYCLE

Sadi Carnot explained clearly with the help of his famous Carnot cylcle, how and to what extent work is obtainable from heat. He designed an engine working in complete cycles where every step was carried out in a reversible manner.

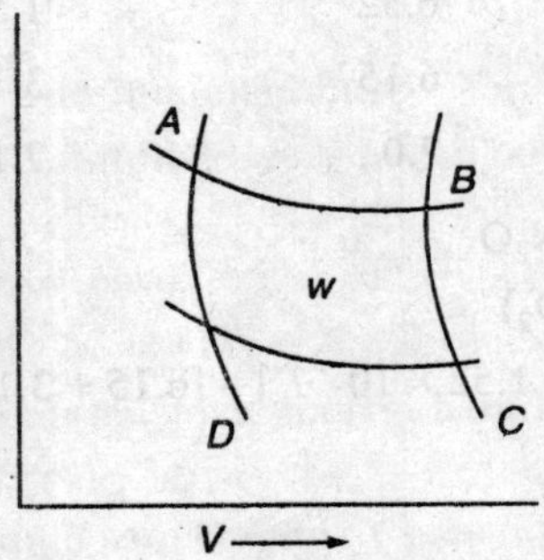

Fig. 5.5 : Carnot cycle.

The Carnot cycle when acting as a heat engine consists of the following steps:

1. *Reversible isothermal expansion of the gas at the "hot" temperature reservoir or source at* T_2 (isothermal heat addition). During this step (A to B on diagram) the expanding gas causes the piston to do work on the surroundings. The gas expansion from volume V_1 to V_2 is propelled by absorption of heat q_2 from the high temperature reservoir. The work done is given by
   $(-w_1) = q_2 = RT_2 \ln (V_2/V_1)$.
2. *Reversible adiabatic expansion of the gas*. For this step (B to C on diagram) we assume that the piston and cylinder are thermally insulated, so that no heat is gained or lost ($q = 0$). The gas continues to expand from volume V_2 to V_3, doing work on the surroundings. The gas expansion causes it to cool to the "cold" temperature, T_1. The work done is given by $(-w_2) = -C_V (T_1 - T_2)$.
3. *Reversible isothermal compression of the gas at the "cold" temperature reservoir or sink at* T_1. (*isothermal heat rejection*). (C to D on diagram) Now the surroundings do work on the gas, causing heat to flow out of the gas to the low temperature reservoir. The gas is compressed from V_3 to V_4. q_1 amount of heat is rejected to the sink. The work done is $(-w_3) = -q_1 = RT_1 \ln (V_4/V_3)$.
4. *Reversible adiabatic compression of the gas* (D to A on diagram). Once again we assume, the piston and cylinder are thermally insulated. During this step, the surroundings do work on the gas, compressing it and causing the temperature to rise to T_2. At this point the gas is in the same state as at the start of step 1. The work done is $(-w_4) = -C_V (T_2 - T_1)$.

   The net work done in the cyclic process,

   $(-w) = (-w_1) + (-w_2) + (-w_3) + (-w_4)$

   $= RT_2 \ln (V_2 / V_1) - C_V (T_1 - T_2) + RT_1 \ln (V_4 / V_3) - C_V (T_2 - T_1)$

   $= RT_2 \ln (V_2/V_1) + RT_1 \ln (V_4/V_3)$.

   Now since AD and BC are two reversible adiabatics.

   $\therefore \; V_1/V_2 = V_4/V_3$.

   $(-w) = R(T_2 - T_1) \ln (V_2/V_1) = q_2 - q_1$.

   Therefore, efficiency of the Carnot cycle is $\eta = (q_2 - q_1)/q_2 = (T_2 - T_1)/T_2$. (5.18)

ENTROPY

The concept of entropy was developed in the 1850s by German physicist Rudolf Clausius who described it as the *transformation-content*, i.e., dissipative energy use of a thermodynamic system or working body of chemical species during a change of state. In contrast, the first law of thermodynamics, formalized through the heat-friction experiments of James Joule in 1843, deals with the concept of energy, which is conserved in all processes. Entropy change has often been defined as a change to a more disordered state at a molecular level. In recent years, entropy has been interpreted in terms of the "dispersal" of energy. Entropy is an extensive state function that accounts for the effects of irreversibility in thermodynamic systems. When a system's energy is defined as the sum of its "useful" energy, (e.g., the energy which is used to push a piston), and "useless energy", i.e., the energy which cannot be used for external work, then entropy may be (most concretely) visualized as the "scrap" or "useless" energy. When heat is absorbed by a gas, the kinetic energy of the gas molecules will be increased and they will move faster and collide with (i) each other (ii) the container and (iii) the face of the piston, thereby the piston will be pushed, Only the third type of collision will bring about the expansion of the gas. Thus a fraction of the total amount of heat absorbed is wasted. This is the unavailable work, and entropy is a measure of this unavailable work. Under an ideal situation, if all the gas molecules were striking the piston perpendicularly, then whole of the heat absorbed would have been converted into work. This can happen if there is a regular and only one possible arrangement. In other situations when the molecules are in random motion, a fraction of work is converted into such a factor which is a measure of the randomness of the molecular system. Entropy is thus a measure of the randomness or disorderness of the molecular system. More the randomness of the system, more is its entropy. Quantitatively, entropy is defined by the differential quantity,

$$dS = đq_{rev}/T \tag{5.19}$$

Combined form of First Law and Second Law of Thermodynamics:

From first law, $dU = đq + đw$ [From equation (5.02)]

From second law, $dS = đq_{rev}/T$ [From equation (5.19)]

Therefore, the combined form may be written as

$$TdS = dU - đw. \tag{5.20}$$

CALCULATION OF ENTROPY

Reversible process

Let us consider an ideal gas performing a reversible change from an initial state T_1, V_1 to final state T_2, V_2. For infinitesimal changes the first law may be written as

$đq_{rev} = dU - đw_{rev}$.

If only mechanical work (i.e., PV work) is considered,

$-đw_{rev} = PdV$.

$đq_{rev} = dU + PdV$.

For an ideal gas $dU = nC_VdT$ and $P = nRT/V$.

Hence, $đq_{rev} = nC_VdT + (nRT/V)dV$.

Now $dS = đq_{rev}/T = nC_V(dT/T) + nR\,(dV/V)$.

Integrating between appropriate limits, we get

$$\Delta S = nC_V \ln (T_2/T_1) + nR \ln (V_2/V_1) \tag{5.21}$$

Irreversible process

Since entropy is state function the entropy change of a system from a given state I to a state II is always the same irrespective of the path under consideration. Thus to calculate the entropy change of a system in an irreversible process we conceive of taking the system from state I to a state II along any reversible path and then divide the heat absorbed at each point by the temperature and sum up the quotients.

Isolated system

For reversible isolated processes the change of entropy is zero whereas for irreversible isolated processes the entropy change is greater than zero. This may be shown as follows:

Let us suppose that small changes are made in the system reversibly and irreversibly. The work done by the system is $(-đw_{rev})$ and $(-đw_{irr})$ respectively.

From first law of thermodynamics,

$dU = đq + đw$.

Since internal energy is a state function it is same whether the process is performed reversibly or irreversibly.

Therefore, $dU = đq_{rev} + đw_{rev} = đq_{irr} + đw_{irr}$.

Now $(-đw_{rev}) > (-đw_{irr})$.

Therefore $đq_{rev} > đq_{irr}$ and $(đq_{rev})/T > (đq_{irr})/T$.

But from second law, $dS = đq_{rev}/T$.

Therefore, $dS > (đq_{irr})/T$

or, $dS > (dU - đw_{irr})/T$.

Also $dS = (dU - đw_{rev})/T$.

In an isolated system there is no exchange of energy and thus $đq = 0$, $dU = 0$ and $đw = 0$.

So it may be concluded that

$dS_{irr} > 0$ and $dS_{rev} = 0$ for isolated system.

An isolated system includes the system under consideration and the surroundings in its immediate vicinity. Therefore, the entropy change for the isolated system actually gives the entropy change of the universe.

Thus $dS_{Universe} > 0$ for irreversible or spontaneous process

and $dS_{Universe} = 0$ for reversible or equilibrium process.

Problem: One mole of an ideal monoatomic gas initially at S.T.P. expands isothermally and irreversibly to 44.8 litres under conditions such that $(-w) = 100$ cal. Calculate the entropy change for gas and surroundings.

Solution: Initial volume at S.T.P. = 22.4 litres

Final volume = 44.8 litres

$\Delta S_{gas} = R \ln (44.8 / 22.4) = 1.987 \ln (44.8 / 22.4) = 1.38$ cal/degree.

Since it is an isothermal expansion, therefore, $\Delta U = 0$.

$q = (-w) = 100$ cal, i.e., the surroundings have given 100 cal heat to the system.

So $\Delta S_{surrounding} = -100 / 273 = -0.37$ cal /degree.

Problem: 5 gram-moles of water initially at 27°C is converted to a final state of vapour at 227°C, the conversion being effected under 1 atmosphere pressure. Assuming the vapour to behave ideally, compute the total change in entropy.

[heat capacity of water = 1 cal/g; heat capacity of water vapour = 0.40 cal/g, latent heat of vapourization of water = 540 cal/g.]

Solution: The process may be divided into three parts.

$$\begin{array}{ccc} \text{Water (27°C)} & \xrightarrow{\Delta S} & \text{water vapour at 227°C} \\ \downarrow \Delta S_1 & & \uparrow \Delta S_3 \\ \text{Water (100°C)} & \xrightarrow{\Delta S_2} & \text{water vapour (100°C)} \end{array}$$

$\Delta S = \Delta S_1 + \Delta S_2 + \Delta S_3$

$\Delta S_1 = 5 \times 18 \times 1 \times \ln (373 / 300) = 19.60$ cal / degree

$\Delta S_2 = (5 \times 18 \times 540) / 373 = 130.29$ cal /degree

$\Delta S_3 = 5 \times 18 \times 0.40 \times \ln (500/ 373) = 10.55$ cal/ degree

So total entropy change $\Delta S = \Delta S_1 + \Delta S_2 + \Delta S_3 = 160.44$ cal/degree.

Entropy of mixing

When two or more gases are brought in contact they mix up in an irreversible fashion leading to an increase in entropy. Let n_A g-moles of gas A and n_B g-moles of gas B be mixed at constant temperature T and pressure P. Before mixing, the entropy of the two gases are as follows:

$$S_A = n_A [C_{P_A} \ln T - R\ln P + S_A^0]$$

$$S_B = n_B [C_{P_B} \ln T - R\ln P + S_B^0]$$

After mixing, the total pressure remaining constant (P) the partial pressures are p_A and p_B such that $P = p_A + p_B$.

The entropies of the two components in the mixture are

$$S_A^m = n_A [C_{P_A} \ln T - R\ln p_A + S_A^0]$$

$$S_B^m = n_B [C_{P_B} \ln T - R\ln p_B + S_B^0].$$

The entropy change due to mixing,

$$\Delta S^m = S_A^m + S_B^m - S_A - S_B = -n_A R\ln (p_A/P) - n_B R\ln (p_B/P).$$

Now $n_A + n_B = n$ moles

Hence, their mol-fractions are $x_A = n_A/n = p_A/P$ and $x_B = n_B/n = p_B/P$.

Therefore, $\Delta S^m = -n_A R\ln x_A - n_B R\ln x_B$

$= -nR [(n_A / n) \ln x_A + (n_B/n) \ln x_B]$

$= -nR [x_A \ln x_A + x_B \ln x_B]$.

So in general for mixing i number of gases entropy of mixing is

$$\Delta S^m = -nR \sum x_i \ln x_i. \qquad (5.22)$$

Problem: Calculate the entropy change when 2.8 litres of oxygen is mixed with 19.6 litres of hydrogen at N.T.P.

Solution: 2.8 litres of oxygen = 2.8 / 22.4 = 1/8 g-mole

19.6 litres of hydrogen = 19.6 / 22.4 = 7/8 g-mole

Total = 1 g-mole

Therefore, mole fraction of oxygen = 1/8

mole fraction of hydrogen = 7/8

Therefore, $\Delta S^m = -1/8 \text{ R ln } 1/8 - 7/8 \text{ R ln } 7/8 = 0.754$ cal/ degree.

JOULE'S EXPERIMENT

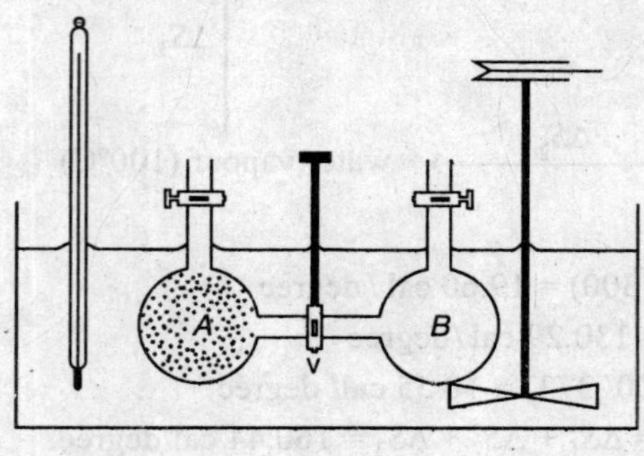

Fig. 5.6 : Joule's Experiment.

Bulb *A* was filled with air under pressure, isolated from a similar evacuated bulb *B* by a valve *V*. The bulbs were immersed in a well-stirred water bath equipped with a sensitive thermometer. After thermal equilibrium had been established, valve *V* was opened to allow the gas to expand into bulb *B*. No change in temperature was detected. In Joule's words, "no change of temperature occurs when air is allowed to expand in such a manner as not to develop mechanical power" (i.e., do no external work, $w = 0$). Since no change of temperature was observed, $q = 0$ and, therefore,

$dU = đq = đw = 0.$

From Joule's experiment it can be shown that

$(\partial U/\partial V)_T = 0.$

JOULE-THOMSON EFFECT

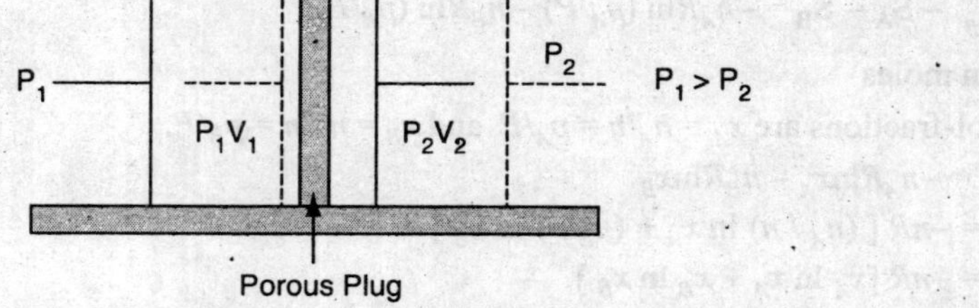

Fig. 5.7 : Joule-Thomson Effect.

Gas was allowed to expand freely through a porous plug from the left hand chamber into the right hand side chamber. The gas expands from a pressure P_1 to pressure P_2 by the throttling action of the porous plug. The entire system is thermally insulated so that the expansion occurs adiabatically. In general, for a real gas it is observed that the final temperature T_2 differs from the initial one, T_1.

The enthalpy change of the system in Joule-Thomson experiment may be shown as follows:

The work done in the left chamber, $w_1 = P_1V_1$.

The work done in the right chamber, $w_2 = -P_2V_2$.

Net work done, $w = w_1 + w_2 = P_1V_1 - P_2V_2$.

Now since the experiment is carried out adiabatically, $q = 0$.

Therefore, from the first law,

$dU = 0 + w$.

or, $U_2 - U_1 = P_1V_1 - P_2V_2$

or, $U_1 + P_1V_1 = U_2 + P_2V_2$

or, $H_1 = H_2$

or, $\Delta H = 0$, i.e., the process is isoenthalpic.

Joule-Thomson Coefficient **(ì $_{JT}$)**

Joule-Thomson coefficient is defined as the change of temperature with pressure at constant enthalpy.

$$\mu_{JT} = (\partial T / \partial P)_H \approx (\Delta T/\Delta P)_H. \quad (5.23)$$

Nature of Joule-Thomson Coefficient

(i) $\mu_{JT} = +ve \Rightarrow$ cooling effect

(ii) $\mu_{JT} = -ve \Rightarrow$ heating effect

(iii) $\mu_{JT} = 0 \Rightarrow$ no temperature change; $\Delta T = 0$.

The temperature at which neither heating effect nor cooling effect is observed is known as the *inversion temperature*.

For any given pressure, a real gas has an inversion temperature, above which expansion at constant enthalpy causes the temperature to rise, and below which expansion at constant enthalpy causes cooling. For most gases at atmospheric pressure, the inversion temperature is fairly high (above room temperature), and so most gases at those temperature and pressure conditions are cooled; hydrogen and helium, however, show heating effect at room temperature.

Explanation

As a gas expands, the average distance between molecules grows. Because of intermolecular attractive forces, expansion causes an increase in the potential energy of the gas. If no external work is extracted in the process ("free expansion") and no heat is transferred, the total energy of the gas remains the same because of the conservation of energy. The increase in potential energy thus means a decrease in kinetic energy and, therefore, in temperature.

A second mechanism has the opposite effect. During gas molecule collisions, kinetic energy is temporarily converted into potential energy. As the average intermolecular distance increases, there is a drop in the number of collisions per time unit, which causes a decrease in average potential energy. Again, total energy is conserved, so this leads to an increase in kinetic energy (temperature). Below the Joule-Thompson inversion temperature, the former effect (work done internally against intermolecular attractive forces) dominates, and free expansion causes a decrease in temperature. Above the inversion temperature, the latter effect (reduced collisions causing a decrease in the average potential energy) dominates, and free expansion causes a temperature increase.

FREE ENERGY AND WORK FUNCTION

None of ΔH, ΔU or ΔS_{syst} alone can decide the direction of a spontaneous process if we focus our attention only on the system. But ΔS_{univ} at constant U and V can be used to decide the direction of a process. However, this is a very odd situation as studying reactions at constant energy condition is a tedious job. Experiments are conveniently performed at constant T and V or at constant T and P. Hence it is necessary to introduce new functions, which can decide the direction of spontaneous processes. The two new functions are defined as

$$G = H - TS, \tag{5.24}$$

$$A = U - TS, \tag{5.25}$$

where G is the Gibbs free energy and A is the Helmholtz free energy.

We can write equation (5.24) as

$$G = U + PV - TS,$$

Differential change in G is given by

$$dG = dU + PdV + VdP - TdS - SdT. \tag{5.26}$$

Combining first and second law of thermodynamics it can be shown that

$$dG = TdS + đw_{rev} + PdV + VdP - TdS - SdT.$$

At constant temperature and pressure $dT = 0$ and $dP = 0$.

Therefore, $(dG)_{T,P} = đw + PdV = đw - (-PdV)$,

where $(-đw)$ represents all types of work and $-PdV$ = mechanical work.

Therefore, $(-dG)_{T,P}$ = non-mechanical work

or, $-\Delta G = (-w_{net})$.

The decrease in free energy is equal to the maximum amount of work, which a system can do isothermally and isobarically over and above the mechanical work.

The differential change in Helmholtz free energy is

$$dA = dU - TdS - SdT. \tag{5.27}$$

Combining first and second laws of thermodynamics it can be shown that

$$dA = TdS + đw_{rev} - TdS - SdT.$$

At constant temperature,

$$dA = dw_{rev}.$$

For finite changes, $-(\Delta A)_T = (-w_{rev})$.

The decrease in Helmholtz free energy is equal to the maximum amount of work, which a system can do isothermally.

MAXWELL'S RELATIONS

We have come across four fundamental thermodynamic relations. These are:

(1) $dU = TdS - PdV$

(2) $dH = TdS + VdP$

(3) $dA = -PdV - SdT$

(4) $dG = VdP - SdT$.

From (1) it follows that

$$(\partial U/\partial S)_V = T \tag{5.28}$$

and $(\partial U/\partial V)_S = -P$. (5.29)

Differentiating equation (5.28) w.r.t. V and equation (5.29) w.r.t. S, we get

$(\partial^2 U / \partial V \partial S) = (\partial T/\partial V)_S$

and $(\partial^2 U / \partial S \partial V) = -(\partial P/\partial S)_V$.

Applying Euler's theorem,

we have $(\partial^2 U / \partial V \partial S) = (\partial^2 U / \partial S \partial V)$.

Therefore, $(\partial T/\partial V)_S = -(\partial P/\partial S)_V$. (5.30)

From (2), we have

$(\partial H / \partial S)_P = T$ (5.31)

and $(\partial H / \partial P)_S = V$. (5.32)

Differentiating equation (5.31) w.r.t. P and equation (5.32) w.r.t. S, we get

$(\partial^2 H / \partial P \partial S) = (\partial T / \partial P)_S$.

and $(\partial^2 H / \partial S \partial P) = (\partial V / \partial S)_P$

Applying Euler's theorem,

we have $(\partial^2 H / \partial P \partial S) = (\partial^2 H / \partial S \partial P)$.

Therefore, $(\partial T / \partial P)_S = (\partial V / \partial S)_P$. (5.33)

From (3), we have

$(\partial A / \partial V)_T = -P$ (5.34)

and $(\partial A / \partial T)_V = -S$. (5.35)

Differentiating equation (5.34) w.r.t. T and equation (5.35) w.r.t. V, we get

$(\partial^2 A / \partial T \partial V) = -(\partial P / \partial T)_V$

and $(\partial^2 A / \partial V \partial T) = -(\partial S / \partial V)_T$

Applying Euler's theorem,

we have $(\partial^2 A / \partial T \partial V) = (\partial^2 A / \partial V \partial T)$.

Therefore, $(\partial P / \partial T)_V = (\partial S / \partial V)_T$. (5.36)

From (4), we have

$(\partial G / \partial P)_T = V$ (5.37)

and $(\partial G / \partial T)_P = -S$. (5.38)

Differentiating equation (5.37) w.r.t. T and equation (5.38) w.r.t. V, we get

$(\partial^2 G / \partial T \partial P) = (\partial V / \partial T)_P$

and $(\partial^2 G / \partial P \partial T) = -(\partial S / \partial P)_T$.

Applying Euler's theorem,

we have $(\partial^2 G / \partial T \partial P) = (\partial^2 G / \partial P \partial T)$.

Therefore, $(\partial V / \partial T)_P = -(\partial S / \partial P)_T$. (5.39)

The equations (5.30), (5.33), (5.36), (5.39) are known as Maxwell's relations and are very useful in dealing with various physico-chemical phenoma.

THERMODYNAMIC EQUATIONS OF STATE

We have seen that $dU = TdS - PdV$.

Dividing this by dV and introducing the condition of constant temperature, we get

$(\partial U / \partial V)_T = T(\partial S / \partial V)_T - P$, but from Maxwell's relations $(\partial S / \partial V)_T = (\partial P/\partial T)_V$

Therefore, $(\partial U/\partial V)_T = T(\partial P/\partial T)_V - P$. (5.40)

Again, $dH = TdS + VdP$.

Dividing this by dP and introducing the condition of constant temperature, we get

$(\partial H / \partial P)_T = T(\partial S / \partial P)_T + V$.

But from Maxwell's relations $(\partial S/\partial P)_T = -(\partial V / \partial T)_P$.

Therefore, $(\partial H/ \partial P)_T = V - T(\partial V/\partial T)_P$. (5.41)

The above two equations (5.40) and (5.41) are called the *thermodynamic equations of state* as they relate to various variables of state.

Problem: Establish the relation,

$$C_p - C_v = \frac{\alpha^2 VT}{\beta}\text{; the terms having their usual significance.}$$

Solution: We know that $S = f(V, T)$.

Therefore, $dS = (\partial S/\partial V)_T dV + (\partial S/\partial T)_V dT$.

Dividing throughout by dT and introducing the condition of constant pressure, we get

$(\partial S/\partial T)_P = (\partial S/\partial V)_T(\partial V/\partial T)_P + (\partial S/\partial T)_V$.

Now $C_P = (\partial q/\partial T)_P = T(\partial S/\partial T)_P$

and $C_V = (\partial q/\partial T)_V = T(\partial S/\partial T)_V$.

Therefore, the equation becomes

$C_P / T = (\partial S/\partial V)_T(\partial V/\partial T)_P + C_V/T$

or, $C_p - C_v = T(\partial S/\partial V)_T(\partial V/\partial T)_P$.

From Maxwell's relations we have $(\partial S/\partial V)_T = (\partial P/\partial T)_V$.

Therefore, $C_p - C_v = T(\partial P/\partial T)_V(\partial V/\partial T)_P$.

The cyclic rule for the relation, $P = f(V, T)$. is

$(\partial P/\partial T)_V(\partial T/\partial V)_P(\partial V/\partial P)_T + 1 = 0$.

This gives $(\partial P/\partial T)_V = -1/[(\partial T/\partial V)_P(\partial V/\partial P)_T] = -(\partial V/\partial T)_P/(\partial V/\partial P)_T$.

Therefore, $C_p - C_v = -T[(\partial V/\partial T)_P]^2/(\partial V/\partial P)_T$.

Now $\alpha = (1/V)(\partial V/\partial T)_P$ and $\beta = -(1/V)(\partial V/\partial P)_T$.

$$\text{Therefore, } C_p - C_v = \frac{\alpha^2 VT}{\beta}.$$

TEMPERATURE DEPENDENCE OF FREE ENERGY: GIBBS-HELMHOLTZ EQUATION

Gibbs-Helmholtz equation relates the free energy change to the enthalpy change and the rate of change of free energy with temperature. This equation may be applied to any change whether physical or chemical that takes place at constant pressure.

$G = H - TS$ [from equation (5.24)]

$= U + PV - TS$.

For an infinitesimal change,

$dG = dU + PdV + VdP - TdS - SdT$ [from equation (5.26)]

Combining the first and the second laws of thermodynamics and considering that the external work is due to volume change only, it can be shown that

$dG = VdP - SdT$. (5.42)

In an isobaric change, $dP = 0$ and equation (5.42) becomes

$(\partial G / \partial T)_P = -S$ [from equation (5.38)]

Substitution of equation (5.38) in equation (5.24) gives

$G = H + T(\partial G / \partial T)_P$

or, $G - T(\partial G / \partial T)_P = H$. (5.43)

This is referred to as Gibbs Helmholtz equation.

On dividing equation (5.43) by T^2 we get

$[G - T(\partial G / \partial T)_P] / T^2 = H / T^2$

or, $[\partial(G/T) / \partial T]_P = -H / T^2$. (5.44)

Hence, $[G/T]_P = -\int H / T^2 \, dT + I_1$. (5.45)

This is another form of Gibbs-Helmholtz equation, where I_1 is an integration constant.

Equation (5.44) can be reset into another form –

$[\partial(G/T) / \partial(1/T)]_P = H$.

For a change from state 1 to state 2 it can be written

$G_1 = H_1 + T(\partial G_1 / \partial T)_P$.

$G_2 = H_2 + T(\partial G_2 / \partial T)_P$.

And $\Delta G = \Delta H + T[\partial(\Delta G) / \partial T]_P$. (5.46)

Therefore, $[\partial(\Delta G/T) / \partial T]_P = -\Delta H / T^2$. (5.47)

Similar relations can be deduced for A.

CONDITION OF SPONTANEITY AND EQUILIBRIUM

The criteria for spontaneity and equilibrium can be expressed in terms of U, H, S, A and G.

We have already seen that for an irreversible system (spontaneous), $TdS > dU + PdV$

And for a reversible system (equilibrium), $TdS = dU + PdV$.

Combining the two we can write, $TdS \geq dU + PdV$.

Criterion in terms of entropy

$dS_{U,V} > 0$ Spontaneity

$dS_{U,V} = 0$ Equilibrium

Criterion in terms of internal energy

$dU_{S,V} < 0$ Spontaneity

$dU_{S,V} = 0$ Equilibrium

Criterion in terms of Enthalpy

We know that $dH = dU + PdV + VdP$.

Therefore,

$dH_{S,P} < 0$ Spontaneity

$dH_{S,P} = 0$ Equilibrium

Criterion in terms of Free energy

$$dG = dH - TdS - SdT$$
$$= dU + PdV + VdP - TdS - SdT$$

Therefore,

$dG_{P,T} < 0$ Spontaneity

$dG_{P,T} = 0$ Equilibrium

Criterion in terms of work function

$$dA = dU - TdS - SdT$$

Therefore,

$dA_{V,T} < 0$ Spontaneity

$dA_{V,T} = 0$ Equilibrium

The criterion in terms of free energy change is the most useful criterion to decide spontaneity of a reaction.

Problem: H_2O (l, –10°C) → H_2O (s, –10°C)

Calculate the change in Gibbs free energy for the process, given vapour pressure of water at –10°C is 2.15 mm of Hg and the vapour pressure of ice at –10°C is 1.95 mm of Hg.

Solution: $\Delta G = RT \ln (P_2 / P_1) = 8.314 \times 263 \ln (1.95 / 2.15) = -213.49$ J/mole.

THERMODYNAMICS OF OPEN SYSTEMS—PARTIAL MOLAR PROPERTIES

The thermodynamic properties U, H, S, A and G are extensive properties as their values depend on the mass or the number of moles of the system. In the discussions so far we assumed the system to be closed, and only changes of temperature and pressure were considered. However, in case of an open system containing two or more components, there can be change in the number of moles of various components as well. In such cases an extensive property X must be a function of the number of moles as well as temperature and pressure.

Let $n_1, n_2, \ldots, n_j$ be the number of moles of the constituents 1, 2,...,j.

Then $X = f(T, P, n_1, n_2, \ldots, n_j)$,

where $n_1 + n_2 + \cdots + n_j = \text{N}$ (total number of moles).

For a small change in temperature, pressure and number of moles of the components,

$$dX = (\partial X/\partial T)_{P,n_1,n_2,\cdots,n_j}\, dT + (\partial X/\partial P)_{T,n_1,n_2,\cdots,n_j}\, dP + (\partial X/\partial n_1)_{T,P,n_2,\cdots,n_j}\, dn_1 + (\partial X/\partial n_2)_{T,P,n_1,n_3,\cdots,n_j}$$
$$dn_2 + \cdots + (\partial X/\partial n_i)_{T,P,n_1,n_3,\cdots,n_j}\, dn_i + \cdots \tag{5.48}$$

The quantity $(\partial X/\partial n_i)_{T,P,n_1,n_3,\cdots,n_j}$ where $j \neq i$ is called the partial molar property of the ith component. It is often represented as x_i. Thus for the ith component:

Partial molar internal energy $= (\partial U/\partial n_i)_{T,P,n_1,n_3,\cdots,n_j}$

Partial molar enthalpy $= (\partial H/\partial n_i)_{T,P,n_1,n_3,\cdots,n_j}$

Partial molar entropy $= (\partial S/\partial n_i)_{T,P,n_1,n_3,\cdots,n_j}$

Partial molar volume $= (\partial V/\partial n_i)_{T,P,n_1,n_3,\cdots,n_j}$

However, the most important one is the partial molar free energy = $\left(\partial G/\partial n_i\right)_{T,P,n_1,n_3,\cdots,n_j}$. It is more popularly designated as the chemical potential (μ_i).

Thus, the chemical potential of a given substance is the change in free energy of the system due to addition of 1 mole of the ith component at constant temperature and pressure when other components remain unaltered.

The Gibbs-Duhem Relation

$$dG=(\partial X/\partial T)_{P,n_1,n_2,\cdots,n_j}dT+(\partial G/\partial P)_{T,n_1,n_2,\cdots,n_j}dP+\left(\partial G/\partial n_1\right)_{T,P,n_2,\cdots,n_j}dn_1+\left(\partial X/\partial n_2\right)_{T,P,n_1,n_3,\cdots,n_j}$$
$$dn_2+\cdots+(\partial G/\partial n_i)_{T,P,n_1,n_3,\cdots,n_j}dn_i+\cdots \quad (5.49)$$

Integrating we have

$$G_{P,T}=\mu_1 n_1+\mu_2 n_2+\mu_3 n_3+\cdots \mu_i n_i+\cdots \quad (5.50)$$

Now complete differentiation of equation (5.50) gives

$$dG_{P,T}=(\mu_1 dn_1+\mu_2 dn_2+\mu_3 dn_3+\cdots+\mu_i dn_i+\cdots)+(n_1 d\,\mu_1+n_2 d\,\mu_2+n_3 d\,\mu_3+\cdots+n_i d\,\mu_i+\cdots). \quad (5.51)$$

Subtracting equation (5.49) from equation (5.51),

$$(n_1 d\,\mu_1+n_2 d\,\mu_2+n_3 d\,\mu_3+\cdots+n_i d\,\mu_i+\cdots)=0. \quad (5.52)$$

These three relations for chemical potentials,

$G_{P,T}=\mu_1 n_1+\mu_2 n_2+\mu_3 n_3+\cdots \mu_i n_i+\cdots=\sum \mu_i n_i$ [from equation (5.50)]

$dG_{P,T}=\mu_1 dn_1+\mu_2 dn_2+\mu_3 dn_3+\cdots \mu_i dn_i+\cdots=\sum \mu_i dn_i$ [from equation (5.49)]

$0=n_1 d\,\mu_1+n_2 d\,\mu_2+n_3 d\,\mu_3+\cdots+n_i d\,\mu_i+\cdots=\sum n_i d\,\mu_i$ [from equation (5.52)]

are commonly known as the Gibbs-Duhem relations.

FUGACITY, ACTIVITY AND ACTIVITY COEFFICIENT

We have seen that the small change in free energy may be expressed as

$dG = VdP - SdT$ [from equation (5.42)]

If we hold temperature constant and vary only the pressure, we can prepare equation (5.42) for integration from pressure P_1 to P_2 as follows:

$dG = VdP$.

Then

$$G(P_2)-G(P_1)=\int_{P_1}^{P_2} VdP$$

or,
$$G(P_2)-G(P_1)=\int_{P_1}^{P_2} VdP. \quad (5.53)$$

Equation (5.53) is quite general and applies to all isotropic substances—solids, liquids, ideal gases and real gases.

Ideal Gases

For ideal gases, $V = nRT/P$.

With this substitution equation (5.53) becomes

$$G(P_2) - G(P_1) + \int_{P_1}^{P_2} (nRT/P)\,dP. \tag{5.54}$$

After integration equation (5.54) becomes

$$G(P_2) = G(P_1) + nRT\ln(P_2/P_1). \tag{5.55}$$

Let P_2 range over the pressures of interest to us and call it just P and P_1 be some standard state pressure and let us call it P^0. Let us now divide the resulting equation by the number of moles of gas, n. With these changes equation (5.55) is written as

$$\frac{G(P_2)}{n} = \frac{G(P_1)}{n} + RT\ln\left(P/P^0\right), \tag{5.56}$$

Now G/n is the chemical potential μ.

Therefore, $\mu = \mu^0 + RT\ln(P/P^0)$,

where μ^0 is the chemical potential at the standard state. In most cases we will set the standard state pressure equal to one atmosphere. Hence,

$$\mu = \mu^0 + RT\ln P. \tag{5.57}$$

Non-ideal Gases

The above equation is not valid for a real gas. For a real gas the evaluation of $\int VdP$ is very complicated. It was G. N. Lewis who devised the following expression:

$$\mu = \mu^0 + RT\ln f, \tag{5.58}$$

where f was named the fugacity of the gas. The fugacity has units of pressure and it is a function of pressure. It contains all the information on the nonideality of the gas. For an ideal gas the fugacity is the same as the pressure. Since all real gases become ideal in the limit as pressure goes to zero we must have

Lt $P \to 0$, $(f/P) = 1$.

At other pressures $f/P = \gamma$ or, $f = \gamma P$,

where γ is called the activity coefficient.

Non-ideal solution

For ideal solutions chemical potential may be expressed as

$$\mu_i = \mu^0_i + RT\ln x_i. \tag{5.59}$$

In case of non ideal solutions the expression becomes

$$\mu_i = \mu^0_i + RT\ln a_i. \tag{5.60}$$

The quantity, a_i, is called the activity of component i. The activity term has no unit. For an ideal solution,

$a_i = x_i$.

For a real solution,

$a_i = \gamma_i x_i$.

In general, the activity coefficient γ_i is a unitless parameter that contains all of the non-ideality of a system.

APPLICATION OF THERMODYNAMICS TO PHASE TRANSITION: CLAPEYRON-CLAUSIUS EQUATION

Let us consider the following phase change:

Liquid $\rightleftharpoons$ Vapour

Let G_1 and G_2 be the free energies of the liquid and vapour, which are in equilibrium.

Let us consider a virtual change of a small amount of liquid vaporizing at equilibrium. The drop in free energy of the liquid is $-\Delta G_1$ and the gain in free energy of the vapour is $+\Delta G_2$. Since the system is in equilibrium, net free energy change $\Delta G = 0$.

Therefore, $-\Delta G_1 + \Delta G_2 = 0$

or, $\Delta G_1 = \Delta G_2$

Now $\Delta G = V\Delta P - S\Delta T$.

Therefore, $V_1\Delta P - S_1\Delta T = V_2\Delta P - S_2\Delta T$

or, $\Delta P / \Delta T = (S_2 - S_1) / (V_2 - V_1) = \Delta S / (V_2 - V_1)$,

i.e., $dP / dT = TdS / T(V_2 - V_1) = L_{Vap} / T(V_2 - V_1)$, (5.61)

where L_{Vap} is the latent heat of vaporization.

This is called the *Clapeyron-Clausius equation*. This equation is applicable to all phase changes.

Now, for vaporization, V_2 (volume of gas) >> V_1 (volume of liquid). Therefore, the above equation may be simplified as follows:

$dP / dT = L_{Vap} / TV_2 = L_{Vap} P / RT^2$ (since $PV_2 = RT$, assuming that the vapour behaves like ideal gas)

or, $dP / P = (L_{Vap} / R)(dT / T^2)$.

Integrating between suitable limits, we get

$\ln (P_2 / P_1) = (L_{Vap} / R) [1/T_1 - 1/T_2]$. (5.62)

Problem

What is the boiling point of water at a place where the atmospheric pressure is 600 mm? (L_{vap} = 540 cal /g)

Solution

From the Clapeyron-Clausius equation,

$\ln (P_2 / P_1) = (L_{Vap} / R) [1/T_1 - 1/T_2]$.

For water, if P_1 = 1 atm = 760 mm, then b.p, T_1 = 373K.

L_{Vap} = 540 × 18 = 9720 cal / mole

P_2 = 600 mm.

So $\ln (600/760) = (9720/1.987) [1/373 - 1/T_2]$

Therefore, T_2 = 366.4 K.

REACTION DYNAMICS

6

INTRODUCTION

The rate law or rate equation for a chemical reaction is an equation which links the reaction rate with concentrations or pressures of reactants and constant parameters (normally rate coefficients and partial reaction orders). To determine the rate equation for a particular system one combines the reaction rate with a mass balance for the system. For a general reaction, $A + B \to C$, the simple rate equation is of the form:

$r = k(T)\,[A]^m\,[B]^n.$

In this equation, $[X]$ expresses the concentration of a given X, usually in mol/litre and $k(T)$ is known as the reaction rate coefficient or rate constant, although it is not really a constant, because it includes everything that affects reaction rate outside concentration, mainly temperature but also ionic strength, surface area of the adsorbent or light irradiation.

The exponents, n and m are called *reaction orders* and depend on the reaction mechanism

ORDER AND MOLECULARITY

The order of reaction with respect to a certain reactant is defined, in chemical kinetics, as the power to which its concentration term in the rate equation is raised.

For example, a given chemical reaction $A + 2B \to C$, with a rate equation, $r = k[A]^1[B]^2$, the reaction order with respect to A would be 1 and with respect to B would be 2, the total reaction order would be $1 + 2 = 3$. It is not necessary that the order of a reaction is a whole number—zero and fractional values, of order are possible—but they tend to be integers. Reaction orders can be determined only by experiment.

The reaction order is not necessarily related to the stoichiometry of the reaction, unless the reaction is elementary. Complex reactions may or may not have reaction orders equal to their stoichiometric coefficients

If the concentration of one of the reactants remains constant (because it is a catalyst or it is in great excess with respect to the other reactants), its concentration can be included in the rate constant, obtaining a pseudo constant. If B is the reactant whose concentration is constant, then

$r = k\,[A]\,[B] = k'\,[A].$

The second order rate equation has been reduced to a *pseudo first order rate equation*. Zero order reactions are often seen for thermal chemical decompositions where the reaction rate is independent of the concentration of the reactant (changing the concentration has no effect on the speed of the reaction).

Molecularity is the number of colliding molecular entities that are involved in a single reaction step. While the order of a reaction is derived experimentally, the molecularity is a theoretical concept and can only be applied to elementary reactions. In elementary reactions, the reaction order, the molecularity and the stoichiometric coefficient are the same, although only numerically, because they are different concepts.

- A reaction involving one molecular entity is called unimolecular.
- A reaction involving two molecular entities is called bimolecular.
- A reaction involving three molecular entities is called termolecular. Termolecular reactions in solutions or gas mixtures are very rare because of the improbability of three molecular entities simultaneously colliding.

FIRST ORDER REACTIONS

A first order reaction depends on the concentration of only one reactant. Other reactants can be present, but each will be zero order. The rate law for a first order reaction is

$r = k[A]$,

k is the first order rate constant that has units of 1/time.

If, and only if, this first-order reaction (1) occurs in a closed system, (2) there is no net build-up of intermediates, and (3) there are no other reactions occurring, it can be shown that

$-d[A]/dt = k[A]$

The integrated first order rate law is

$\ln [A] = -kt + \ln [A]_0$

or, $[A] = [A]_0 e^{-kt}$.

A plot of ln $[A]$ vs. time t gives a straight line with a slope equal to the reaction rate constant.

The rate expression for the first order reaction may be expressed in terms of product formed.

	A	$\rightarrow$	B
Time $t = 0$	a		0
Time $t = t$	$a - x$		x

If 'a' moles per litre is the initial concentration of A and 'x' moles per litre be the concentration of the product formed after time 't', then the rate expression is given by

$-d(a-x)/dt = k(a-x)$

or, $dx/dt = k(a-x)$

or, $dx/(a-x) = kdt$.

Integrating between limits 0 to x and 0 to t respectively, we get

$\ln a/(a-x) = kt$,

the unit of k is time^{-1}.

The half life or the time required to reduce the concentration of the reactant to half of its original value, can be determined for a first order reaction as follows:

$t_{½} = (1/k) \ln a/\{a - (a/2)\} = (\ln 2)/k = 0.693/k$.

Characteristic of a first order reaction is that all the half-lives are equal.

Examples of first order reactions:

$2H_2O_2$ (l) $\rightarrow$ $2H_2O$ (l) + O_2 (g)

SO_2Cl_2 (l) $\rightarrow$ SO_2 (g) + Cl_2 (g).

These are all first order with respect to the reactant.

SECOND ORDER REACTIONS

A second order reaction depends on the concentrations of one second order reactant, or two first order reactants.

For a second order reaction, its reaction rate is given by

$r = k [A]^2$ or $r = k [A] [B]$.

There may be two possibilities in case of reaction of the second order:

(i) the rate may be proportional to the product of two equal initial concentrations

(ii) the rate may be proportional to the product of two different initial concentrations.

CASE 1

	$2A$	$\rightarrow$	Product	or,		A	+	B	$\rightarrow$	Product
Time $t = 0$	a		0		Time $t = 0$	a		a		0
Time $t = t$	$a - x$		x		Time $t = t$	$a - x$		$a - x$		x

If 'a' moles per litre is the initial concentration of A and 'x' moles per litre be the concentration of the product formed after time 't' then the rate expression is given by

$-d(a - x)/dt = k(a - x)^2$

or, $dx/dt = k(a - x)^2$

or, $dx/(a - x)^2 = kdt$.

Integrating between limits 0 to x and 0 to t respectively, we get

$x/a(a - x) = kt$

Half life, $t_{1/2} = 1/ak$.

CASE 2

	A	+	B	$\rightarrow$	Product
Time $t = 0$	a		b		0
Time $t = t$	$a - x$		$b - x$		x

Here 'b' moles per litre is the initial concentration of B.

$-d[A]/dt = -d[A]/dt = k[A][B]$

$= -d(a - x)/dt = k(a - x)(b - x)$

or, $dx/dt = k(a - x)(b - x)$

or, $dx/\{(a - x)(b - x)\} = kdt$.

Integrating between limits 0 to x and 0 to t respectively, we get

$1/(a - b) \ln \{b(a - x)\}/\{a(b - x)\} = kt$.

Unit of k is $\text{conc}^{-1}\ \text{time}^{-1}$ for second order reactions.

Example of a second order reaction:

$2NO_2\,(g) \rightarrow 2NO + O_2\,(g)$.

THEORIES OF REACTION RATES

Temperature and reaction rates: The Arrhenius equation:

It was found that in homogeneous thermal reactions the rate of reaction doubled or trebled for every ten degree rise in temperature. The ratio called the temperature coefficient is $(k_{t+10})/k_t \approx 2$ to 3.

The Arrhenius equation is an expression that shows the dependence of the rate constant k of chemical reactions on the temperature T (in Kelvin) and activation energy E_a. He gave the empirical formulation,

$d\ln k/dt = E_a/RT^2$

or, $k = Ae^{-Ea/RT}$,

where A is the pre-exponential factor or simply the prefactor and R is the gas constant. The units of the pre-exponential factor are identical to those of the rate constant and will vary depending on the order of the reaction. If the reaction is first order it has the units s^{-1}, and for that reason it is often called the frequency factor or attempt frequency of the reaction. Taking the natural logarithm of the Arrhenius equation yields:

$$\ln k = \ln A - E_a/RT.$$

Differentiating w.r.t. temperaure,

$$d(\ln k)/dT = E_a/RT^2.$$

So, when a reaction has a rate constant which obeys the Arrhenius equation, a plot of ln(k) versus T^{-1} gives a straight line, whose slope and intercept can be used to determine E_a and A. This procedure has become so common in experimental chemical kinetics that practitioners have taken to using it to *define* the activation energy for a reaction. That is the activation energy is defined to be ($-R$) times the slope of a plot of ln(k) vs. ($1/T$) :

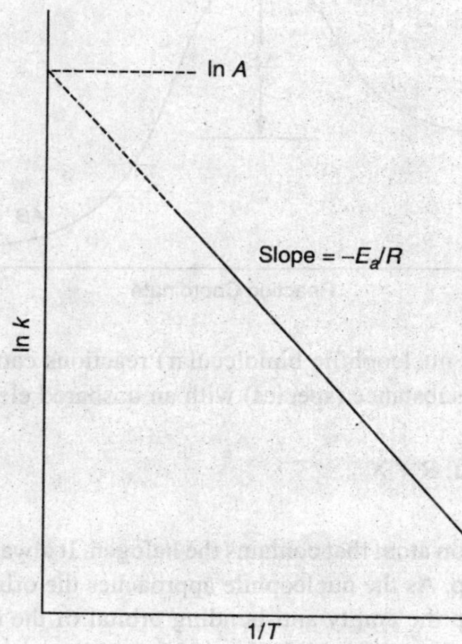

Arrhenius argued that in order for reactants to be transformed into products, they first needed to acquire a minimum amount of energy, called the activation energy E_a. At an absolute temperature T, the fraction of molecules that have a kinetic energy greater than E_a is proportional to $e^{-Ea/RT}$. The concept of activation energy explains the exponential nature of the relationship.

Transition State Theory

Transition state theory provides an approach to explain the temperature and concentration dependence of the rate law. For example, for the elementary reaction,

$$A + BC \rightarrow AB + C.$$

The rate law is $-r_A = kC_AC_{BC} = Ae^{-Ea/RT}\,C_AC_{BC}$.

In transition state theory (TST) an activated molecule is formed during the reaction at the transition state between forming products from reactants.

$$A + BC \rightleftharpoons A - B - C^{\#} \xrightarrow{\nu_1} AB + C.$$

The rate of reaction is equal to the product of the frequency, ν_I, of the activated complex crossing the barrier and the concentration of the transition state complex,

$-r_A = \nu_I C_{ABC}\#.$

The transition state molecule $A - B - C^{\#}$ and the reactants are in pseudo-equilibrium at the top of the energy barrier

$K_C^{\#} = C_{ABC}\# / C_A C_{BC}.$

Combining,

$-r_A = \nu_I K_C^{\#} C_A C_{BC}.$

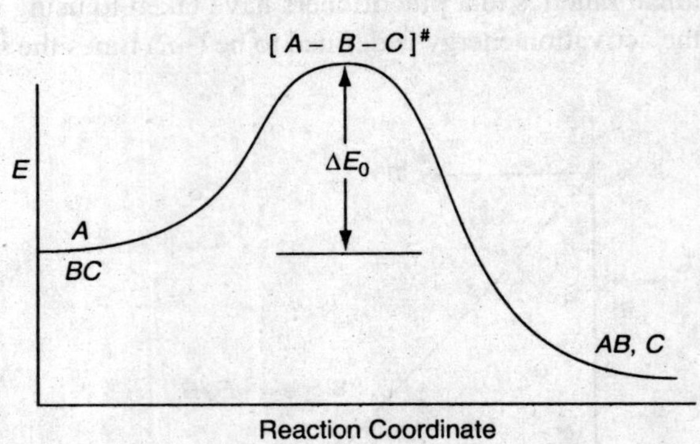

Many of the S_N2 (substitution nucleophilic bimolecular) reactions can be described by transition state theory. A nucleophile ($Nu:^-$) is a substance (species) with an unshared electron. It is a species that seeks a positive centre.

$$Nu^- + RX \rightarrow Nu{:}\,R + X^-,$$

where X is a halogen.

The nucleophile seeks the carbon atom that contains the halogen. It always approaches from the backside, directly opposite the leaving group. As the nucleophile approaches the orbital that contains the nucleophile electron pairs, it begins to overlap the empty anti-bonding orbital of the carbon atom bearing the leaving group (Solomon, T.W.G, *Organic Chemistry*, 6/e Wiley 1996, p.233).

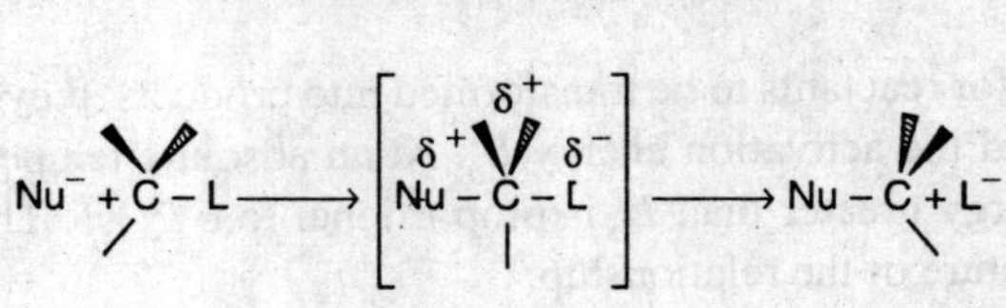

7

ELECTROCHEMISTRY

INTRODUCTION

At the atomic level all chemical processes involve electrical interactions. Thus, in a sense all chemistry may be encompassed within the sphere of electrochemistry. Strictly speaking, electrochemistry is the study of solutions of electrolytes and of phenomenon occurring at the electrodes dipped in these solutions. William Gilbert, Queen Elizabeth's physician, was the first to coin the term 'electric' and applied it to the phenomenon of attraction of small bits of paper or pith to surfaces rubbed with fur. However, it was Michael Faraday whose researches laid down the foundations of electrochemistry. Owing largely to him, the elegant nomenclature used in today's electrochemistry was devised—electrode, electrolyte, ion, anion, cation.

CONDUCTANCE OF ELECTROLYTIC SOLUTIONS

Metallic conductors are known to follow Ohm's Law,

$$V = IR, \quad (7.01)$$

where I is the current, V is the potential difference between the terminals of the conductor and R is the resistance. In a metallic conductor of length (l) and cross section (a) the resistance (R) is given as

$$R = \rho l / a. \quad (7.02)$$

where ñ is the specific resistance or resistivity of the conductor. Specific resistance may be defined as the resistance of a unit length of conductor of unit cross section.

Reciprocating equation (7.02), we get

$$1 / R = a / (\rho l). \quad (7.03)$$

Here $1/R$ or the reciprocal of resistance is called the conductance (G). It is a measure of how easily electricity flows along a certain path through an electrical element. The SI derived unit of conductance is the siemens (formerly referred to as the reciprocal ohm or *mho*). The reciprocal of specific resistance ($1/\rho$) is called the specific conductance or conductivity κ. If the electrical conductance between opposite faces of a 1-metre cube of material is 1 siemens, then the material's electrical conductivity is 1 siemens per metre. Thus

$$\kappa = G \times (l/a), \quad (7.04)$$

where (l/a) is called the *cell constant*.

In Chemistry we are concerned with solutions and not metallic conductors. A modified definition of specific conductance or conductivity may be stated as: the conductance of a solution enclosed between two electrodes of unit area of cross section, unit distance apart. Now the conductance of the solution is dependent on the number of ions present or the concentration of the solution. Hence, terms that can demonstrate this dependence have to be introduced.

Equivalent conductance (Λ): The conductance of a solution containing 1g-equivalent of dissolved electrolyte placed between two electrodes, unit distance apart, is called its *equivalent conductance*. Direct experimental determination of equivalent conductance entails the use of electrodes of enormous sizes so the specific conductance is first determined experimentally followed by the determination of equivalent conductance from it using the following formula:

$$\Lambda = a/l \times \kappa. \tag{7.05}$$

If we take a solution having concentration c g-equivalent per litre, then the volume of solution containing 1g-equivalent of dissolved electrolyte would be $1/c$ litre. If this volume of solution is placed between two electrodes, unit distance apart, then area of cross section, $a = 1/c$ sq m.

Therefore, $\Lambda = \kappa/c$. (7.06)

Molar conductance (μ): it is defined as the conductance of a solution containing 1g-mole of dissolved electrolyte placed between two electrodes, unit distance apart,

$$\mu = \kappa/c', \tag{7.07}$$

where c′ is the molar concentration.

VARIATION OF CONDUCTANCE WITH CONCENTRATION

Specific conductance of an ionic solution increases with increasing concentration. For strong electrolytes, the increase in specific conductance with increase of concentration is sharp. However, for weak electrolytes, the increase in specific conductance is more gradual. In both cases, the increase is due to an increase in the number of ions per unit volume of the solution. For strong electrolytes, which are completely ionized, the increase in specific conductance is almost proportional to the concentration. In weak electrolytes, however, the increase in specific conductance is not large due to the low ionization of the electrolytes and consequently the specific conductance does not go up so rapidly as in the case of strong electrolytes.

Equivalent conductance has been found to increase with increase in dilution of the electrolytic solution. However the trend is different for strong and weak electrolytes.

Strong electrolyte: The change in equivalent conductance of strong electrolytes with dilution is relatively small and ultimately tends to reach a limiting value. The accepted idea is that strong electrolytes are completely ionized in solution even in ordinary concentrations, so the number of ions will remain the same in all concentrations. With dilution, the speed of the ions increases. This is because the density of ions in strong electrolytes is very high; each ion is surrounded by oppositely charged ions resulting in interionic attraction. The ions may even give rise to ion pairs of the type ($A^+ B^-$). Such interionic attraction effectively reduces the speed of the ions and hence equivalent conductance of the solution. With increasing dilution the ions move further apart thereby reducing the interionic attraction and thus move more freely. At infinite dilution, the interionic attraction practically disappears and the limiting value of equivalent conductance is reached.

Weak electrolyte: The equivalent conductance of weak electrolytes is low at ordinary concentration owing to its small extent of ionization. With increase in dilution the ionization increases considerably and with the increase in number of ions the equivalent conductance also increases. In case of weak electrolyte interionic attraction is not of any importance.

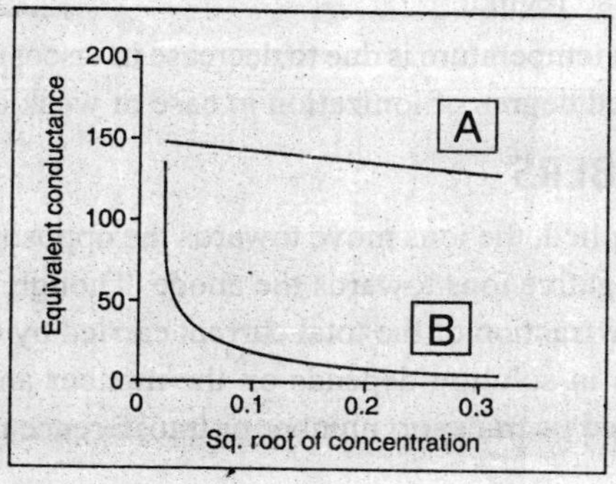

Fig. 7.1 : Equivalent conductance vs. $\sqrt{c}$.

EQUIVALENT CONDUCTANCE AT INFINITE DILUTION

An important relation can be obtained by extrapolating the curve for strong electrolyte to $c \to 0$ where all interionic attractions vanish. The limiting value thus obtained by extrapolation is called equivalent conductance at infinite dilution. The following empirical relation was suggested by Kohlrausch for the variation of equivalent conductance of strong electrolyte with dilution.

$$\Lambda = \Lambda_0 - b\sqrt{c}, \qquad (7.08)$$

where b = a constant for the given electrolyte.

Λ_0 = equivalent conductance at infinite dilution.

The experimental realization of equivalent conductance at infinite dilution for weak electrolyte is not possible owing to the steep increase in Ë at high dilutions.

KOHLRAUSCH'S LAW OF INDEPENDENT MIGRATION OF IONS

The law states that at infinite dilution, where all electrolytes are completely ionized and where all interionic effects are absent, each ion migrates independently of its co-ion and contributes a definite share to the total equivalent conductance of the electrolyte.

$$\Lambda_0 = \lambda_0^+ + \lambda_0^-, \qquad (7.09)$$

where λ_0^+ and λ_0^- are the conductance of cation and anion respectively.

In case of weak electrolytes where direct determination of Λ_0 is not possible Kohlrausch's law may be applied. For example, we can calculate Λ_0 for acetic acid in the following manner—

$$\Lambda_0^{HAc} = \Lambda_0^{NaAc} + \Lambda_0^{HCl} - \Lambda_0^{NaCl}$$

$$= \lambda_0^{Na^+} + \lambda_0^{Ac^-} + \lambda_0^{H^+} + \lambda_0^{Cl^-} - \lambda_0^{Na^+} - \lambda_0^{Cl^-} = \lambda_0^{Ac^-} + \lambda_0^{H^+}.$$

EFFECT OF TEMPERATURE ON CONDUCTANCE

The conductance of all electrolytes increases with increasing temperature. The variation of equivalent conductance at infinite dilution with temperature is given by an empirical relation—

$$\Lambda_0^t = \Lambda_0^{25}\,[\,1 + x\,(\,t - 25)], \qquad (7.10)$$

where Λ_0^t and Λ_0^{25} are the equivalent conductances at t°C and 25°C and x is a constant for each electrolyte.

For strong electrolytes even at appreciable concentration, the above equation holds true. For weak electrolytes the variation is not so regular.

The rise in conductance with temperature is due to decrease in viscosity of the solution resulting in increase in speed of ions and an increased degree of ionization in case of weak electrolytes.

TRANSFERENCE NUMBERS

When a potential gradient is applied, the ions move towards the opposite electrodes – the positive ions move towards the cathode and the negative ions towards the anode. Though the total charge deposited in the two electrodes remains constant, the fraction of the total current carried by the ions is not necessarily equal. The current transported by the ions in solution depends on the number and speed of the ions. The fraction of current carried by an ion is called its transport number or transference number.

$$t_+ = I_+ / I \quad (7.11)$$
$$= n_+ z_+ eu / n_+ z_+ e(u+v) \quad (7.12)$$
$$= u/(u+v) \quad (7.13)$$
$$= \lambda^0_+ / \Lambda_0. \quad (7.14)$$
$$t_- = I_- / I \quad (7.15)$$
$$= n_- z_- ev / n_- z_- e(u+v) \quad (7.16)$$
$$= v/(u+v) \quad (7.17)$$
$$= \lambda^0_- / \Lambda_0. \quad (7.18)$$

Therefore, $t_+ + t_- = 1$ (7.19)

and $t_+ / t_- = u/v.$ (7.20)

Here

t_+ & t_- = transference number of cation and anion respectively

n_+ & n_- = number of cations and anions per cc respectively

z_+ & z_- = charge of each cation and anion respectively

I = current passing through the solution

I_+ & I_- = current carried by cations and anions respectively

e = quantity of electricity associated with unit charge

u = velocity of cation

v = velocity of anion

λ^0_+ & λ^0_- = ion conductance of cations and anions respectively

Λ_0 = equivalent conductance at infinite dilution.

Problem: The resistance of a conductivity cell filled with 0.02 NKCl at 18°C is 17.60 ohms, filled with 0.1 N CH_3COOH, the resistance is 91.8 ohms. If specific conductance for 0.02 N KCl is 2.399×10^{-3} $ohm^{-1}cm^{-1}$ and ion conductances at infinite dilution of H^+ and CH_3COO^- are 315 $ohm^{-1}cm^2 \ equiv^{-1}$ and 35 $ohm^{-1}cm^2 \ equiv^{-1}$, then calculate degree of dissociation of CH_3COOH.

Solution: For KCl solution R = 17.60 ohms; $\kappa = 2.399 \times 10^{-3}$ $ohm^{-1}cm^{-1}$.

So cell constant = $17.60 \times 2.399 \times 10^{-3} = 0.042 \ cm^{-1}$.

For CH_3COOH solution R = 91.8 ohms.

So $\kappa = 0.042 / 91.8 = 4.575 \times 10^{-4}$ $ohm^{-1}cm^{-1}$.

$\Lambda(CH_3COOH) = (4.575 \times 10^{-4} \times 10^3)/0.1 = 4.575\ ohm^{-1} cm^2\ equiv^{-1}$.

$\Lambda^0(CH_3COOH) = \lambda_0(CH_3COO^-) + \lambda_0(H^+) = 315 + 35 = 350.$

So degree of dissociation $\alpha = \Lambda/\Lambda^0 = 4.575/350 = 0.013.$

IONIC MOBILITY

The velocity with which an ion would move under a potential gradient of 1 volt / cm in a solution is called its ionic mobility.

$$u = \lambda^0_+ / F \tag{7.21}$$

$$= t_+ \Lambda_0 / F \tag{7.22}$$

$$v = \lambda^0_- / F \tag{7.23}$$

$$= t_- \Lambda_0 / F, \tag{7.24}$$

where F = one Faraday.

THE INTERIONIC ATTRACTION THEORY

We have already seen that though the number of ions remains constant for a strong electrolytic solution at all concentrations, yet the equivalent conductance is found to increase with dilution. This is because at ordinary concentrations the ions are close together and hence the interionic forces retard their motion. As dilution increases the ions move further apart thereby decreasing the interionic force and as a result increase the equivalent conductance. The interionic forces results due to four different types of forces:

(i) The electric force acting on the ion due to the external electric field.

(ii) Frictional or viscous force which retards the motion of the ions.

(iii) As a result of electrostatic attraction between oppositely charged ions, each ion in solution is surrounded by an ionic atmosphere of opposite charge. In absence of an electric field this ionic atmosphere is spherically and symmetrically distributed. When a potential is applied across the solution, ions move towards oppositely charged electrodes. For example, under the influence of the electric field when a cation moves towards the cathode, its negatively charged ion atmosphere moves towards the anode and thus lags behind. As the number of negative ions is greater behind than in front of the central ions the motion of the cation is retarded. The retarding force acts in the opposite direction to that of the motion of the central ion and this dragging effect is known as the *asymmetry* effect or *relaxation effect*.

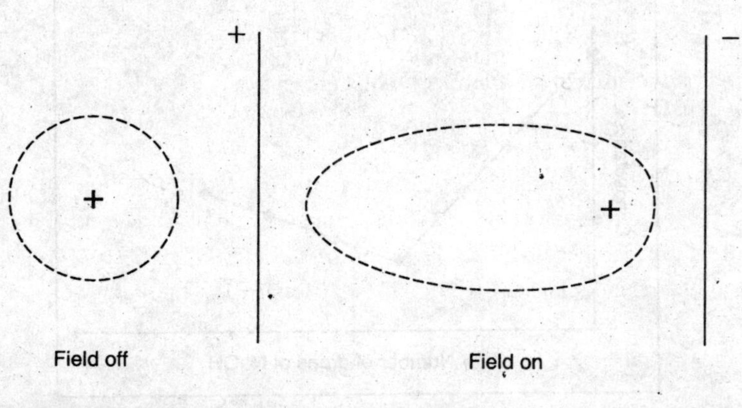

Fig. 7.2 : Ionic atmosphere in presence and absence of electric field.

(iv) The ionic atmosphere associated with the solvent molecules also moves in the opposite direction to that of the ions. Thus the cations have to swim upstream against the flow of the solvent. This phenomenon is known as *electrophoretic effect.*

HYDRATION OF IONS

The ion conductance depends on charge and speed of the ions. So one might expect that lighter and smaller ions having similar charge will have greater speed. However, it was found that the ion conductance of the cations increased with atomic weight from Li^+ to Cs^+. This may be attributed to the fact that the Li^+ ion being smaller, the electric field in its neighborhood is larger and is associated with a larger number of water molecules. This greater degree of hydration in Li^+ as compared to K^+ or Cs^+ gives Li^+ a larger dimension thereby reducing its speed.

CONDUCTOMETRIC TITRATION

The conductance of an electrolytie solution depends primarily on the concentration and types of the ions. Conductivity measurement can thus provide a sensitive measure of the changes taking place in ionic composition in the course of chemical reaction occurring in the solution and this is used in conductometric titration.

Let us consider a solution of a strong acid, hydrochloric acid, HCl for instance, to which a solution of a strong base, sodium hydroxide, (NaOH) is added. The reaction

$$H^+Cl^- + Na^+OH^- \rightarrow H_2O + Na^+Cl^-$$

occurs. For each amount of NaOH added equivalent amount of hydrogen ions is removed. Effectively, the faster moving H^+ cation is replaced by the slower moving Na^+ ion, and the conductivity of the titrated solution as well as the measured conductance of the cell fall. This continues until the equivalence point is reached, at which we have a solution of sodium chloride, NaCl. If more base is added, an increase in conductivity or conductance is observed, since more ions are being added and the neutralization reaction no longer removes an appreciable number of any of them. Consequently, in the titration of a strong acid with a strong base, the conductance has a minimum at the equivalence point. This minimum can be used instead of an indicator dye to determine the end-point of the titration. Conductometric titration curve, that is a plot of the measured conductance or conductivity values against the number of millilitres of NaOH solution, is shown in Fig.7.3.

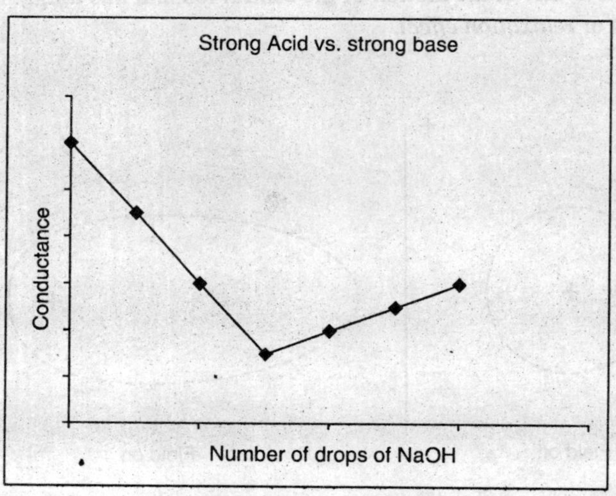

Fig. 7.3 : Conductometric titration of strong acid with strong base.

The position of the equivalence point may be localized precisely as the point of intersection of two straight lines both determined using experimental readings obtained. It makes the conductometric titration more objective and independent of a nature of an indicator used in the end-point method. This is one of the advantages of the instrumental method.

Now let us consider the titration of solution of weak acid, such as acetic acid, CH_3COOH, using a solution of strong base, NaOH. As we know, the weak acids, as well as other weak electrolytes, are dissociated into very small extent and they exist in solution essentially in the form of the neutral acid molecules. When a solution of NaOH is added the reaction occurs

$$CH_3COOH + Na^+OH^- \rightarrow H_2O + CH_3COO^-Na^+$$

and, as is seen, the undissociated molecules of acetic acid are transformed into dissociated molecules of sodium acetate. And this transformation increases the conductivity of the solution.

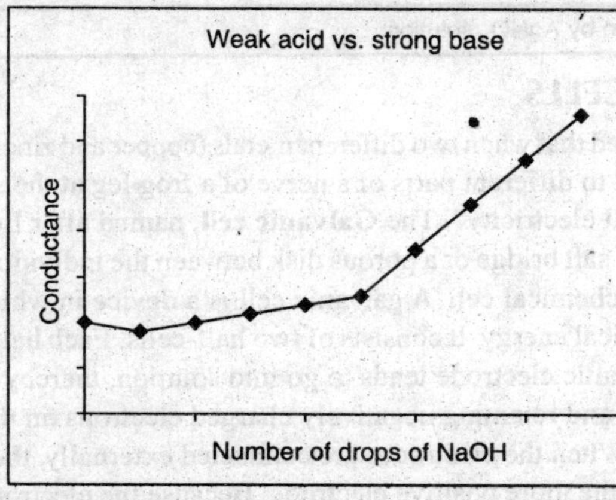

Fig. 7.4 : Conductometric titration curve for acetic acid titrated using solution of sodium hydroxide.

It should be noted, however, that an initial decrease in conductivity of the solution may be observed after addition of the first drops of titrant due to common ion effect. Thus, a mild increase in conductivity of a titrated solution is observed until the equivalence point is reached, at which we have a solution of sodium acetate, CH_3COONa. If an excess of titrant, that is the sodium hydroxide solution, is added a sharp increase in conductivity is observed. This distinct difference in a rate of increase is related to the fact that the excess OH^- anions, as well as the protons, exhibit particular mechanisms of charge migration. More detailed inspection of the conductometric titration curve presented in Fig.7.4. indicates that the equivalence point is less sharp than that observed for the strong acid. Thus, it should be localized as the intersection point of two lines determined by two section of the conductometric curve.

Precipitation titrations may also be studied conductometrically. In a reaction of the type

$$K^+Cl^- + Ag^+NO_3^- \rightarrow AgCl + K^+NO_3^-$$

On addition of titre Ag NO_3 the Cl^- ions are replaced by NO_3^- ions. The ion conductances of Cl^- and NO_3^- are not much different. Hence, the conductance of the solution with successive additions of Ag NO_3 does not alter much until equivalence point is reached. On addition of excess of Ag NO_3 there will be a sharp rise in conductance owing to the unused Ag^+ and NO_3^- ions. The intersection of the two lines gives the end-point.

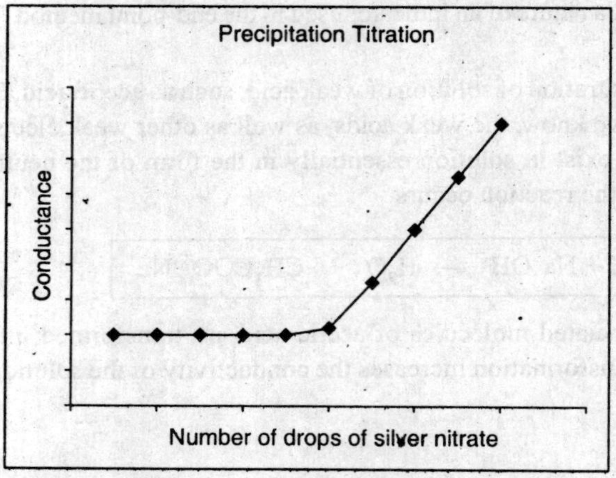

Fig. 7.5 : Titration of KCl solution by $AgNO_3$ solution.

ELECTROCHEMICAL CELLS

In 1780, Luigi Galvani discovered that when two different metals (copper and zinc, for example) were connected together and then both touched to different parts of a nerve of a frog leg at the same time, they made the leg contract. He called this "animal electricity". The **Galvanic cell**, named after Luigi Galvani, consists of two different metals connected by a salt bridge or a porous disk between the individual half-cells. It is also known as a *voltaic cell* and an electrochemical cell. A galvanic cell is a device in which free energy of a chemical process is converted into electrical energy. It consists of two half-cells. Each half-cell has an electrode and an electrolyte. The metal of a metallic electrode tends to go into solution, thereby releasing positively charged metal ions into the electrolyte, and retaining negatively charged electrons on the electrode. Thus each half-cell has its own half-reaction. When the electrodes are connected externally, the electrons tend to flow from the more negative electrode to the more positive electrode. Because the electrons have negative charge, this produces an electric current which is opposite to the electron flow. At the same time, an equal electric current flows through the electrolyte. By definition, the anode is the electrode where oxidation (removal of electrons) takes place and the cathode is the electrode where reduction (gain of electrons) takes place. At the surface of separation of the electrode and electrolyte there exists a potential difference called the *electrode potential*. Electricity is generated due to potential difference between two electrodes. If there are no other potential differences in the cell the EMF of the cell is taken to be equal to the algebraic sum of the two electrode potentials, allowance being made for the direction of the potential difference when asssessing its sign. The **standard electrode potential**, abbreviated E^0, is the measure of individual potential of any electrode at standard ambient conditions, which is at a temperature of 298K, solutes at a concentration of 1 M, and gases at a pressure of 1 bar.

FREE ENERGY AND EMF OF A CELL REACTION

If the EMF of a voltaic cell is E volts, and the process taking place is accompanied by the passage of n faradays i.e, nF coulombs, then the work done by the system in the cell is nFE volt-coulombs or joules. Now if the cell is a reversible one, then this work represents the maximum work. Since electrical work does not involve mechanical work resulting from a volume change, it may be taken as equal to the change of free energy accompanying the cell reaction.

Therefore, $-\Delta G = nFE$. (7.25)

Criteria of spontaneity in terms of ΔG and E:

Reaction	ΔG	E
Equilibrium	0	0
Spontaneous	–ve	+ve
Non-spontaneous	+ve	–ve

Entropy and Enthalpy changes from EMF:

From Gibbs-Helmholtz equation,

$\Delta G = \Delta H + T\,[\partial(\Delta G) / \partial T]_P$. (7.26)

Now, $\Delta G = -nFE$. (7.27)

Therefore, differentiating ΔG w.r.t. temperature at constant pressure gives

$[\partial(\Delta G) / \partial T]_P = -nF\,(\partial E / \partial T)_P$. (7.28)

From equations (7.26) and (7.28), we get

$\Delta G = \Delta H - nFT\,(\partial E / \partial T)_P$. (7.29)

Also $\Delta G = \Delta H - T\,\Delta S$. (7.30)

Comparing equations (7.29) and (7.30), we get

$\Delta S = nF\,(\partial E / \partial T)_P$ (7.31)

Again, from equations (7.27) and (7.29), we get

$-nFE = \Delta H - nFT\,(\partial E / \partial T)_P$. (7.32)

Therefore, $\Delta H = nF\,[T\,(\partial E / \partial T)_P - E]$. (7.33)

NERNST EQUATION

Nernst equation gives the electrode potential *(E)*, relative to the standard electrode potential, (E^0), of the electrode couple or, equivalently, of the half-cells of a battery.

For the reaction,

$Ox + ne^- \rightarrow red$

$E = E^0 + (RT/nF)\ln(a_{Ox} / a_{red})$. (7.34)

The activities of pure solid or liquid phases are taken as unity. For a solution at room temperature (25°C) replacing activities by concentration:

$E = E^0 + (0.0591/n)\log([Ox] / [red])$. (7.35)

CALCULATION OF STANDARD ELECTRODE POTENTIALS

To overcome the difficulty of measuring individual potential of an electrode, an electrode of unknown reduction potential can be paired with a reference electrode of a known potential. The ultimate reference is the standard hydrogen electrode (SHE) whose potential is defined to be exactly zero volt.

The SHE consists of a small platinum sheet coated with platinum black and placed in a solution where the hydrogen ion activity is unity. The electrode is partly immersed in the solution and partly exposed to an atmosphere of hydrogen gas. Pure hydrogen gas at 1 atmosphere is continuously bubbled through the solution. The electrode is represented as

$Pt \mid H_2\ (P = 1\ atm) \mid H^+\ (a = 1)$

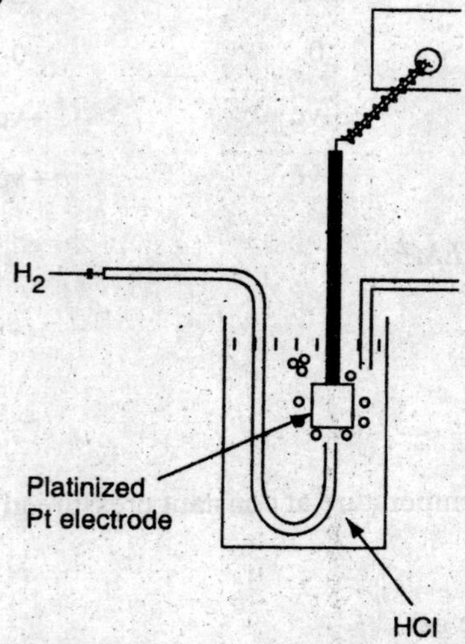

Fig. 7.6 : Standard Hydrogen Electrode.

For example, to measure the standard reduction potential of zinc metal electrode, an electrochemical cell can be built with a zinc metal electrode (e.g., a zinc electrode immersed in 1 M $ZnSO_4$ solution) as the anode. The anode half-reaction is then:

$Zn(s) \rightarrow Zn^{2+}(aq,\ 1\ M) + 2e^-$.

The SHE is used as the cathode, and overall cell can be written in shorthand form:

$Zn(s) \mid Zn^{2+}(aq, 1\ M) \parallel H^+(aq, 1\ M) \mid H_2(g,\ 1\ atm)$,

Since the reduction half-reaction has a potential of zero, the emf of the cell, E^0_{cell}, corresponds to the potential of the zinc metal electrode, because

$E^0_{cell}(0.76V) = E^0_{2H^+(aq)\rightarrow H_2(g)}(0V) + E^0_{Zn(s)\rightarrow Zn^{2+}(aq)}(0.76V)^0$ where the superscript $(^0)$ denotes that standard states are employed.

Since the electrode potentials are conventionally defined as reduction potentials, the sign of the potential for the metal electrode being oxidized must be reversed when calculating the overall cell potential. The electrode potentials are independent of the number of electrons transferred and so the two electrode potentials can be simply combined to give the overall *cell* potential even if different numbers of electrons are involved in the two electrodes.

TABLE 7.01: *Standatd Reduction Potentials at* 25°C

Half-Reaction	E^0 (V)
$Li^+_{(aq)} + e^- \longrightarrow Li_{(s)}$	−3.05
$K^+_{(aq)} + e^- \longrightarrow K_{(s)}$	−2.93
$Ba^{2+}_{(aq)} + 2e^- \longrightarrow Ba_{(s)}$	−2.90

contd.

$Sr^{2+}_{(aq)} + 2e^- \longrightarrow Sr_{(s)}$	−2.89
$Ca^{2+}_{(aq)} + 2e^- \longrightarrow Ca_{(s)}$	−2.87
$Na^+_{(aq)} + e^- \longrightarrow Na_{(s)}$	−2.71
$Mg^{2+}_{(aq)} + 2e^- \longrightarrow Mg_{(s)}$	−2.37
$Be^{2+}_{(aq)} + 2e^- \longrightarrow Be_{(s)}$	−1.85
$Al^{3+}_{(aq)} + 3e^- \longrightarrow Al_{(s)}$	−1.66
$Mn^{2+}_{(aq)} + 2e^- \longrightarrow Mn_{(s)}$	−1.18
$2\ H_2O + 2e^- \longrightarrow H_{2(g)} + 2OH^-_{(aq)}$	−0.83
$Zn^{2+}_{(aq)} + 2e^- \longrightarrow Zn_{(s)}$	−0.76
$Cr^{3+}_{(aq)} + 3e^- \longrightarrow Cr_{(s)}$	−0.74
$Fe^{2+}_{(aq)} + 2e^- \longrightarrow Fe_{(s)}$	−0.44
$Cd^{2+}_{(aq)} + 2e^- \longrightarrow Cd_{(s)}$	−0.40
$PbSO_{4(s)} + 2e^- \longrightarrow Pb_{(s)} + SO_4{}^{2-}_{(aq)}$	−0.31
$Co^{2+}_{(aq)} + 2e^- \longrightarrow Co_{(s)}$	−0.28
$Ni^{2+}_{(aq)} + 2e^- \longrightarrow Ni_{(s)}$	−0.25
$Sn^{2+}_{(aq)} + 2e^- \longrightarrow Sn_{(s)}$	−0.14
$Pb^{2+}_{(aq)} + 2e^- \longrightarrow Pb_{(s)}$	−0.13
$2\ H^+_{(aq)} + 2e^- \longrightarrow H_{2(g)}$	0.00
$Sn^{4+}_{(aq)} + 2e^- \longrightarrow Sn^{2+}_{(aq)}$	+0.13
$Cu^{2+}_{(aq)} + e^- \longrightarrow Cu^+_{(aq)}$	+0.13
$SO_4{}^{2-}_{(aq)} + 4H^+_{(aq)} + 2e^- \longrightarrow SO_{2(g)} + 2\ H_2O$	+0.20
$AgCl_{(s)} + e^- \longrightarrow Ag_{(s)} + Cl^-_{(aq)}$	+0.22
$Cu^{2+}_{(aq)} + 2e^- \longrightarrow Cu_{(s)}$	+0.34
$O_{2(g)} + 2\ H_2 + 4e^- \longrightarrow 4\ OH^-_{(aq)}$	+0.40
$I_{2(s)} + 2e^- \longrightarrow 2\ I^-_{(aq)}$	+0.53
$MnO_4{}^-_{(aq)} + 2\ H_2O + 3e^- \longrightarrow MnO_{2(s)} + 4\ OH^-_{(aq)}$	+0.59

$O_{2(g)} + 2H^+_{(aq)} + 2e^- \longrightarrow H_2O_{2(aq)}$	+0.68
$Fe^{3+}_{(aq)} + e^- \longrightarrow Fe^{2+}_{(aq)}$	+0.77
$Ag^+_{(aq)} + e^- \longrightarrow Ag_{(s)}$	+0.80
$Hg_2^{2+}{}_{(aq)} + 2e^- \longrightarrow 2\,Hg_{(l)}$	+0.85
$2Hg^{2+}_{(aq)} + 2e^- \longrightarrow Hg_2^{2+}{}_{(aq)}$	+0.92
$NO_3^-{}_{(aq)} + 4H^+_{(aq)} + 3e^- \longrightarrow NO_{(g)} + 2\,H_2O$	+0.96
$Br_{2(l)} + 2e^- \longrightarrow 2\,Br^-_{(aq)}$	+1.07
$O_{2(g)} + 4\,H^+_{(aq)} + 4e^- \longrightarrow 2H_2O$	+1.23
$MnO_{2(s)} + 4\,H^+_{(aq)} + 2e^- \longrightarrow Mn^{2+}_{(aq)} + 2H_2O$	+1.23
$Cr_2O_7^{2-}{}_{(aq)} + 14\,H^+_{(aq)} + 6e^- \longrightarrow 2\,Cr^{3+}_{(aq)} + 7\,H_2O$	+1.33
$Cl_{2(g)} + 2e^- \longrightarrow 2\,Cl^-_{(aq)}$	+1.36
$Au^{3+}_{(aq)} + 3e^- \longrightarrow Au_{(s)}$	+1.50
$MnO_4^-{}_{(aq)} + 8H^+_{(aq)} + 5e^- \longrightarrow Mn^{2+}_{(aq)} + 4\,H_2O$	+1.51
$Ce^{4+}_{(aq)} + e^- \longrightarrow Ce^{3+}_{(aq)}$	+1.61
$PbO_{2(s)} + 4H^+_{(aq)} + SO_4^{2-}{}_{(aq)} + 2e^- \longrightarrow PbSO_{4(s)} + 2\,H_2O$	+1.70
$H_2O_{2(aq)} + 2\,H^+_{(aq)} + 2e^- \longrightarrow 2H_2O$	+1.77
$Co^{3+}_{(aq)} + e^- \longrightarrow Co^{2+}_{(aq)}$	+1.82
$O_{3(g)} + 2\,H^+_{(aq)} + 2e^- \longrightarrow O_{2(g)} + H_2O$	+2.07
$F_{2(g)} + 2e^- \longrightarrow F^-_{(aq)}$	+2.87

Since the values are given in their ability to be reduced, the bigger the standard reduction potentials, the easier they are to be reduced. In other words, they are simply better oxidizing agents. For example, F_2 has 2.87V and Li^+ has –3.05V. F_2 reduces easily and is, therefore, a good oxidizing agent. In contrast, Li^+ would rather undergo oxidation (hence a good reducing agent). Thus Zn^{2+} whose standard reduction potential is–0.76V can be oxidized by any other electrode whose standard reduction potential is greater than –0.76V (e.g., H^+(0V), Cu^{2+}(0.16V), F_2(2.87V)) and can be reduced by any electrode with standard reduction potential less than –0.76V (e.g., H_2(–2.23V), Na^+(–2.71V), Li^+(–3.05V)).

CELL CONVENTION

Let us take the examle of the Daniell cell consisting of zinc electode dipped in $ZnSO_4$ solution and copper electrode dipped in $CuSO_4$ solution. The standard reduction potentials of the two electrodes are

$E^0_{Zn^{2+}/Zn} = -0.763$ V

$E^0_{Cu^{2+}/Cu} = +0.337$ V.

Thus in this cell Cu^{2+} will act as an oxidizing agent and itself gets reduced.

Cell representation

Anode (–) Cathode (+)

$$Zn(s) \mid ZnSO_4(aq) \parallel CuSO_4(aq) \mid Cu(s)$$

Oxidation Reduction

At the anode,

$$E_{Zn^{2+}/Zn} = E^0_{Zn^{2+}/Zn} - (RT/2F)\ln a_{Zn}/a_{Zn^{2+}}$$

$$= E^0_{Zn^{2+}/Zn} - (0.0591/2)\log a_{Zn^{2+}} \quad (\text{since } a_{Zn} = 1)$$

Similarly, at the cathode,

$$E_{Cu^{2+}/Cu} = E^0_{Cu^{2+}/Cu} - (0.0591/2)\log a_{Cu^{2+}}$$

$$E_{Cell} = E_{red}\,(\text{R.H.S.}) - E_{red}\,(\text{L.H.S.})$$

$$= E_{Cu^{2+}/Cu} - E_{Zn^{2+}/Zn}.$$

Half-cell reactions;

R.H.S. : $Cu^{2+} + 2e^- \rightarrow Cu$

L.H.S. : $Zn(s) \rightarrow Zn^{2+} + 2e^-$

Overall raction : $Cu^{2+} + Zn \rightarrow Zn^{2+} + Cu$.

Problem: For the following cell, write down the cell reactions and calculate the EMF at 25°C:

$Ag(s) \mid Ag^+ (a = 0.1) \parallel Zn^{2+} (a = 0.1) \mid Zn$

Given, $E^0_{Ag^+/Ag} = 0.799$ volt; $E^0_{Zn^{2+}/Zn} = -0.763$ volt. Is the cell reaction spontaneous?

Solution: *Cell reaction:*

At the R.H.E. $Zn^{2+} + 2e^- \rightarrow Zn(s)$

At the L.H.E. $(Ag \rightarrow Ag^+ + e^-) \times 2$

Overall reaction: $Zn^{2+} + 2Ag \rightarrow Zn + 2Ag^+$

$$E_{cell} = E^0_{cell} - (RT/2F)\ln\left(a_{Ag^+}\right)^2/\left(a_{Zn^{2+}}\right)$$

$$= E^0_{Zn^{2+}/Zn} - E^0_{Ag^{2+}/Ag} = -0.763 - 0.799 = -1.533 \text{ volt.}$$

As E_{cell} is negative, so the reaction is not spontaneous.

THE CALOMEL REFERENCE ELECTRODE

The standard hydrogen electrode is often inconvenient to use as a reference electrode experimentally. It has to have freshly made platinum electrodes, a solution of HCl with a very aocurately determined concentration corresponding to unit activity of hydrogen ions, and the safe provision of a hydrogen gas source at 1 atmosphere pressure.

It is generally much more convenient to use a **calomel electrode** as a secondary reference electrode.

It consists of mercury at the bottom, over which a paste of mercury – mercurous chloride, is placed. A solution of KCl saturated with mercurous chloride is placed over the paste. A platinum wire is used for making electrical contact. There may be three types of electrodes depending on the concentration of the KCl solution.

TABLE 7.2: *Different types of Calomel Electrode*

Type	Concentration of KCl solution	Potential (at 25°C)
Saturated Calomel Electrode (SCE)	Saturated	0.2415 V
Normal Calomel Electrode (NCE)	1(N)	0.2800 V
Decinormal Calomel Electrode (DNCE)	0.1(N)	0.3338 V

The electrode reaction when the electrode acts as a cathode may be written as

$$\tfrac{1}{2}Hg_2Cl_2\,(s) + e^- \rightarrow Hg + Cl^-\left(a_{Cl^-}\right)$$

The potential is given as

$$E_{cal} = E^0_{cal} - (RT/F)\ln a_{Cl^-}$$

Of the three electrodes SCE is the most popular one. To obtain the potential of any other electrode it is combined with the calomel electrode, and EMF of the resulting cell is measured.

DETERMINATION OF pH

The solution whose pH is to be determined is taken in a vessel and a Pt–wire carrying a Pt-foil coated with platinum black is immersed in the solution. Pure H_2-gas at 1 atmosphere is continuously bubbled so that the Pt black and the solution are saturated and at equilibrium. This half cell is then coupled with a reference electrode (a SHE or SCE) through a KCl salt bridge.

For SHE Cell $Pt|H_2O(p=1\,atm.)|H^+\left(a_{H^+}\right)||H^+\,(a=1)|H_2\,(p=1\,atm.)|Pt$

$$E_{Cell} = E_{SHE} - E^0_{H^+/\frac{1}{2}H_2} + RT/F \ln p^{1/2}_{H_2}/a_{H^+}$$

$$= 0 - 0 - RT/F \ln a_{H^+} = -0.0591 \log a_{H^+}.$$

Now $pH = -\log a_{H^+}$ $\therefore E_{Cell}/0.0591 = pH$

For SCE, Cell $Pt|H_2\,(p=1\,atm.)/H^+(a_{H^+})||Cl^-|Hg_2Cl_2\,(S)|Hg$

$$E_{Cell} = E_{SHE} - E^0_{H^+/\frac{1}{2}H_2} + RT/F \ln p^{1/2}_{H_2}/a_{H^+}$$

$$= 0.2415V - 0 - RT/F \ln a_{H^+}$$

$$\therefore (E_{Cell} - 0.2415)/0.0591 = pH.$$

STRUCTURE AND REACTIVITY OF ORGANIC MOLECULES

8

INTRODUCTION

Chemical reactivity of a substance is determined by the ease with which bonds are broken in reactants and new ones being formed in products. Whenever a reagent attacks a molecule, displacement of electrons takes place and chemical reaction occurs. During the course of reaction, reactants are either directly converted into products (one step reaction) or some intermediates are formed whose number determines the steps of a chemical reaction. For example, in two-step reactions one intermediate is formed and in three-step reactions two intermediates are formed. The kind, shape and structure of these intermediate species determine which product is formed and in what amount starting with the reactants. In this chapter a brief discussion on electron displacement effects, shapes and structures of intermediate organic species and types of organic reactions is made.

ELECTRONEGATIVITY

It is defined as the power of an atom in a molecule to attract the bonded electrons towards itself. It is not a fixed quantity. It depends on the following factors:

(i) ***Valence state of the atoms***: The electronegativity of atom increases with its positive oxidation state, since the tendency to attract an electron will increase with the positive charge of the cation. It also decreases with the negative oxidation state of an atom, since the tendency to attract an electron will decrease with the negative charge of the anion.

For example: electronegativity of Fe^{3+} > electronegativity of Fe^{2+}.

(ii) ***Hybridization***: The electronegativity of carbon increases with the increasing *s*-character of its hybrid orbital. This is due to high penetration power of *s*- orbital electrons, since they are nearer to the nucleus and hence exert stronger attractive power.

For example:

Hybridization	*sp*	sp^2	sp^3
s-character in the hybrid orbitals	50%	33%	25%
Electronegativity	3.29	2.75	2.48

Whenever a reagent (called attacking agent) attacks a molecule (called substrate) having covalent bond, chemical reaction takes place provided partial or complete displacement of the covalent bond electrons in molecule takes place.

Since an attacking reagent always bears either a positive charge or a negative charge for a reaction to take place on the covalent bond, the latter must possess oppositely charged centres. In other words, the substrate molecule although as a whole is electrically neutral, it must develop polarity on some of its carbon atoms or substituents linked together which is made possible by displacement (partial or complete) of the bonding electrons. The electronic displacement in a molecule in turn may be due to certain effects, some of which are permanent (e.g., inductive effect and mesomeric effect) and others are temporary (e.g., electrometric effect and inductometric effect). The former effects are permanently operating in the real molecule and are known as polarization effect while the latter are brought into play by the attacking reagent and as soon as the attacking reagent is removed, the electronic displacement disappears. Such effects are called polarizability effects. Sometimes s and d subscripts are used to represent static (permanent) and dynamic (temporary), i.e., time variable effects respectively. Such changes or effects involving displacement of electron in the substrate molecule are often called electron displacement effect.

FACTORS INFLUENCING A CHEMICAL REACTION

A reaction takes place only when a reagent attacks a molecule. So, the electron density in the substrate molecule will govern the nature of the attacking reagent. Substrate although as a whole electrically neutral, it must develop polarity on some of its carbon atoms and substituents linked together for reaction to occur. Polarity in a substrate develops through the electronic displacements in a molecule. The displacement of electrons in a substrate is due to certain effects, some of which are permanent, and others are temporary.

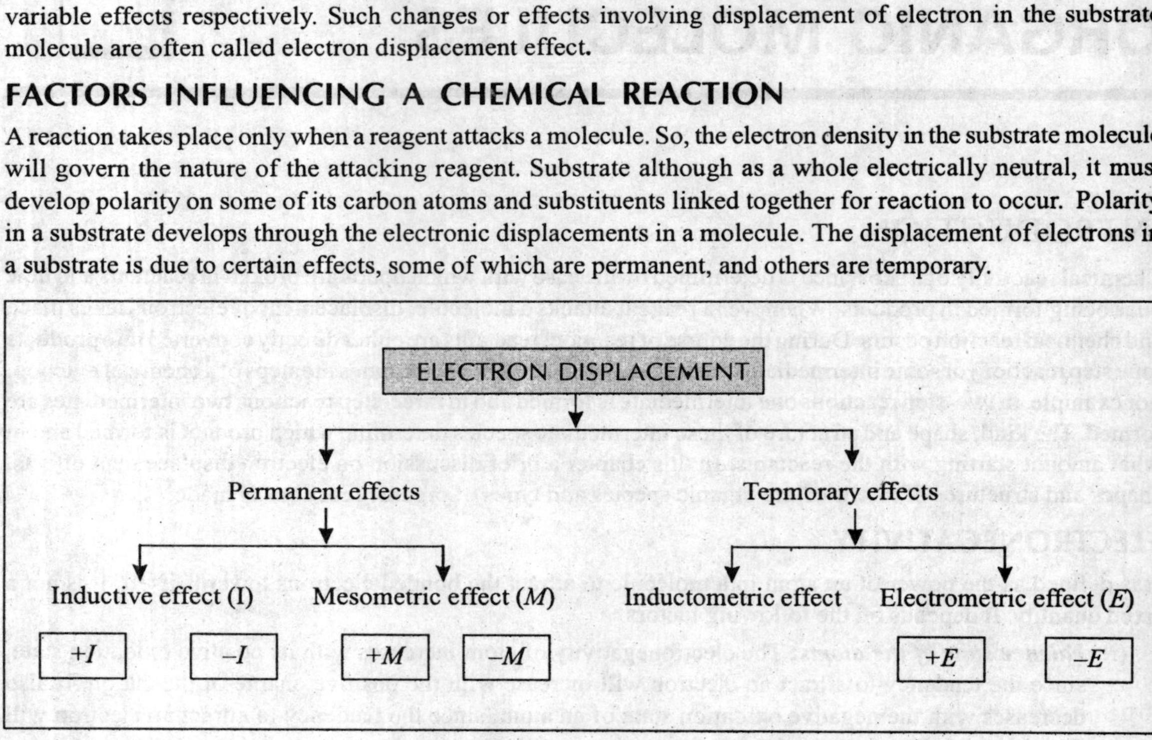

Permanent effects are permanently operating in the real molecule and are known as polarization effects or static effects. Temporary effects are brought into play by the attacking reagent and as soon as the attacking reagent is removed, the electronic displacement of substrate disappears. Such effects are also known as polarizability effects or dynamic (time variable) effects.

The following types of electron displacement effects operate in electronic displacement:

1. Inductive effect
2. Mesomeric effect
3. Inductometric effect
4. Electrometric effect

INDUCTIVE EFFECT

The inductive effect is defined as the permanent displacement of electrons forming a covalent bond towards the more electronegative element or group. This introduces a certain degree of polarity in the bond. The more electronegative atom acquires a small negative charge (∂^{-}) and the less electronegative atom acquires a small positive charge (∂^{+}). Assuming electronegativity of X > C > Y, then the electron pair forming the

C–X bond is somewhat more attracted towards the atom X with the result X attains a partial negative charge (∂^-) while the carbon attains a partial positive charge (∂^+).

$$C:X => \overset{\partial^+}{C} \longrightarrow \overset{\partial^-}{X}$$

Similarly, the electron pair forming the C–Y bond can be represented as

$$C:Y => \overset{\partial^-}{C} \longleftarrow \overset{\partial^+}{Y}$$

When the electronegative atom joins by ó bond to a chain of carbon atoms, the positive charge on the carbon atom is relayed to the other carbon atoms. But the partial positive charge is gradually decreasing away from polar bond and at third or fourth carbon atom it is almost zeros for all practical purposes.

$$\overset{\partial^{++++}}{C_4} \longrightarrow \overset{\partial^{+++}}{C_3} \longrightarrow \overset{\partial^{++}}{C_2} \longrightarrow \overset{\partial^{+}}{C_1} \longrightarrow \overset{\partial^{-}}{X}$$

This process of electron shift along a chain of atoms due to the presence of some polar bond in the molecule is termed as inductive effect.

Inductive effect may be due to the atom or groups and is represented by the symbol → the arrow pointing towards the more electronegative element. Relative inductive effect has been measured with reference to hydrogen. The order of electron withdrawing effect (−I effect) is

$-NO_2 > -CN > -COOH > -F \sim -COOH > -Cl > -Br > -I > -OH \sim -OR > -C_6H_5 > -H$ and the order of electron releasing (+ I effect) is

$-H > Me_3C \rightarrow Me_2CH \rightarrow MeCH_2 \rightarrow -C_2H_5 > -CH_3$

The alkyl groups are less electron withdrawing than the hydrogen and are, therefore, considered as electron releasing. The effect is additive, and greater the number of electron withdrawing group the stronger is the effect.

Applications

Inductive effect is useful in correlating structure with reactivity

(a) **Acid strength of aliphatic carboxylic acid**: The strength of acid depends on the extent of ionization– the greater the ionization the stronger is the acid. The strength of acid is denoted by the numerical value of pK_a ($pK_a = -\log K_a$), where K_a is the acid dissociation constant. Smaller the numerical value, stronger is the acid.

$$CH_3-\overset{O}{\overset{\|}{C}}-OH \longrightarrow CH_3-\overset{O}{\overset{\|}{C}}-O^- + H^+ \quad (pKa = 4.76)$$

Acetic acid

$$H-\overset{O}{\overset{\|}{C}}-OH \longrightarrow H-\overset{O}{\overset{\|}{C}}-O^- + H^+ \quad (pKa = 3.77)$$

Formic acid

Formic acid is stronger than acetic acid due to +I effect of CH_3- group present in the latter.

(b) **Acid strength of aromatic carboxylic acid**: The α-carbon of benzoic acid is sp^2 hybridized. Hence, benzoic acid is a stronger acid than its saturated analogue cyclohexane carboxylic acid. Electron withdrawing group substituted at *o*- and *p*- position enhances the acid strength.

(c) **Dipole moment**: Greater the inductive effect more will be the dipole moment in the molecule since inductive effect leads to a dipolar character in the molecule. For example, dipole moments $CH_3-F > CH_3-Cl > CH_3-Br > CH_3-I$, since –I effect of F > Cl > Br > I.

(d) **Bond length**: Greater the inductive effect, smaller will be the bond length in the molecule since inductive effect leads to ionic character in the bond. For example, bond length increases in the order $CH_3-F < CH_3-Cl < CH_3-Br < CH_3-I$, since the –I effect decreases in the order F > Cl > Br > I.

(e) **Reactivity of alkyl halides**: Due to the presence of C–X polar bond in alkyl halides, alkyl halides are more reactive than the corresponding alkanes. Reactivity is directly related to C–X bond polarity.

(f) **Basic strength of amines**: Amines are basic because they are electron pair donor or proton acceptor. The readiness with which it happens determines the relative strength of amines. Groups which have (+I) effect, when attached to nitrogen atom make it simpler for the lone pair of electrons on nitrogen to be easily available hence they increase the basic strength.

But the groups with (–I) effect makes it difficult for the easy availability of lone pair of electrons on nitrogen; hence, they decrease the basic strength.

The increasing order of basic strength is as follows:

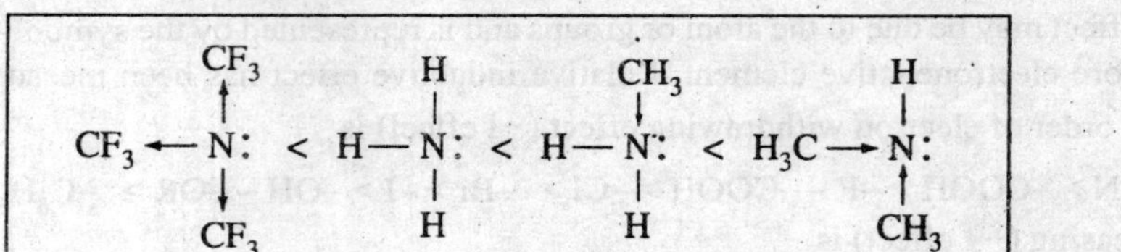

MESOMERIC EFFECT

The permanent polarization of a group conjugated with a π-bond or a set of alternate π-bond is transmitted through the π electrons of the system resulting in a different distribution of electrons in the unsaturated chain. This kind of electron redistribution in unsaturated compound conjugated with electron releasing or electron withdrawing groups or atoms is called the mesomeric effect.

For example, when the carbonyl group is conjugated with a carbon chain of alternate single and double bond the positively charged carbonyl carbon exerts electron transfer towards itself via the π electrons. Thus, the polarization of the carbonyl group is transmitted via π electrons of the carbon chain.

$$CH_2=CH-CH=CH-\underset{\displaystyle H}{\underset{|}{C}}=O \longleftrightarrow C^+H_2-CH=CH-CH=\underset{\displaystyle H}{\underset{|}{C}}-O^-$$

Similarly, the lone pair on the nitrogen atom repels the π electron of the conjugated chain of the amino compound.

$$H_2N-CH=CH-CH=CH_2 \leftrightarrow H_2N^+=CH-CH=CH-CH_2^-$$

The π electrons of molecule are delocalized due to the mesomeric effect resulting in number of resonance structures, which give the stability to the ions.

This is a permanent effect in the ground state of the molecule, which is indicated by the dipole moment.

Mesomeric effect takes place in unsaturated and specially conjugated system via their π-orbitals (π bond). In this effect, complete charge separation takes place and full charges are created.

Mesomeric effect is of two types:

When the transference of electron pair is away from the atom or group it is known as electron repelling mesomeric effect and is indicated by the + M effect. For example,

+M effect group – OH, – OR, – NR_2, – $\dot{N}$HR, – SR, – X, etc.

When the transference of electron pair is towards the atom or group it is known as electron attracting mesomeric effect and is indicated by the –M effect. For example,

–M effect group – CHO, > C = O, – CN, – NO_2, – SO_3H, etc.

The acidity of phenol is due to the +M effect of –OH group. The mesomeric transfer of the lone pair on the oxygen atom of phenol to the π electrons of the benzene ring results in several resonating structures with a positive charge on the oxygen atom. This helps the hydrogen atom of -OH group to leave as proton.

Applications of Mesomeric effect

(a) *Dipole moment*: Dipole moment of $CH_3 - CH_2 - Cl > CH_2 = CH - Cl$ is due to resonance of vinyl chloride.

(b) *Bond length*: Benzene has three double bonds and three single bonds within the ring. So, it is written as

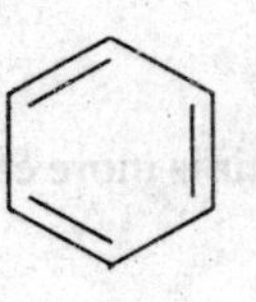

It appears that there should be two different types of bond lengths corresponding to single (1.54 Å) and double bonds (1.33 Å).

But the fact is that all the bond lengths in the benzene ring are equal. This can be explained by writing the following resonating structures of benzene:

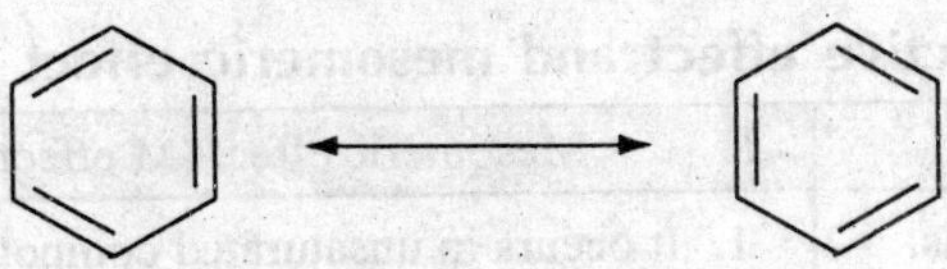

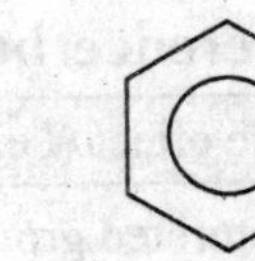

With the net result, resonance hybrid has the structure where the position of single and double bonds are delocalized and hence each bond has partial single and double bond character and hence equal bond length.

(c) *Strength of acids*: Acid strength of phenol R – C(=O) – OH > 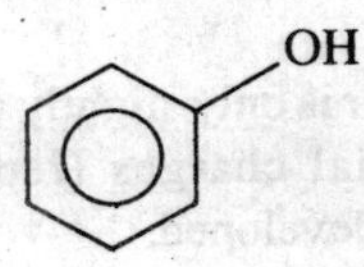 > ROH

Because the stability of corresponding conjugate bases in the order

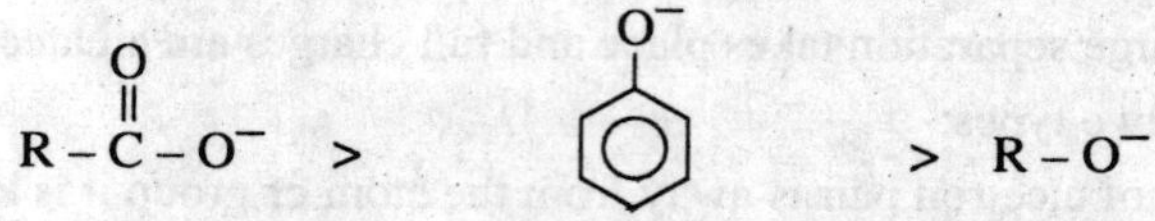

carboxylate anion phenoxide ion alkoxide ion

This is due to the more resonance stabilization of carboxylate anion compared to phenoxide ion

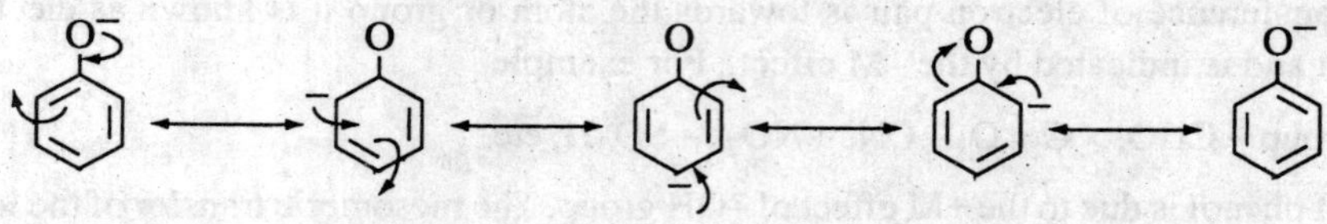

though the number of canonical forms are more in case of phenoxide ions.

This is because in case of the phenoxide ions the negative centre resides on the more electropositive carbon atom compared to the carboxylate ion where the negative charge resides on the more electronegative oxygen atom. Alkoxide ion is not stabilized at all by resonance.

(d) *Strength of bases*:

Basic strength of $RNH_2 > NH_3 > C_6H_5NH_2$

It is due to +I effect of R in RNH_2. Thus, lone pair is more easily available. But in $C_6H_5NH_2$, phenyl group has –M effect.

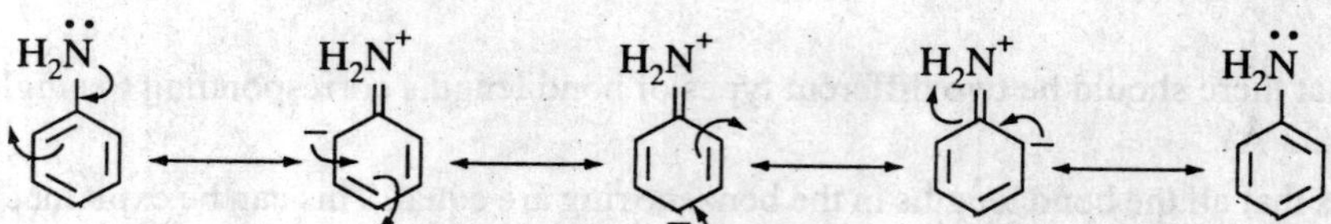

The lone pair present at the nitrogen in aniline is delocalized in the ring by resonance; hence, it is not free for donation leading to decrease of basic strength of aniline.

The essential difference between inductive effect and mesomeric effect

Inductive effect (I effect)	Mesomeric effect (M effect)
1. It operates in saturated group of compounds.	1. It occurs in unsaturated compounds having conjugated system
2. It involves electrons of ó-bond.	2. It involves electrons of π-bond and lone pair.
3. It is transmitted over a short distance. The effect dies out rapidly since the ó-bond electrons are tightly held.	3. It is transmitted from one end to other of the unsaturated chain since the effect involves delocalized π- electrons.
4. The electron pair is only slightly displaced and thus only partial charges (represented by symbols ∂) are developed.	4. The electron pair is transferred completely resulting full positive and negative charges.

INDUCTOMETRIC EFFECT

Inductive effect is always present in a molecule. It may be increased temporarily and prior to some reactions or change by the approach of a charged species (attacking reagent). Such a temporary effect is known as inductometric effect. This effect is time dependent and it always aids the reaction, never inhibits it. For example,

$$O_2N \leftarrow \underset{H}{\overset{H}{\overset{\delta+}{C}}} - \overset{\delta++}{H} \xrightarrow{OH^-} O_2N - \underset{H}{\overset{H}{C}} + H_2O$$

ELECTROMERIC EFFECT

In electromeric effect, electron displacement occurs by the complete transfer of shared pair of π electrons of a multiple bond (double or triple bond) to one of the atoms in the presence of reagent (attacking species).. Thus, electromeric effect produces polarity in a multiple bonded compound when the reagent approaches towards it. When multiple bond is present between two different atoms, electromeric shift towards more electronegative atom takes place. It involves the π electrons.

Consider the attack of a nucleophile to the carbonyl group of aldehydes or ketones. In presence of the nucleophile, the π electrons of the $>C=O$ group gets transferred to the oxygen atom. This effect is temporary and exists so long as the attacking reagent is present.

$$>C=O \text{ –attacking reagent} \rightarrow >C^+ - O^-$$

$$>C=O + CN^- \rightarrow >C^+ - O^- + CN^- \rightarrow >\underset{CN}{C} - O^- + \text{E effect}$$

This type of electron displacement is also possible in case of carbon–carbon double bond in presence of reagent.

$$>C=C< + H^+ \rightarrow >\underset{H}{C} - O^- \qquad \text{–E effect}$$

If the transfer of electrons takes place towards the attacking reagent, it is called +E effect and if it goes away from the attacking reagent, then it is called –E effect.

When multiple bond is present between two similar atoms (for example, in alkenes and alkynes, multiple bond is present between two carbon atoms), then there are two possibilities:

(i) *Symmetric alkenes or alkynes*: Here, electromeric shift can take place in any direction.

(ii) *Asymmetric alkenes or alkynes*: Here, direction of electron shift is governed by the inductive effect or electromeric effect (conjugate effect) of the group present at doubly or triply bonded atom. Since conjugate effect is stronger, hence when both inductive as well as conjugate effects are operating in unsaturated molecule, conjugate effect usually overcome inductive effect.

On the close approach of a reagent, the electronic system of an unsaturated molecule is deformed. When the reagent is removed without allowing the reaction to take place, the electronic system reverts to the original ground state of the molecule. This kind of polarizability of multiple bonds is known as the *electromeric effect.*

RESONANCE

The phenomenon, in which several structures involving identical position of atoms can be assumed to contribute to the true structure of a molecule but none of them can be said to represent it completely, the molecule is referred to the resonance hybrid and the phenomenon is called resonance.

The various structures are called as resonating structures or contributing structures. It is to be noted that resonating structures have no real existence. They are useful because they allow us to describe molecules, radicals or ions for which a single Lewis structure is not sufficient. The actual structure lies somewhere in between the various possible resonating structures and is called resonance hybrid. The difference in energy between the resonance hybrid and the most stable resonating structure is called as *resonance energy*.

The different structures of a compound devised by different method of pairing electron in a fixed atomic skeleton are called the *resonating* or *canonical structures*. The actual structure of the compound is then a combination of all these structures and hence, the compound is called the *resonance hybrid*. A hybrid is more stable than any one of the contributing structures. The contributing structure is denoted by the double-headed arrow ($\leftrightarrow$) indicating that the real structure involves both the way of pairing electron.

For example: 1, 3-butadiene

$$\overset{4}{C}H_2 = \overset{3}{C}H - \overset{2}{C}H = \overset{1}{C}H_2$$

The bonds between $C_1 - C_2$ and $C_3 - C_4$ are found to be longer than the carbon-carbon double bond and bond between $C_2 - C_3$ is slightly shorter than the carbon-carbon single bond. To explain this, several other structures may be devised by shifting a pair of π electrons.

$$CH_2 = CH - CH = CH_2(1) \rightarrow C^{\delta-}H_2 - C^{\delta+}H - CH = CH_2(2) \rightarrow CH_2 = CH - C^{\delta+}H - C^{\delta-}H_2(3)$$
$$\leftrightarrow C^{\delta-}H_2 - CH = CH - C^{\delta+}H_2(4) \rightarrow C^{\delta+}H_2 - CH = CH - C^{\delta-}H_2(5)$$

The structures 2 and 3 show partial single bond character between $C_3 - C_4$ and $C_1 - C_2$ bond and structure 4 and 5 show the partial double bond character between $C_2 - C_3$ bond. This explains the observed anomalies in the bond distance in the hybrid.

Resonance stability increased with increased number of resonance structures for the larger number of nuclei are brought closer by the binding effect of the loose electron.

Rules of resonance

1. All the resonance structures must conform to Lewis structure, e.g., carbon cannot be pentavalent.
2. The nuclear position of all the atoms in resonance structures must be same.
3. All the resonance structures must have the same number of paired electrons. For example, $\dot{C}H_2 - CH = CH - \dot{C}H_2$ is not considered a resonance structure of butadiene.
4. Resonance structures formed from greater number of covalent bonds are more stable.
5. Resonance structures with similar charge on adjacent atom are insignificant due to electrostatic repulsion and instability, e.g., $C^+H_2 - C^-H - C^-H - C^+H_2$ does not contribute to the stability of the hybrid.
6. Structures with the positive charge on electronegative element are unimportant, e.g., $>C^- - O^+$ of carbonyl group is insignificant.
7. Resonance stabilization is greatest when there are at least two equivalent canonical structures, e.g., $C^+H_2 - CH = CH_2 \rightarrow CH_2 = CH - C^+H_2$
8. Charge separation structures are less important than those in which the charge is delocalized. This is due to the electrostatic attraction between unlike charges.

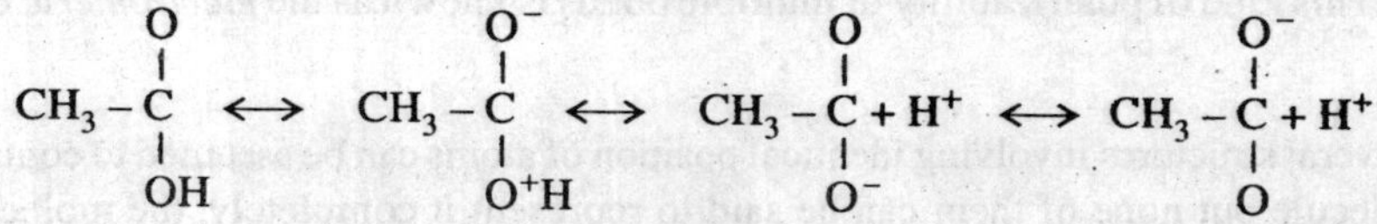

Less stable More stable

9. All the atoms present in a molecule, which are taking part in the resonance, should be coplanar. This is necessary for the effective overlap of the *p*-orbital and delocalization of the electron.

HYPERCONJUGATION

The inductive effect of alkyl group is found to be in the order $Me_3C^- > Me_2CH^- > MeCH_2^- > CH_3^-$ when attached to a saturated carbon. But, when they are attached to an unsaturated carbon (C = C) the order is reversed. In this case, an electron releasing effect operates by different methods than in inductive effect.

The electron releasing effect is explained by assuming that the ó orbital of the $\alpha - C - H$ bond overlaps with the adjacent π-orbital. The displacement of the electron pair of the $\alpha - C - H$ bond causes a partial positive charge on the hydrogen atom without the actual proton release.

This ability of the σ-bond electron of $\alpha - C - H$ bond to undergo conjugation with the adjacent π electron is called *hyperconjugation*.

The other example:

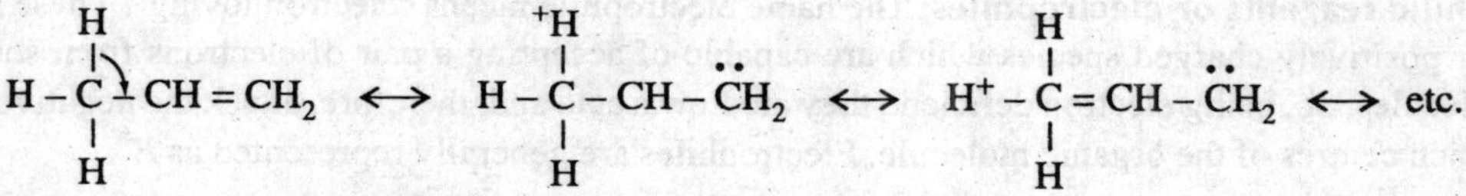

Thus, all the three C–H bond of a methyl group in propylene can polarize the adjacent π-bond. The order of this effect is $CH_3^- > RCH_2^- > R_2CH^- > R_3C^-$.

Thus, the adjacent π bond can polarize all the three C – H bonds of a methyl group in propylene.

The order of this effect is $CH_3^- > RCH_2^- > R_2C^- > R_3C^-$.

The ortho and para directing effect of **methyl** group is similarly attributed to electron release by hyperconjugation.

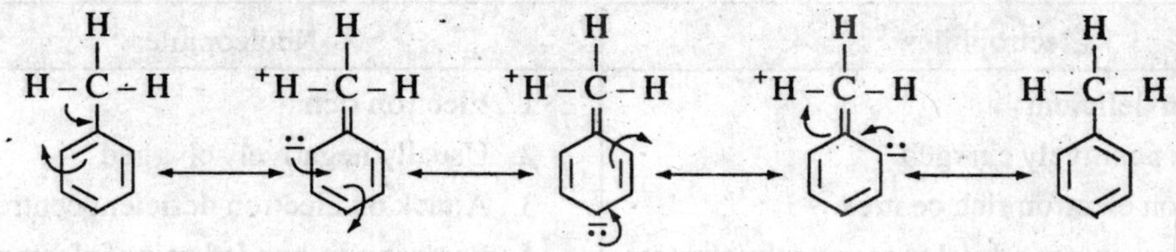

ORGANIC REACTIONS AND THEIR MECHANISMS

Organic reactions involve the transformation of molecule from one structure into another by change in bonds or rearrangement of atoms. In a chemical reaction, a molecule, which undergoes the change or undergoes attack, is called the *substrate*, and the attacking species, which is responsible for the change, is the *reagent*. The substrate and the reagent when interact form a new compound called the *product*.

Substrate + Reagent → Products

The bonds in the substrate molecule are broken or cleaved to give fragments, which are very reactive and constitute transitory intermediates. At once they may react with other species present in the system to form new bonds to give the products.

This conversion of substrate into products may be a single step or multi-step process. The steps of an organic reaction showing the breaking and making of new bonds of carbon atoms in the substrate leading to the formation of the final products through transitory intermediates are often referred to as reaction mechanism.

Substrate → Intermediate (Transitory) → Products

So, organic reaction proceeds by the attack of certain active species on the reactant molecule (generally called substrate). The active species are called *attacking species* or *reagent*. Based on the reaction types, they are classified into the following classes:

(a) Nucleophilic reagents or nucleophiles
(b) Electrophilic reagents or electrophiles
(c) Free radical.

(a) **Nucleophilic reagents or nucleophiles:** The term nucleophile means 'nucleus loving'. These are either neutral or negatively charged species which are capable of donating a pair of electrons to some other molecules. So, they are electron rich species and act as Lewis bases. They, therefore, attack the electron deficient centres of the organic molecule. They are generally represented as Nu^-. Examples of nucleophiles:

(i) Charged nucleophiles: Cl^-, OR^- CN^-, NO_2^-, $RCOO^-$, SH^-, OH^-, NH_2^-, etc.

(ii) Neutral nucleophiles (Lewis bases): H_2O, ROH, NH_3, RNH_2, R_2NH, R_3N, etc.

(b) **Electrophilic reagents or electrophiles:** The name electrophile means 'electron loving'. These are neutral or positively charged species which are capable of accepting a pair of electrons from some other molecules. So, being electron deficient, they are Lewis acid and, therefore, attack the negative or electron rich centres of the organic molecule. Electrophiles are generally represented as E^+.

Examples of electrophiles

(i) Charged electrophiles: NO_2^+, Br^+, Cl^+, etc.

(ii) Neutral electrophiles (Lewis acids): BF_3, $ZnCl_2$, $FeCl_2$, etc.

(c) **Free radical:** These are neutral species having an unpaired electron.

Some examples are $Cl^{\bullet}$, $Br^{\bullet}$, $RCH_3^{\bullet}$, $OR^{\bullet}$.

The main differences between nucleophiles and electrophiles are shown in a tabular form.

Electrophiles	Nucleophiles
1. Electron deficient	1. Electron rich
2. Usually positively charged	2. Usually negatively charged
3. Attack on electron rich centres	3. Attack on electron deficient centres
4. Possess an empty orbital to receive the electron pair from the nucleophiles.	4. Possess non-bonded pair of electrons, which they donate to electrophiles.

CLEAVAGE OR FISSION OF COVALENT BONDS

A covalent bond between two atoms of the given compound may be broken in different ways depending on

(i) the nature of given organic compound;
(ii) the nature of the attacking reagent and
(iii) the reaction conditions.

Fission of covalent bond leads to the generation of intermediate organic species, which are unstable, and so new compounds are formed.

Organic reactions take place with the breaking (fission) of bonds and formation of new bonds. The fission of covalent bonds can take place in two ways:

1. Homolytic cleavage or Homolytic fission
2. Heterolytic cleavage or Heterolytic fission.

Homolytic cleavage or homolytic fission

In this type of cleavage, the shared electron pair of a covalent bond gets equally distributed between the fragments. Thus, this type of cleavage results in the formation of species having odd or unpaired electrons. Such species are called free radicals, which are electrically neutral. For example, $A:B \rightarrow A\bullet + B\bullet$

Homolytic cleavage takes place when the bond is non-polar in nature, high temperature or pressure or presence of high-energy radiation. It generally takes place in gas or vapour phase and is initiated by heat, light or organic peroxides.

Heterolytic cleavage or heterolytic fission

In this type of cleavage, both the shared electrons of the covalent bond are taken away by one of the fragments while the other fragment does not get any of the shared electrons. For example,

$A\!:\!B \rightarrow A^{+} + :B^{-}$ (when electronegativity of B > A)

or, $A^{-}: + B^{+}$ (when electronegativity of B < A)

So, this type of cleavage results in the formation of ions. When the bonding elements differ widely in electronegativity, i.e., the bond is polar in nature, the heterolytic fission occurs generally in presence of polar solvents and usually at lower temperatures.

REACTION INTERMEDIATES

Reaction intermediates are the species formed during the reaction (homolytic or heterolytic bond fission), but are lost before the reaction is complete. These are short-lived fragments and are very reactive. Some of the important reaction intermediates based on carbon are

Carbocation and Carboanion (Formed during heterolytic fission)

Free radical (Formed during homolytic fission)

Carbenes (Formed by photolysis).

CARBOCATION OR CARBONIUM ION

A group of atoms containing a positively charged carbon atom is called *carbonium ion* or *carbocation*. It contains only six electrons in its valence shell and is formed due to heterolytic bond fission where the leaving group takes the shared pair of electrons

$$-\overset{|}{\underset{|}{C}}-X \longrightarrow -\overset{|}{\underset{|}{C}}{}^{+} + X^{-}$$

Carbonium ions are classified as primary (1°), secondary (2°) and tertiary (3°) depending upon the carbon having the positive charge. If the positive charge is on the primary carbon, then the carbocation is primary; if it is on secondary carbon, then it is called secondary carbocation and so carbonium ions are named after the parent alkyl groups and adding the words carbonium ions.

Stability of carbocation

Carbocations are stabilized by the dispersal of charge. Positive charge of carbocation can be dispersed either by inductive effect (+I effect) of the alkyl groups or by resonance or by hyperconjugation. They are able to reduce the positive charge on the carbon atom. Greater is the number of alkyl groups on the carbon carrying positive charge, greater would be the stability of the carbonium ion due to greater dispersal of charge. The decreasing order of stability of carbocation is 3° > 2° >1°.

Carbocations are highly reactive species because it has only six electrons in its valence shell and has tendency to complete its octet. So, order of reactivity of carbocation is the reverse order of stability, i.e., order of reactivity 1° > 2° >3°.

CARBANION

A group of atoms containing a negatively charged carbon atom is called carbanion. Carbanions are formed by heterolytic fission where the shared electron pair remains with the carbon atom.

Like the carbocations, they are also classified as primary (1°), secondary (2°) and tertiary (3°) depending on the nature of the carbon atom bearing the negative charge. They are also named after the parent alkyl group and then adding the word carbanion.

$$-\overset{|}{\underset{|}{C}}-X \longrightarrow -\overset{|}{\underset{|}{C}}\!:^{-} + X^{+}$$

Stability of carbanion

Carbanions are stabilized by the dispersal of negative charge. Negative charge of carbanions can be dispersed either by the electron releasing inductive effect (+ I effect) of bonded atoms or groups of atoms or by resonance

Decreasing order of the stability of carbanion is 1° > 2° >3°. Presence of electron releasing group (+I effect) makes the carbanion less stable. So, greater the number of alkyl groups on the carbon bearing negative charge, greater would be the negative charge on the carbon, hence, instability is produced.

$\bar{C}H_3-$	> $CH_3\bar{C}H_2-$	> $CH_3-\bar{C}H-CH_3$	> $CH_3-\underset{\underset{CH_3}{|}}{\bar{C}}-CH_3$
	Primary (1°)	Secondary (2°)	Tertiary (3°)
Methyl carbanion ion	Ethyl carbanion ion	Isopropyl carbanion ion	t-butyl carbanion ion

Presence of electron attracting groups (-I effect) like $-\overset{+}{N}O_2$, $\overset{+}{Br}$, etc will stabilize carbanion by withdrawing the negative charge on carbon partially.

Carbanions are highly reactive species because the carbon atom is electron rich and can donate its non-bonding pair of electrons to some other group for sharing. They behave as nucleophiles and the order of reactivity of carbanions is reverse the order of its stability, i.e., 3° > 2° >1°.

FREE RADICALS

A free radical is a species, which has one or more unpaired electrons. It is paramagnetic and can be detected by electron spin resonance spectroscopy.

A free radical may be defined as an atom or group of atoms containing unpaired electron. They are formed by homolytic cleavage of a covalent bond and are neutral. If the unpaired electron is on the carbon atom, then it is called *carbon-free radical* or *alkyl-free radical*. Alkyl free radicals are classified as primary (1°), secondary (2°) and tertiary (3°) depending upon the nature of carbon atom carrying the unpaired electron.

$H_3\dot{C}$	$CH_3\dot{C}H_2$	$CH_3-\dot{C}H-CH_3$	$CH_3-\underset{\underset{CH_3}{|}}{\dot{C}}-CH_3$
	Primary (1°)	Secondary (2°)	Tertiary (3°)
Methyl-free radical	Ethyl-free radical	Isopropyl-free radical	t-butyl-free radical

The decreasing order of stability of alkyl-free radicals are in the order 3° > 2° > 1°, i.e., of the same order of stability of carbonium ions. Since, free radicals contain one unpaired electron they are highly unstable and reactive because they have the tendency to share with some other atom or group to complete the octet. So, order of reactivity of alkyl-free radicals is reverse the order of stability 1° > 2° > 3°.

CARBENES

A carbene is a group of atoms which contains a carbon atom with only six electrons in its valence shell, out of which two electrons are unshared. These are neutral species formed by the removal of two univalent groups or a divalent group along with one of the two bonding pairs of electrons. They are unstable and highly reactive in nature. They contain a carbon, which is deficient in electrons and hence, can accept a pair of electrons to complete their outer shell. Therefore, carbenes act as electrophiles. Being highly reactive, carbenes are generally produced in situ. They can be generated in a number of ways—

1. By photolysis of diazomethane or ketone:

(a) $\underset{\text{Diazomethane}}{CH_2 = \overset{+}{N} = \overset{-}{N}} \xrightarrow{\text{UV light}} \underset{\text{carbene}}{:CH_2} + N_2$

(b) $CH_2 = \dot{C} = O \xrightarrow{\text{UV light}} :CH_2 + CO$

2. By reacting chloroform in presence strong alkali:

$$CHCl_3 + OH^- \rightarrow :CCl_2 + Cl^- + H_2O$$

Carbenes exist in two different states–singlet and triplet. In singlet state the unshared electrons are paired while in triplet state they are not paired. Triplet carbene is more stable than the singlet carbene.

Carbene undergoes addition reaction with carbon-carbon double bond to form cyclic structure.

$$\underset{\text{Propene}}{CH_3-CH=CH_2} \quad + \quad \underset{\text{Diazomethane}}{CH_2N_2} \longrightarrow \underset{\text{Methylcyclopropane}}{CH_3CH\overset{CH_2}{\text{—}}CH_2}$$

Intermediate organic species based on carbon and their properties are summarized in the following table

Serial number	Intermediate	Hybridization	Shape	Structure	Formal charge	Stability
1.	Carbocation/ Carbonium ion	sp^2	Trigonal planar	$>\overset{+}{C}-$	+1	3° > 2° > 1°
2.	Carbonion	sp^3	Pyramidal	$-\overset{..}{\underset{\mid}{C}}-$	–1	1° > 2° > 3°
3.	Free radical	sp^2 sp^3	Trigonal planar Pyramidal	$>\dot{C}-$ $-\overset{\cdot}{\underset{\mid}{C}}-$	Zero —	3° > 2° > 1° —
4.	Carbene Singlet	sp^2	Bent	$>C:$	Zero	—
5.	Carbene Triplet	sp	Linear	$-\overset{\bullet}{\underset{\bullet}{C}}-$	Zero	—

TYPES OF ORGANIC REACTIONS

Organic reactions can be classified into four general types

(a) Substitution reaction

(b) Elimination reaction

(c) Addition reaction

(d) Rearrangement reaction

(a) SUBSTITUTION REACTIONS

Reactions involving the replacement of an atom or group from a molecule by some other atom or group without changing the rest part of the molecule are called substitution reactions. The product formed by the replacement of a group by a new one is called *substituted product*. For example:

(i) $CH_3–CH_2–OH + PCl_5 \rightarrow CH_3–CH_2–Cl + POCl_3 + HCl$

(ii) $C_6H_6 + HNO_3\ (conc.H_2SO_4) \rightarrow C_6H_5NO_2 + H_2O$

(iii) $R\text{-}I + aq.\ KOH \rightarrow ROH + KI$

Substitution reactions may be initiated by an electrophile, nucleophile, or free radical.

Depending on the nature of the substituting species there are three types of substitution reactions:

(i) Electrophilic substitution reactions
(ii) Nucleophilic substitution reactions
(iii) Free radical substitution reactions.

(i) Electrophilic substitution reactions

These reactions are mostly found with aromatic systems where the high electron density attracts electrophiles. In such reactions, electrophile is the attacking species and the leaving group departs without the pair of bonding electrons.

When a substitution reaction takes place by the attack of an electrophile, it is called *electrophilic substitution reaction*., e.g., bromination or nitration of benzene.

$$\text{Benzene} + Br_2 \xrightarrow{FeBr_3} C_6H_5Br + HBr$$

$$\text{Benzene} + HNO_3 \xrightarrow{H_2SO_4} C_6H_5NO_2 + H_2O$$

Here, the electrophile $\overset{+}{Br}$ ion is formed by the reaction of $FeBr_3$ with bromine, and electrophile $\overset{+}{NO_2}$ ion is formed by the reaction of H_2SO_4 with HNO_3.

$$FeBr_3 + Br_2 \rightarrow \overset{+}{Br} + FeBr_4^- \text{ and}$$

$$H_2SO_4 + HNO_3 \rightarrow \overset{+}{NO_2} + H_3SO_5^- \rightarrow NO_2^- + HSO_4^- + H_2O$$

(ii) Nucleophilic substitution reactions

When a substitution reaction takes place by the attack of a nucleophile, the reaction is called *nucleophilic substitution reaction*, e.g., hydrolysis of alkyl halides in presence of a base.

$$R-X + NaOH \rightarrow R-OH + NaX$$

Nucleophilic substitution reactions are of two types:

(i) S_N1 (Unimolecular Nucleophilic substitution) and
(ii) S_N2 (Bimolecular Nucleophilic substitution)

S_{N1} reaction

S_N1 is the abbreviation used for unimolecular nucleophilic substitution reaction. Here, the rate of the reaction is first order, i.e., depends only on the concentration of the substrate, i.e., the alkyl halide. Generally, tertiary alkyl halides undergo this type of reaction and the reaction proceeds in two steps.

(a) Formation of carbocation intermediate by the heterolytic cleavage of substrate molecule

$R-X \rightarrow [R\cdots X] \rightarrow R^{(+)} + X^{(-)}$, (slow). Hence, it is rate determining step. Rate α [alkali halide].

(b) Combination of carbocation intermediate with the nucleophile to yield product.

$$R^{(+)} + OH \rightarrow R-OH \quad \text{(fast)}$$

The term unimolecular is used here since the rate-determining step involves only one molecule.

S_N2 reaction

S_N2 is the abbreviation used for bimolecular nucleophilic substitution reaction. Here, the rate of the reaction depends on the concentration of the alkyl halide as well as the base. So, the order of the reaction is two.

$$\text{Rate } \alpha \text{ [alkali halide] [base]}$$

Generally primary alkyl halides undergo this type of substitution reaction and it is a single step process.

(iii) Free radical substitution reactions

Chlorination of alkane is an example of free radical substitution reactions and takes place in presence of heat or light.

$$CH_4 + Cl_2 \xrightarrow{\text{heat / light}} CH_3Cl + HCl$$

Mechanism of this reaction involves free radical chain reaction.

Step (a): Chain initiating step: Energy from heat or light is absorbed; homolytic cleavage of chlorine into atoms takes place, which are reactive particles for subsequent steps.

$$Cl_2 \xrightarrow{\text{heat / light}} 2Cl^{\bullet}$$

Step (b): Chain propagation steps: Here the reaction keeps on propagation by the simultaneous consumption and formation of reactive particles, free radicals.

$$CH_4 + Cl^{\bullet} \longrightarrow HCl + CH_3^{\bullet}$$

$$Cl_2 + CH_3^{\bullet} \longrightarrow CH_3Cl + Cl^{\bullet}$$

Step (c): Chain termination step: In such steps, reactive free radicals are consumed and not generated.

$$Cl^{\bullet} + Cl^{\bullet} \rightarrow Cl_2$$

$$CH_3^{\bullet} + CH_3^{\bullet} \rightarrow CH_3CH_3$$

$$CH_3^{\bullet} + Cl^{\bullet} \rightarrow CH_3Cl$$

(b) ELIMINATION REACTIONS

These reactions involve the elimination or removal of two or more atoms /groups from the adjacent carbon atoms of same organic molecule under suitable conditions to form a product with multiple bond. Elimination can be considered as reverse of addition reaction. For example,

(i) Acid catalyzed dehydration of alcohol

$$\underset{H}{\underset{|}{CH_2}} - \underset{OH}{\underset{|}{CH_2}} \xrightarrow{\text{conc. } H_2SO_4 \text{ at } \sim 170\ ^\circ C} CH_2 = CH_2 + H_2O$$

(ii) Base catalyzed dehydrohalogenation of alkyl halide

$$\underset{H}{\underset{|}{CH_2}} - \underset{Br}{\underset{|}{CH_2}} \xrightarrow{\text{alcoholic KOH}} CH_2 = CH_2 + H_2O + KBr$$

(iii) Dehalogenation of alkyl dihalide

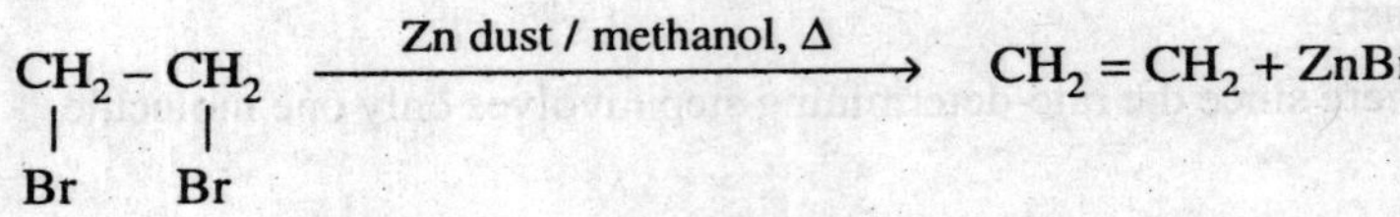

Like substitution reactions, elimination reactions are also classified as

1. E_1 reactions and
2. E_2 reactions.

E_1 reaction

E_1 represents unimolecular elimination reaction. The rate of the reaction depends only on the concentration of the substrate. Rate α [substrate]. Therefore, it is a first order reaction, which proceeds through two steps.

In the first step, the substrate undergoes slow heterolysis to form a carbocation intermediate and a halide ion X^-. This is the slow, rate-determining step and it involves only substrate. Thus, the rate of an E_1 reaction is independent of base concentration.

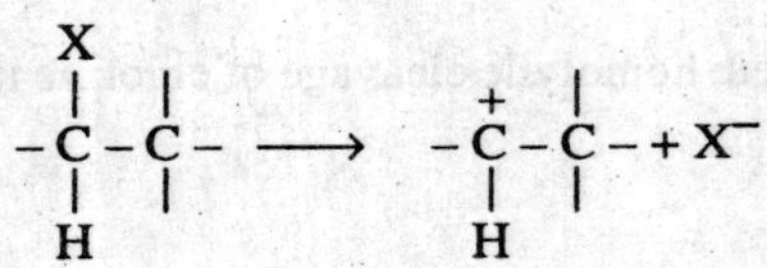

slow rate determining step, rate = k [substrate] = k [RX]

In the second step, the intermediate carbocation rapidly loses a proton to the base and forms the alkene.

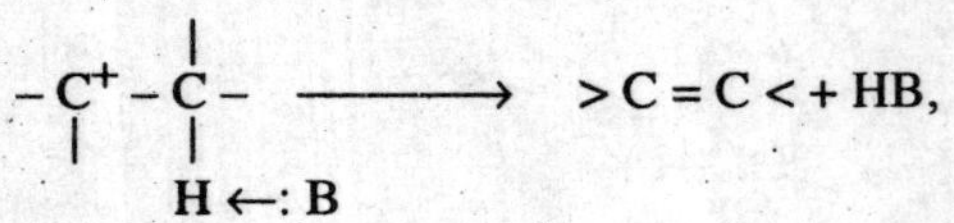

fast step

E_2 reaction

E_2 represents bimolecular elimination. In E_2 reaction, base : B pulls a proton from the substrate; simultaneously leaving group, X departs with its electron pair but hydrogen leaves its electron pair behind to form the double bond. Everything is happening simultaneously, in a single step, via a single transition state.

The rate is found to be dependent on the concentration of the substrate as well as on the concentration of reagent. So, it is bimolecular elimination reaction and follows second order rate law. Rate α [substrate] [Reagent]

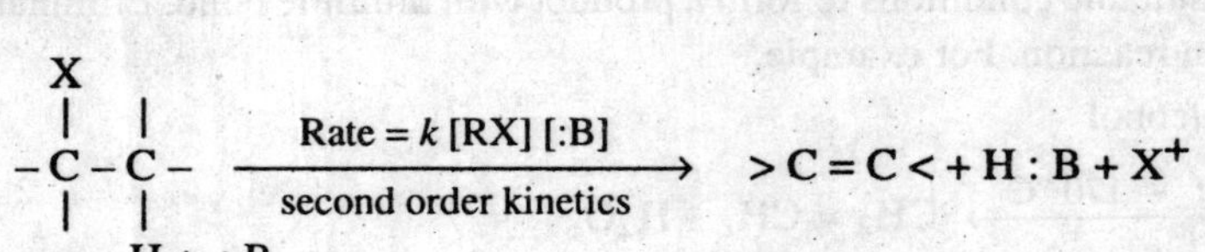

(c) ADDITION REACTIONS

These reactions are those where atom or group of atoms are simply added to a molecule containing multiple bond. They involve the combination of two molecules to form a single molecule without the elimination of any atom or group. In these reactions generally a bond breaks up forming two bonds, which are more stable.

The product formed is known as addition product or adducts. So, addition reactions are generally the characteristics of unsaturated compounds. For example,

(i) 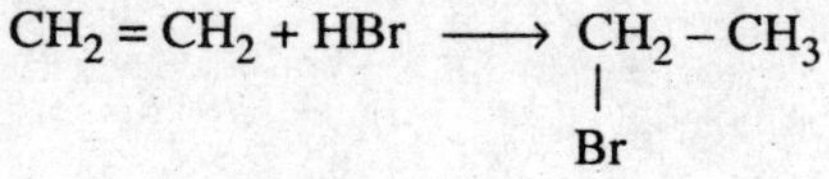

(ii) $CH \equiv CH + 2H_2 \rightarrow \text{Ni at 575K} \rightarrow CH_3 - CH_3$

(iii) 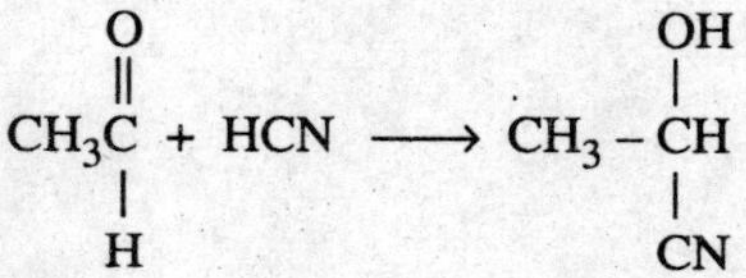

Addition reactions are also of two types:

1. Electrophilic addition reaction
2. Nucleophilic addition reaction.

1. Electrophilic addition reaction

When addition reaction takes place by the initial attack of an electrophile to a multiple bonded compound, it is called *electrophilic addition reaction*. Here, carbocations are formed as intermediates due to the reaction between electrophilic reagents and unsaturated reactant molecule. Compounds containing carbon–carbon double and triple bond undergo such reaction. Example: addition of HBr to an alkene

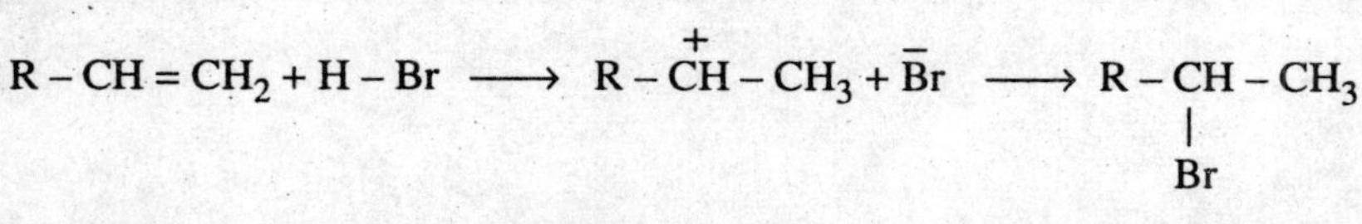

Here, the initial attack is by the proton (electrophile) to the $- C = C -$ double bond and then the bromide ion (nucleophile) adds on to the carbonium ion forming the addition product. The complete reaction takes place in multiple steps.

2. Nucleophilic addition reaction

When the addition reaction proceeds by the initial attack of a nucleophile it is called nucleophilic addition reaction. Compounds containing carbon–oxygen double bond (carbonyl compounds) undergo such reactions. Example: addition of HCN to acetaldehyde.

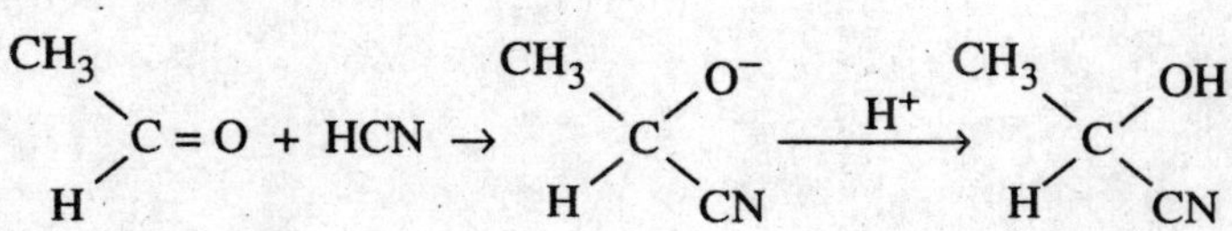

Here, the nucleophile CN^- attacks the carbonyl carbon and in the next step the electrophile (H^+) gets joined with the negatively charged oxygen ion.

(d) REARRANGEMENT REACTIONS

In this type of reaction, migration of group or atoms takes place from one position to the other within a molecule. The product formed is the structural isomer of the original one and possesses different properties. For example,

(i) 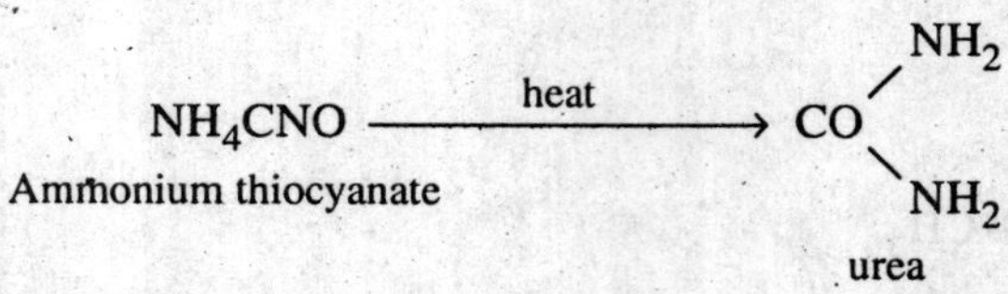

(ii) 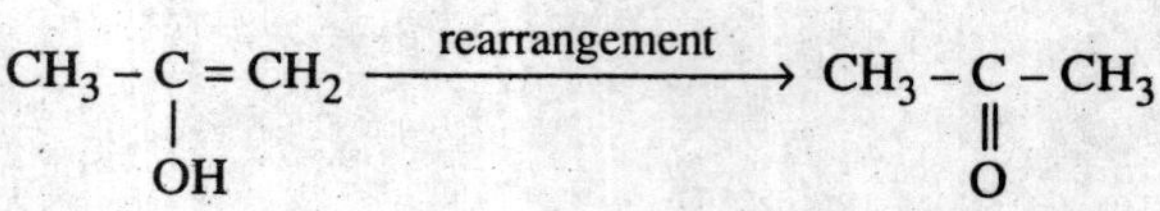

9

POLYMER

INTRODUCTION

In this modern world, polymers are an integral part of everyone's life style. They have very diverse structures, and applications are ranging from domestic articles to sophisticated scientific and medical instruments. These materials have penetrated into almost all the applications like transport, agriculture, construction, education, etc. These materials are used as fibres, rubbers, adhesives and paints, etc. In fact, the very existence of life is virtually the formation, transformation and decomposition of biopolymers, viz., carbohydrates, proteins and nucleic acids. Therefore, in view of their importance, a proper understanding of polymeric materials is very essential.

POLYMERS AND THEIR DEFINITIONS

Polymers are high molecular weight compounds whose structures are made up of a large number of simple repeating units, usually obtained from simple compounds of low molecular weight and are formed by joining together of thousands of smaller molecular units by chemical bonds. Due to their large size, polymers are also sometimes called as *macromolecules*.

The chemical process leading to the formation of polymer is known as *polymerization* and the number of monomeric units contained in the polymer is known as *degree of polymerization*. Smaller simple molecules, which combine with each other acting as repeating unit to form polymer molecules, are known as Monomers.

For example,

$$\underset{\substack{\text{Ethylene or ethene}\\ \text{(Monomer)}}}{n(CH_2 = CH_2)} \longrightarrow \underset{\substack{\text{Polyethylene or polythene}\\ \text{(Polymer)}}}{-\!\!\left(CH_2 - CH_2 \right)\!\!-_n}$$

Where n is the degree of polymerization, it can be 10^4 or more.

Polymers, which are synthesized from only one kind of monomer, are called *homopolymers* whereas polymers which are prepared from more than one kind of monomer are called *copolymers*.

SHORTHAND FOR WRITING POLYMER REACTIONS

Polymerization reactions are time consuming to write. Chemists have a shorthand way of writing the structural formula for a polymer molecule and the equation for polymerization reaction. The monomer is shown on the left hand (reactants) side of the equation. On the right hand (products) side, the repeating polymer unit is enclosed in brackets and the line passes through the brackets to show that the chain continues in both directions. The subscripts n shows how many times the repeating unit appears in the polymer.

$$\underset{\text{Ethylene}}{n(CH_2 = CH_2)} \xrightarrow{\text{polymerization}} \underset{\text{Polyethylene}}{-\!\!\left[CH_2 - CH_2 \right]\!\!-_n}$$

Frequently, the end groups of polymers are unknown – they may arise from impurities present in the reaction mixture. The properties of a polymer are governed almost entirely by the bulk of the polymer molecule rather than the end groups. To emphasize the basic structure of the polymer, it is customary not to include the end groups in the formula.

FUNCTIONALITY

Functionality means the number of bonding sites present in a monomer. The double bond in ethylene monomer ($CH_2 = CH_2$) can be considered as a site for two free valences. When the double bond is broken, two single bonds become available for combination.

For example, $$CH_2 = CH_2 \longrightarrow -CH_2-CH_2-$$

Thus, ethylene can be considered as bifunctional because it has two reactive or bonding sites. For a substance to act as a monomer, it must have at least two reactive or bonding sites.

POLYMER CLASSIFICATION

Depending on the structures, physical properties, thermal characteristics, etc, polymers can be classified into several ways.

1. *Based on physical state*: Polymer materials may be amorphous (e.g., LDPE, Rubbers, etc.) or semi-crystalline (e.g., HDPE, Nylon, Polyesters, etc).
2. *Based on structure/shape*: Polymer molecule may be linear, branched or crossed linked. Their structures and illustrative examples are shown below:

Polymer	Structure	Example
Linear	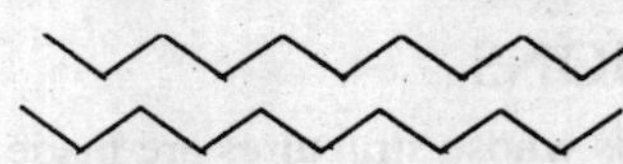	HDPE (High density polythene)
Branched	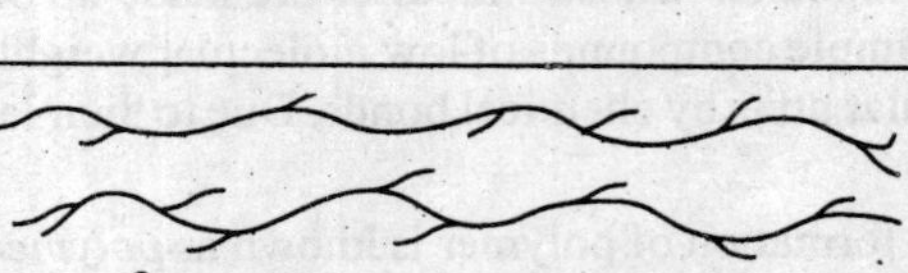	LDPE (Low density polythene)
Cross-linked	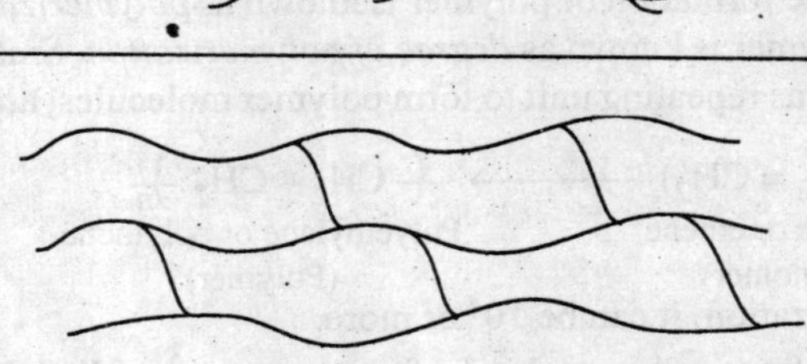	XLPE (Cross-linked polythene)Rubbers, Thermosets

3. *Based on configuration or Tacticity*: According to the stereochemistry **(orientation)** of monomeric, units in a macromolecule can be classified into the following categories:
   (i) Atactic polymers in which the side groups are oriented at random around the backbone chain, e.g., polypropylene.

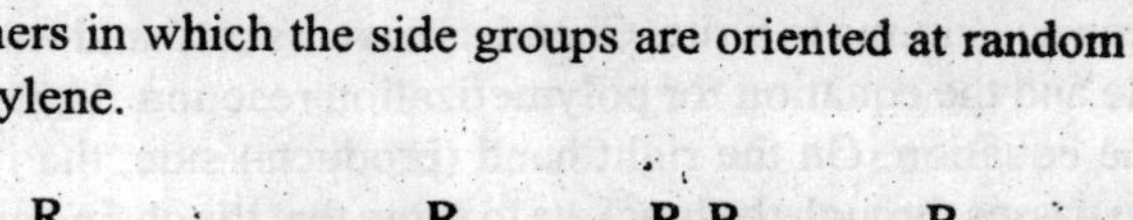

(ii) Isotactic polymers in which the side groups are all on the same side of the main chain, e.g., natural rubber.

(iii) Syndiotactic polymers in which the side groups are arranged in alternating fashion around the main chain. Polystyrene made by metallocene catalyzed polymerization is crystalline with a melting point 270°C is an example of syndiotactic polymer.

4. *Based on end use*: Depending on the end use, a polymer may be classified into various ways: (i) *Plastics*: A polymer which on treatment of heat and pressure, can be shaped into hard and tough articles known as plastics, e.g., PVC, PS, etc. (ii) *Elastomers* are those polymers, which are vulcanized rubber products having good strength and flexibility, e.g., natural rubber, silicone rubber and polyurathene, etc. (iii) *Fibres* are those polymers whose chains are held by strong intermolecular forces like hydrogen bonding. They are crystalline in nature and of high tensile strength, due to strong intermolecular forces. For example, nylons, polyesters, etc. (iv) *Films* (e.g., PP, LDPE, HDPE, PVC and PET, etc.), (v) *Adhesives* (e.g., PVA, epoxy resin, etc.), (vi) *Paints* (e.g., epoxy, etc.), (vii) *Membranes* (e.g., polyacetylene, polyaniline, etc.)
5. *Based on origin/source*: According to the origin, polymers may be grouped as natural or synthetic. Polymers isolated from naturally occurring materials are called natural polymers (e.g., cotton, silk, wool, rubber, etc). Polymers synthesized chemically from monomers are called *synthetic polymers* (e.g., polythene, PVC, nylon, polyester, etc).
6. *Based on conductance*: Polymers are mostly insulator but some are conducting (e.g., polyaniline, polypyrole, etc).
7. *Based on type of monomers*: Polymers can be homopolymer (e.g., PE, PP, PVC, etc.) or copolymer (e.g., styrene acrylonitrile, styrene butadiene rubber, etc.) when the types of monomers are same or different respectively.
8. *Based on character of monomer*: Polymers can be polar (e.g., PET, Nylon, etc) or non-polar (e.g., PE, PP, etc)
9. *Based on thermal behaviour*: According to thermal behaviour polymer is divided into two types: *thermoplastic* and *thermosetting polymers* (thermosets). Thermoplastic polymers are those, which softens readily on heating but on cooling they stiffen again reversibly, that is why they can be moulded, reshaped and reclaimed from wastes. Thus, they can be processed again and again. These polymers are usually soluble in suitable solvents. Most addition polymers and some of condensation polymers are thermoplastic (e.g., polyethylene, polypropylene, polystyrene, polyamides, nylons, PVC, polyesters, PET, PMMA, PTEE, etc). Thermoplastics are amorphous or semi-crystalline and have either liner or branched structures.

Thermosetting polymers are those polymers, which during moulding (by heat) become hardened and once they have solidified cannot be softened, i.e., they are permanent setting polymer. Such polymers are prepared in two steps. The first step is the formation of long chain molecules, which are capable of further reaction with each other. The second step is the application of heat, which causes a reaction to occur between the chains; thus producing a complex three-dimensional cross-linked network structure; so thermosetting polymers cannot be remoulded, reused or reclaimed. These polymers are hard, strong and brittle, and are generally insoluble in solvent. Examples: PF, UF, MF, vulcanized rubber, epoxy, etc.

10. *Based on eco-friendly nature*

    Polymer materials are generally durable (mostly all) but can be biodegradable (e.g., starch based polyethylene).

TABLE 9.1: *Summary of classifications of polymers*

CLASSIFICATION OF POLYMERS	
Based on structure/shape	Linear; Branched; Cross-linked
Based on physical state	Amorphous; Semi-crystalline
Based on response to heat	Thermoplastics; Thermosets
Baset on end-use	Fibres; Plastics; Elastomers; Films; Adhesives; Paints; Membranes
Based on tacticity	Isotactic; Syndiotactic; Atactic
Based on origin	Natural; Synthetic
Based on conductance	Insulators; Conducting
Based on environment friendly nature	Biodegradable; Durable
Based on type of monomers	Polar; Non-polar
Based on number of monemers	Homopolymer; Copolymer
Based on type of atoms present in the polymer	Organic polymers; Inorganic polymers
Based on growth mechanism of polymerization	Step growth polymer; Chain growth polymer

POLYMERIZATION

Polymerization is a process by which simple (monomer) molecules join together to form very large molecules with high molecular weight. Hence, the synthesis of large molecular weight polymers is termed as *polymerization*. There are three different ways for doing polymerization:

(i) By opening a double bond, e.g.,

$$n\,(CH_2=CH_2) \longrightarrow \left[CH_2-CH_2 \right]_n$$

Ethene — Polythene

(ii) By opening a ring, e.g.,

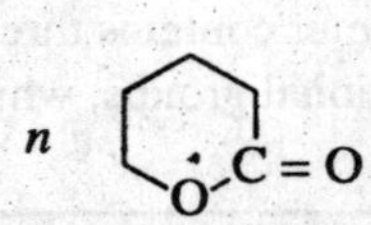

$$\longrightarrow \left[O-(CH_2)_4-\overset{O}{\overset{\|}{C}} \right]_n$$

δ-valerolactone — Polyester

(iii) By using molecules having two functional groups, e.g.,

$$n\ HOOC-(CH_2)_4-COOH + n\ H_2N-(CH_2)_6-NH_2 \longrightarrow$$

Adipic acid — Hexamethylenediamine

$$\left[\overset{O}{\overset{\|}{C}}-(CH_2)_4-\overset{O}{\overset{\|}{C}}-\underset{H}{\underset{|}{N}}-(CH_2)_6-\overset{H}{\overset{|}{N}} \right]_n$$

Nylon 6, 6

Polymerization reaction can be classified into following three types:

1. Addition Polymerization or Chain Growth Polymerization
2. Condensation Polymerization or Step-Growth Polymerization
3. Co polymerization

Addition polymerization

The process in which the polymers are formed from the corresponding monomer without losing any materials and the polymer formed is an exact multiple of monomer, is called *addition polymerization*.

$$n\ CH_2=\underset{X}{\underset{|}{CH}} \longrightarrow \left[CH_2-\underset{X}{\underset{|}{CH}} \right]_n$$

where X may be H, CH_3, Cl, CN, C_6H_5, etc.

Peroxides or acids catalyze these reactions. The reaction requires high pressures and temperatures. Much lower temperature and pressure can be used with so-called Ziegler-Natta catalysts, which consist of a trialkyl aluminium and titanium tetrachloride in an inert solvent.

Condensation polymerization

This type of polymerization takes place between two bi- or poly-functional monomers with elimination of small molecules likes water, HCl, etc. For example, hexamethylene diamine and adipic acid condense to form a polymer, nylon 6:6.

$$\underset{\text{Hexamethylenedimine Diamine}}{H_2N-(CH_2)_6-NH_2} + \underset{\text{Adipic acid Diacid}}{HOOC-(CH_2)_4-COOH} \xrightarrow{-2n\,H_2O}$$

$$\underset{\text{Polyhexamethylene adipate (Polyamide) (Nylon 6:6)}}{\left[HN-(CH_2)_6-NH-CO-(CH_2)_4-CO\right]_n}$$

Thus, condensation polymerization is an intermolecular combination, and it takes place through the different functional groups (in the monomer) having affinity for each other. When monomer contains three functional groups, it may give rise to a cross-linked structure polymer. The types of functional groups, which are most important in the condensation reactions, are:

1st group	2nd group	Product	Example
Hydroxyl, –OH	Carboxyl, –COOH	Polyester, –OOC–	Polyethylene terephthalate (Terylene)
Amine, $-NH_2$	Carboxyl, –COOH	Polyamide, –NH–CO–	Polyhexamethylene adipate (Nylon - 6:6)
Hydroxyl, –OH	Isocyanate, OCN–	Polyurethane, –CONH–	Spandex fibre

The differences between addition and condensation polymerization are summarized in Table 2

TABLE 9.2: *Differences between Addition and condensation polymerization*

Addition Polymerization	*Condensation Polymerization*
1. It requires the presence of double bond in the monomer.	1. It requires two reactive functional groups to be present at both ends of the monomer. There should be at least two different bi- or poly-functional monomers having functional groups with affinity for each other.
2. No by-product is formed.	2. Generally a by-product is formed.
3. Homo-chain polymer, generally a thermoplastic is obtained.	3. Hetero-chain polymer either thermoplastic or thermoset can be obtained.
4. The growth of chain is at one active centre.	4. The growth of chain occurs at minimum of two active centres.
5. **Examples of Addition Polymerizations** (i) Polymerization of ethylene to form polyethylene (PE): $\underset{\text{Ethylene}}{n\ CH_2=CH_2} \longrightarrow \underset{\text{Polyethylene}}{\left[CH_2-CH_2\right]_n}$	5. **Examples of Condensation Polymerization** (i) Condensation polymerization of hexamethylene diamine and adipic acid to form nylon 6:6 (polyamide) $\underset{\text{Hexamethylene diamine}}{n\ H_2N-(CH_2)_6-NH_2} + \underset{\text{Adipic acid}}{n\ HOOC-(CH_2)_4-COOH}$ $\downarrow$ $\underset{\text{Nylon 6:6}}{\left[\underset{H}{\underset{\vert}{N}}-(CH_2)_6-\underset{H}{\underset{\vert}{N}}-\underset{O}{\underset{\Vert}{C}}-(CH_2)_4-\underset{O}{\underset{\Vert}{C}}\right]_n}$

contd.

Addition Polymerization	*Condensation Polymerization*
(ii) Polymerization of vinyl chloride to form polyvinylchloride (PVC): $n\ CH_2=CHCl \longrightarrow \left[CH_2-CHCl \right]_n$ Vinyl chloride Polyvinyl chloride	(ii) Condensation polymerization of caprolactum to form nylon-6. 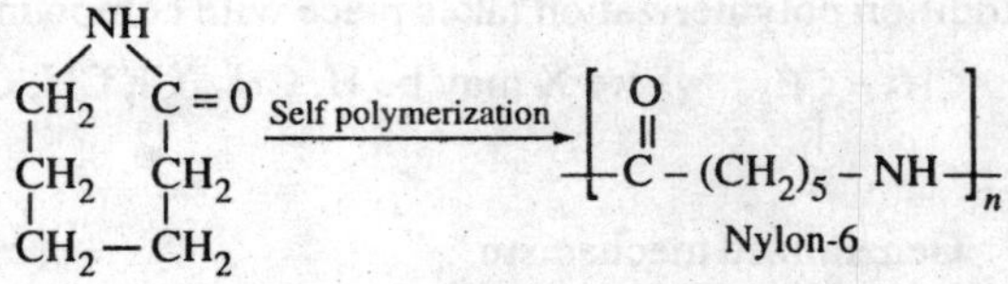 Nylon-6
(iii) Polymerization of styrene to form polystyrene (PS): 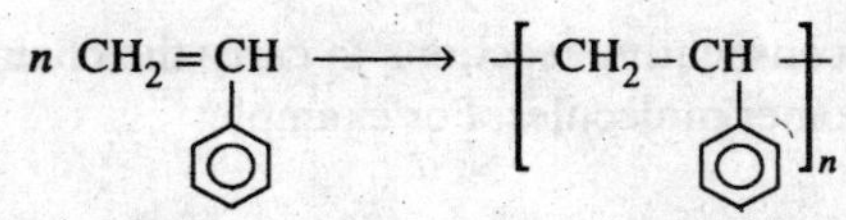 Styrene Polystyrene	(iii) Condensation polymerization of terephthalic acid and ethylene glycol to form polyethylene terephthalate. 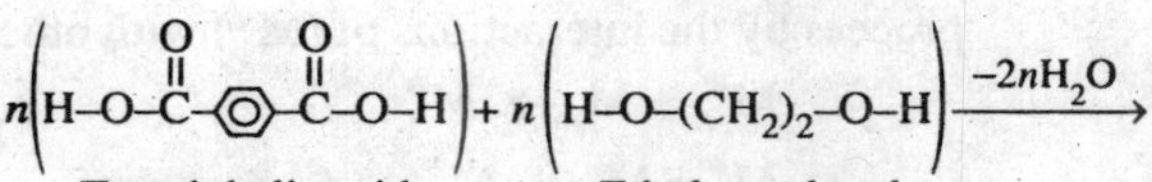 Terephthalic acid Ethylene glycol 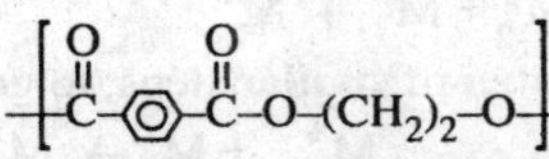 Polyethylene terephthalate

TABLE 9.3: *Comparison between Addition Polymerization and Condensation Polymerization*

Addition Polymerization	Condensation Polymerization
1. Only growth reaction, adds repeating units one at a time to the chain.	1. Any two molecular species present can react
2. Number of units decreases steadily throughout the reaction.	2. Monomer disappears early in reaction.
3. High molecular mass polymer is formed at once.	3. Polymer molecular mass (or DP) rises steadily throughout the reaction.
4. Longer reaction times have a little effect on molecular weight, but gives higher yields.	4. To obtain high molecular weight, longer reaction time is essential.
5. The reaction mixture contains only monomers, high polymers and about 10^{-8} part of growing chains.	5. All types of molecular species are present at any stage.

A polymer with a given molecular structure can be prepared by either addition or condensation method. For example: The simplest polymer, polyethylene can be made by

(a) Addition polymerization:

$$n\,(CH_2=CH_2) \longrightarrow \left[CH_2-CH_2 \right]_n \quad (I)$$

(b) $$n\,Br(CH_2)_{10}Br \xrightarrow[-2nNaBr]{2nNa} \left[(CH_2)_{10} \right]_n \quad (II)$$

Polymer (I) is very hard, tough solid with melting point 115–135°C.

Polymer (II) is low molecular weight, waxy solid having melting point 87–105°C.

In the above example, same polyethylene polymer shows different properties. This is due to the difference in mechanisms of the polymerization reactions, which determines the molecular size of the polymer either by free radical mechanism or ionic mechanism depending upon the reagents used.

Mechanism of addition polymerization

Addition polymerization takes place with compounds having the multiple bonds.

$CH_2=CH-X$ where X may be H, CH_3, Cl, CN, C_6H_5, etc.

Generalized mechanism

The polymerization proceeds in three stages:

(i) *Chain initiation steps*: Interaction of monomer (M) with initiator (In) for its activation and generation of reactive species (M*) occurs in this step.

(ii) *Chain propagation steps*: It consists of a series of consecutive reactions to carry the chain growth process by the interactions of (M*) with other monomer molecules. For example:

$$M^* + M \rightarrow M_1^*$$
$$M_1^* + M \rightarrow M_2^*$$
$$M_2^* + M \rightarrow M_3^*$$

(n–1) number of similar steps, to generate M^*_{n-1}

$$M^*_{n-1} + M \rightarrow M^*_n$$

(iii) *Chain termination steps*: The chain growth is finally terminated, i.e., arrested in this step.

$M^*_n \rightarrow M_n$ where M^*_n is activated growing polymer chain and M_n is inactive polymer molecule.

The reactive particle M* can be a free radical, cation or anion, and accordingly addition polymerization can be classified broadly as

(i) free radical polymerization
(ii) ionic polymerization (a) cationic polymerization and (b) anionic polymerization
(iii) coordination polymerization.

These are briefly discussed below:

1. Free radical polymerization

In free radical polymerization, monomer is activated by its transformation into radical by the action of light, heat, ionizing radiation or by adding chemicals known as *initiator*. Initiators are compounds, which readily dissociate homolytically to generate free radicals and the monomer molecules can interact with these free radicals for their activation. Example of well-known initiator is benzoyl peroxide. It can generate free radicals according to the following reactions:

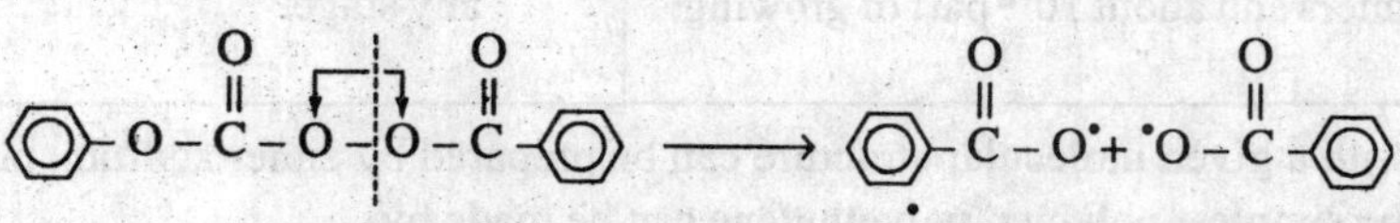

Benzoylperoxide $\longrightarrow 2\dot{C}_6H_5 + 2\,CO_2$

and $H_3C-C(CH_3)(CN)-N=N-C(CH_3)(CN)-CH_3 \longrightarrow 2\,H_3C-\dot{C}(CH_3)(CN) + N_2$

Azobis iso butyronitrile (AIBN)

Such reactions can be written in general as $In \rightarrow 2\dot{R}$ where In is an initiator molecule and $\dot{R}$ is free radical.

Free radical mechanism

Chain initiation

It involves two steps

(i) Formation of free radical by hemolytic dissociation of the initiator, In $\rightarrow$ 2R

(ii) Formation of the reactive monomer intermediate.

$$\dot{R} + CH_2 = \underset{\displaystyle X}{\underset{|}{CH}} \longrightarrow R - CH_2 - \underset{\displaystyle X}{\underset{|}{\dot{C}H}}$$

(*a*)

Chain propagation

It is accomplished by addition of monomer molecules to (a) of chain initiation reaction leading to the formation of macromolecular radicals. In general,

$$R - CH_2 - \underset{\displaystyle X}{\underset{|}{\dot{C}H}} + n\,(CH_2 = \underset{\displaystyle X}{\underset{|}{CH}}) \longrightarrow R \left[CH_2 - \underset{\displaystyle X}{\underset{|}{CH}} \right]_n CH_2 - \underset{\displaystyle X}{\underset{|}{\dot{C}H}}$$

The chain propagation steps determine

(i) the rate of polymerization, (ii) the molecular weight of the polymer, (iii) the structure of the polymer chain, (iv) the mode of monomer addition, as the reactive particle can attack an asymmetric vinyl monomer at either carbon atom of the double bond.

Chain termination

The growing polymer (macromolecular) chain can be terminated by

(i) *Recombination*, which consists of combination of two macro radicals

$$-CH_2 - \underset{\displaystyle X}{\underset{|}{\dot{C}H}} + \underset{\displaystyle X}{\underset{|}{\dot{C}H}} - CH_2 - \longrightarrow -CH_2 - \underset{\displaystyle X}{\underset{|}{CH}} - \underset{\displaystyle X}{\underset{|}{CH}} - CH_2 -$$

$$-CH_2 - \underset{\displaystyle X}{\underset{|}{\dot{C}H}} + \dot{R} \longrightarrow -CH_2 - \underset{\displaystyle X}{\underset{|}{CH}} - R$$

At T = 60 °C, polystyrene and acrylonitrile chains terminate mainly by recombination.

(ii) *Disproportionation*, which involves the transfer of a hydrogen atom from one macro radical to another to form two macromolecules, one of them with a double bond at its end.

$$-CH_2 - \underset{\displaystyle X}{\underset{|}{\dot{C}H}} + \underset{\displaystyle X}{\underset{|}{\dot{C}H}} - CH_2 - \longrightarrow -CH_2 - \overset{\displaystyle H}{\overset{|}{\underset{\displaystyle X}{\underset{|}{CH}}}} + \overset{\displaystyle H}{\overset{|}{\underset{\displaystyle X}{\underset{|}{C}}}} = CH -$$

At $T \geq 60$ °C, PMMA and vinyl acetate chains terminate mainly by disproportionation.

(iii) *Reaction with polymerization inhibitors*

Inhibitors can also terminate the growing polymer chain. Inhibiters are the substances, which quench the chain propagation process by reacting with active free radicals to form stable product or inactive free radicals. Hydroquinone, nitrobenzene, dinitrobenzene and trinitrobenzene, etc. are well known inhibitors. Oxygen is also a powerful inhibitor as it reacts with active radicals to form relatively unreactive peroxy radical ($M_n^* + O_2 \rightarrow M_n{-}O{-}O^{\bullet}$).

That is why radical polymerization is carried out in inert atmosphere.

(iv) *Reaction with the solvent molecules* like carbon tetrachloride:

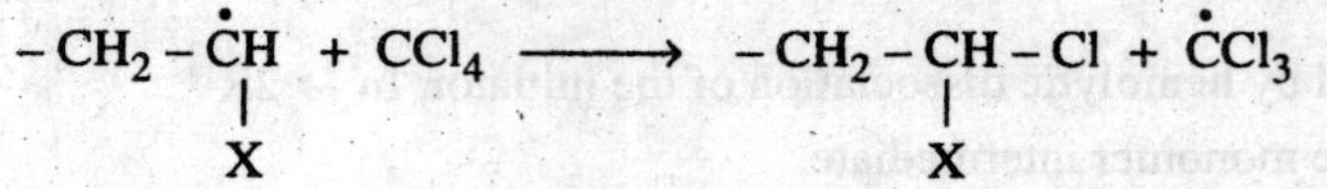

Then, $\dot{C}Cl_3 + \dot{C}Cl_3 \rightarrow CCl_3 - CCl_3$

(v) *Chain-transfer reactions* in free radical polymerization.

In these reactions, the original growing free radical chain is terminated by the reaction with the monomer molecule and a new chain is initiated.

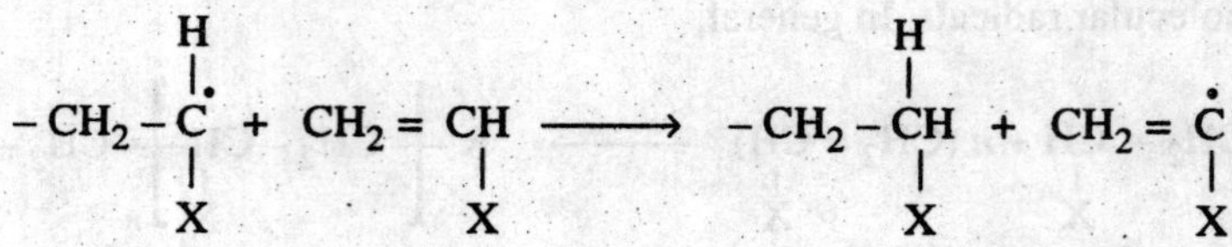

2. Ionic Polymerization

Addition or chain polymerization can also be brought out by ions instead of free radicals. Ionic polymerization depends on the formation and propagation of the ionic species. This polymerization generally proceeds in presence of catalysts. Depending on the charge of the ion formed, the polymerization process is of two types:

(A) Cationic polymerization and

(B) Anionic polymerization.

(A) Cationic polymerization

Monomers with electron releasing substituents (such as $-OCH_3$, $-OC_2H_5$, $-C_6H_5$) undergo cationic polymerization in presence of Lewis acids or Friedel Craft catalysts like $AlCl_3$, BF_3, $SnCl_4$, etc. They add on to the monomer forming a carbonium ion. An essential feature of cationic addition polymerization is that it takes place at very high rates, even at very low temperatures. This helps to suppress termination, transfer and other chain breaking reactions, which destroy the propagating centres.

Chain initiation

$SnCl_4 + H_2O \longrightarrow [SnCl_4 . OH]^{(-)} + H^+$ $\qquad H_2SO_4 \longrightarrow HSO_4^- + H^+$

$BF_3 + H_2O \longrightarrow [BF_3 : OH]^{(-)} + H^+$ $\qquad HF \longrightarrow F^- + H^+$

$AlCl_3 + H_2O \longrightarrow [AlCl_3 . OH]^{(-)} + H^+$

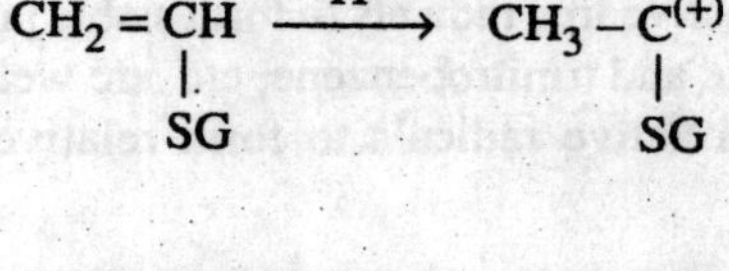

Chain propagation

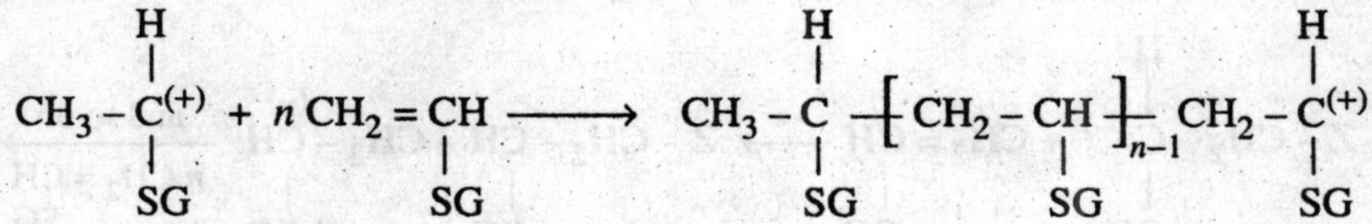

Chain termination

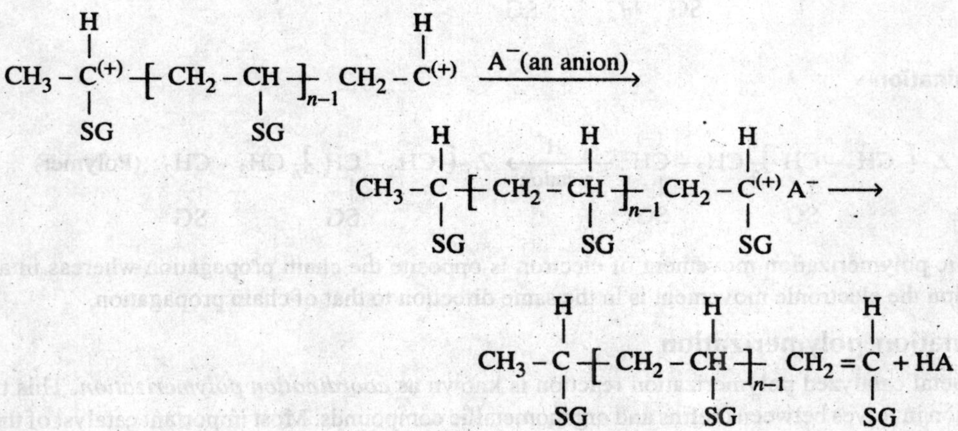

(B) Anionic polymerization

Monomers with electron attracting substituents (such as –CN, –$COOCH_3$ and –C_6H_5, etc) undergo anionic polymerization in the presence of $NaNH_2$, KNH_2, alkali metals and *Grignard* reagent as catalysts. Thus, strong bases are typical catalysts.

In anionic polymerization, the catalyst interacts with the monomer to generate a carbanion as the active centre for chain growth. As the electron withdrawing substituents stabilizes the intermediate carbanion, hence, anionic polymerization is favourable with the monomers having such substituents.

Chain initiation

A new anion (carbanion)

Chain propagation

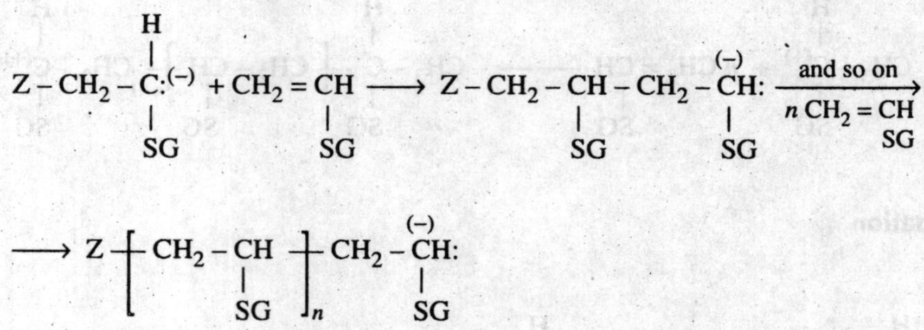

Chain termination

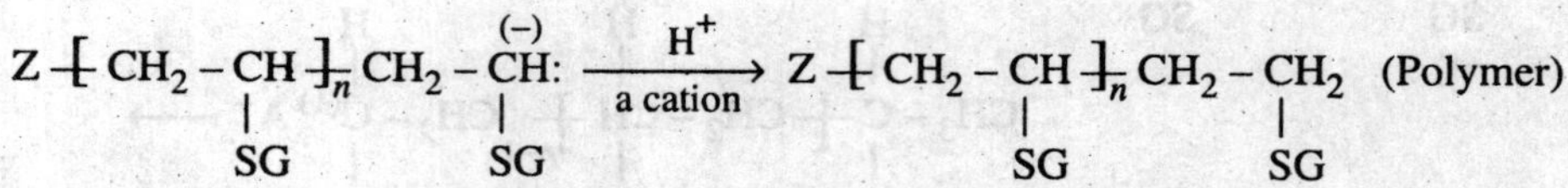

In cationic polymerization movement of electron is opposite the chain propagation whereas in anionic polymerization the electronic movement is in the same direction to that of chain propagation.

3. Coordination polymerization

Transition metal catalyzed polymerization reaction is known as *coordination polymerization*. This type of polymerization involves between olefins and organometallic compounds. Most important catalyst of this type is Ziegler-Natta catalyst. Ziegler (1953) and Natta (1955) discovered that in the presence of a combination of a transitional metal halide like $TiCl_4$ or $TiCl_3$ or halides of V, Zr, Cr, Mo and W with an organometallic compound like triethylaluminium, trimethylaluminium, stereo specific polymerization can be carried out. Here, organo aluminium acts as electron pair donor, and titanium chloride acts as electron pair acceptor. The catalyst complex is insoluble in solvents hence it is a heterogeneous catalyst. Mechanism of coordination polymerization may be represented as follows:

Chain initiation

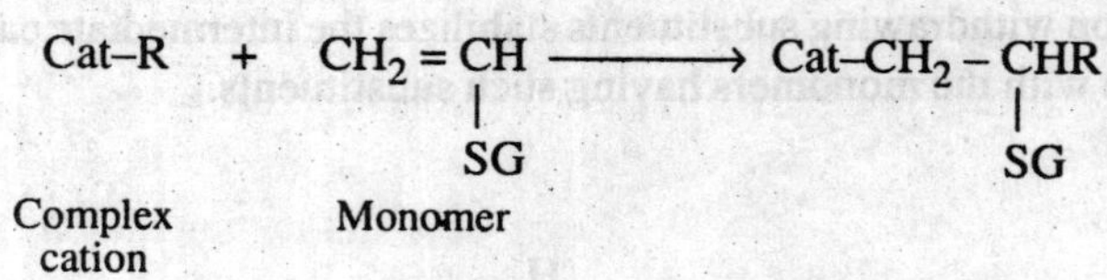

Chain propagation

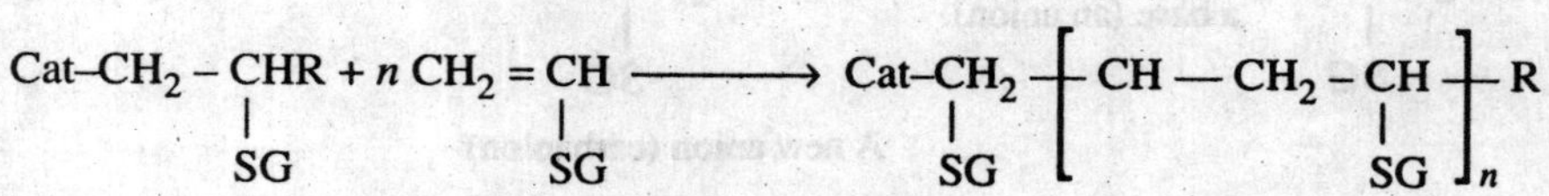

Chain termination (with active hydrogen compound)

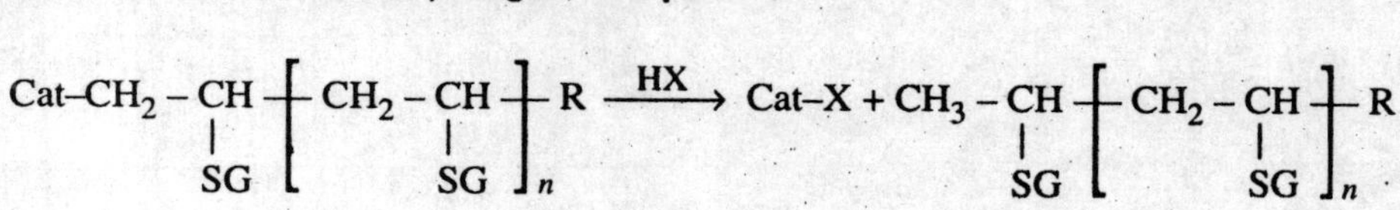

Significance of coordination polymerization

(i) This process leads to preferential formation of isotactic polymer but conventional polymerization gives only atactic polymer.

(ii) This process always results to the linear polymer free from any branching.

(iii) Using this catalytic process selective synthesis of polymer is possible. For example, polymerization of isoprene using Ziegler-Natta catalyst leads to the formation of 1, 4-polyisoprene, which is identical with natural rubber.

Copolymerization

Copolymerization is the joint polymerization of two or more monomeric species. High molecular weight compounds obtained by copolymerization are called *copolymers*. For example, butadiene and styrene copolymerize to yield SBR rubber.

$$\underset{\text{Butadiene (75\%)}}{x[CH_2=CH-CH=CH_2]} + \underset{\text{Styrene (25\%)}}{n[H_2C=CH-C_6H_5]} \xrightarrow{\text{copolymerization}}$$

$$\left[CH_2-CH=CH-CH_2\right]_x \rightarrow \left[CH_2-CH-C_6H_5\right]_n$$

Poly butadiene–co-styrene
(Styrene–Butadiene Rubber, SBR)

SOME IMPORTANT THERMOPLASTIC POLYMERS

(i) Polyethylene or polythene (PE)

Polythene is the simplest hydrocarbon polymer; it is mainly of two types, e.g., low density (LDPE) and high density (HDPE) polyethylene. It is obtained by polymerizing ethylene at high pressure of 1000-5000 atmospheres and at 250 °C in the presence of free radical initiators like oxygen, peroxides, hydro peroxides, azo compounds, etc.

$$\underset{\text{Ethylene}}{n(CH_2=CH_2)} \xrightarrow[\text{1000-5000 atmospheres}]{\text{polymerization at 250 °C}} \underset{\text{Polyethylene}}{\left[CH_2-CH_2\right]_n}$$

High density polyethylene is generally prepared in two ways:

(i) Using Ziegler-Natta catalyst $[TiCl_4 + Al(C_2H_5)_3]$, ethylene is polymerized under 6-7 atmospheric pressure at 60-70°C.

(ii) Using supported metal oxide catalyst, such as chromium or molybdenum oxide, supported over alumina or silica, ethylene is polymerized at 35 atmospheric pressure and a temperature of 60-200°C.

Polyethylene has been produced commercially since 1943. It is used in the manufacture of house wares such as buckets and dustbins, pipes, carpet backing, packaging materials and cable insulation. Here, the monomer (ethylene) contains a double bond and the polymer (polyethylene) does not. The electrons of the monomer π-bond have moved and are used to link one monomer unit to another by sigma bonds as indicated by extended lines in abbreviated polymer formula. The backbone of the polymer consists of the carbon atoms that originally formed the double bonds. Nothing is lost. The monomers simply add to each other.

TABLE 9.41: *Differences between LDPE and HDPE*

Characteristics	LDPE	HDPE
➡ Structure	Branched	Linear
➡ Density	Low (0.91-0.925g/cm³)	High (0.941-0.965g/cm³)

contd.

Characteristics	LDPE	HDPE
➡ Crystallinity	Low (~55%) Thugh and flexible	High (80-90%) Soft and hard
➡ Mechanical property	85-136 atm	204-313 atm
➡ Tensile strength	High	Low
➡ Permeability (water/gas)	Low (115°C)	High (125-135°C)
➡ Soften temperature		
➡ Applications		
(i) Mouldings	Toys, mugs, squeeze bottle particularly for detergents, ink tubes for pens, etc.	Crates, bottles for milk, household chemicals and drug packaging, food tubs, industrial containers, overhead tanks, domestic water and gas piping, etc.
(ii) Films	Carrier bags, general packaging, etc.	Films for wrapping food products.

(ii) Polypropylene (PP)

It is obtained by polymerizing propylene in presence of Ziegler-Natta catalyst at 20-120°C under low pressure in inert solvent like heptanes. Polypropylene is used in the manufacture of house wares (pans, drinking glasses), sterilizable hospital equipments, toys, electronic components, tubes, water-pipes and fibres, etc.

$$\underset{\text{Propylene}}{n\,(CH_3-CH=CH_2)} \xrightarrow{\text{polymerization}} \underset{\text{Polypropylene}}{\left[\overset{CH_3}{\overset{|}{CH}} - CH_2 \right]_n}$$

(iii) Polyvinyl Chloride (PVC)

It is obtained by heating a water emulsion of vinyl chloride containing water, in presence of a small amount of benzoyl peroxide or hydrogen peroxide in an autoclave maintained at 40-45°C under pressure. It is used in the manufacture of imitation leather sheets, floor covering corrugated roofing materials, moulding articles, insulators and gramophone records, etc.

$$\underset{\text{Vinyl chloride}}{n\,(CH_2=\overset{Cl}{\overset{|}{CH}})} \xrightarrow{\text{polymerization}} \underset{\text{Polyvinyl chloride}}{\left[CH_2-\overset{Cl}{\overset{|}{CH}} \right]_n}$$

Vinyl chloride is easily obtained from acetylene by treatment with HCl at 60-80 °C under 1-1.5 atmosphere pressure in presence of mercuric chloride as catalyst.

$$CH \equiv CH + HCl \xrightarrow{HgCl_2} CH_2=CHCl$$

(iv) Orlon

It is obtained by polymerizing vinyl cyanide (acrylonitrile). Orlon is used as a synthetic textile fibre in clothing and carpeting.

$$\underset{\text{Vinyl cyanide}}{n\,(CH_2=\overset{\overset{\large CN}{|}}{C}H)} \xrightarrow{\text{polymerization}} \underset{\text{Orlon}}{\left[CH_2-\overset{\overset{\large CN}{|}}{C}H \right]_n}$$

Vinyl cyanide can be prepared either by treating acetylene with HCN in presence of barium cyanide or by catalytic oxidation of propylene in presence of ammonia.

$$CH \equiv CH + HCN \xrightarrow{Ba(CN)_2} CH_2=CHCN$$

$$CH_3-CH=CH_2+3O_3 \xrightarrow{2NH_3 \text{ at } 450°C} CH_2=CHCN+6H_2O$$

(v) Teflon (PTFE)

It is obtained by polymerizing tetrafluoroethylene under pressure in presence of benzoyl peroxide as catalyst. Teflon is familiar because of its use as non-stick coating particularly for cooking utensils (Nirlep). Because of its low chemical reactivity, excellent toughness, electrical and heat resistance, Teflon is used as insulation for electrical items and in the manufacture of gaskets and valves.

$$\underset{\text{Tetrafluoroethylene}}{n\,(CF_2=CF_2)} \xrightarrow{\text{polymerization}} \underset{\text{Teflon}}{\left[CF_2-CF_2 \right]_n}$$

Tetrafluoroethylene is obtained by treating chloroform with HF.

$$CHCl_3+2HF \longrightarrow CHClF_2+2HCl$$

$$2CHClF_22Cl \xrightarrow{\text{at } 900\ °C} CF_2=CF_2+2HCl$$

(vi) Polystyrene (PS)

It is manufactured by polymerization of styrene (dissolved in ethyl benzene) in presence of benzoyl peroxide catalyst. Polystyrene is transparent, light (density 1.05-1.07 g/cm^3), highly electric insulting and excellent resistant to moisture, acids and chemicals. Polystyrene is used in the manufacture of food containers, cosmetic bottles, television cabinets, plastic cups, packaging and toys, etc.

$$\underset{\text{Styrene}}{n\;CH_2=\underset{\underset{\large C_6H_5}{|}}{C}H} \xrightarrow{\text{polymerization}} \underset{\text{Polystyrene}}{\left[CH_2-\underset{\underset{\large C_6H_5}{|}}{C}H \right]_n}$$

(vii) Polyamides (Nylons)

Polyamides are synthetic polymers, which have recurring amide groups. Nylons belong to this class. They are used mostly for making fibres, which find use in making socks, ladies hoses, undergarments, dresses, carpets, etc. They are also used for making filaments, ropes, bristles tooth-brush, films, tires, cords, etc.

Nylon 6:6 is obtained by the polymerization of hexamethylene diamine with adipic acid.

$$\underset{\substack{\text{Hexamethylenedimine}\\ \text{Diamine}}}{H_2N-(CH_2)_6-NH_2} + \underset{\substack{\text{Adipic acid}\\ \text{Diacid}}}{HOOC-(CH_2)_4-COOH} \xrightarrow{-2n\,H_2O}$$

$$\underset{\substack{\text{Polyhexamethylene adipate (Polyamide)}\\ \text{(Nylon 6:6)}}}{\left[HN-(CH_2)_6-NH-CO-(CH_2)_4-CO \right]_n}$$

Preparation of nylon 6:6 from 1, 3-butadiene can be carried out as follows

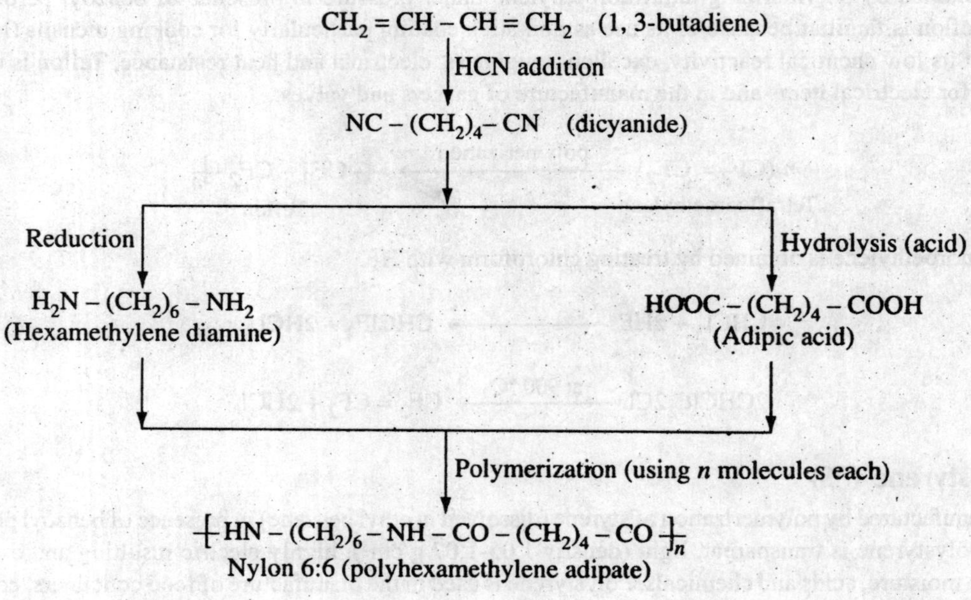

SOME IMPORTANT THERMOSETTING RESINS

(i) Phenolic resins (Bakelite)

Phenolic resins are condensation polymerization products of phenolic derivatives (like phenol, resorcinol) with aldehydes (like formaldehyde, furfural). Most important member of this class is bakelite or phenol-formaldehyde resin. It is prepared by condensing phenol with formaldehyde in presence of acidic or alkaline catalyst. The initial reaction results in the formation of *o*- and *p*-hydroxy methyl phenol, which react to form linear polymer, novolac. Thus,

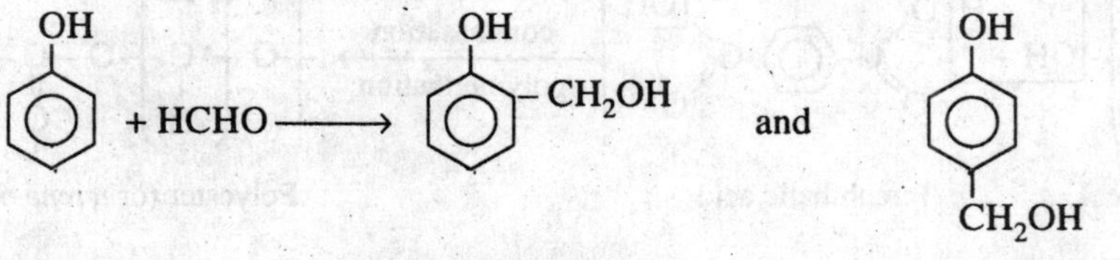

Phenol   *o*-Hydroxy methyl phenol   *p*-Hydroxy methyl phenol

Novolac, when heated with hexamethylenetetramine, which provides formaldehyde, is converted into a three dimensional cross-linked polymer, known as *Bakelite*. In presence of excess formaldehyde under alkaline condition similar methylene bridge compound is obtained, known as resole.

Bakelite is rigid hard, scratch-resistant, infusible, water resistance, insoluble solid, which is resistant to non-oxidizing acids, salts and many organic solvents, but is attacked by alkalis. It possesses excellent electrical insulating character.

Uses

(a) For making electrical insulator parts like switches, plugs, switch boards, heater handles, etc.
(b) For making moulded articles like telephone parts, cabinets for radio and television.
(c) For impregnating fabrics, wood and paper.
(d) Alcoholic solution of resole is used in paint and varnish.

(ii) Amino resins (Urea-formaldehyde resins)

Amino resins are condensation products, obtained by the reaction of urea or melamine with formaldehyde. Commercially important amino resin is urea-formaldehyde, which is prepared by the reaction between 2 parts of urea and 1 part of formaldehyde, in basic medium in a stainless vessel at about 50°C. Primary products are mono- and dimethylol urea.

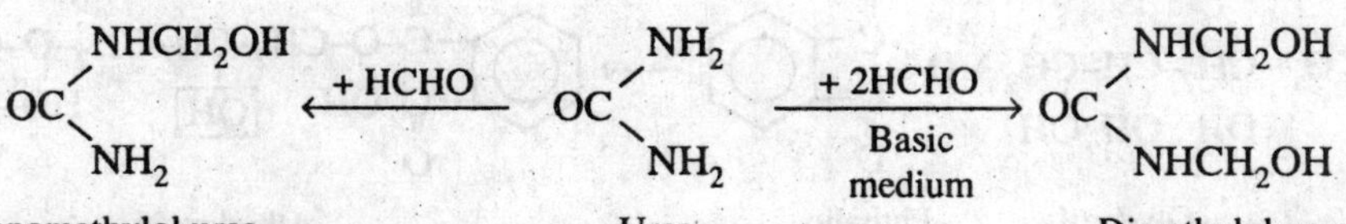

Monomethylol urea   Urea   Dimethylol urea

For moulding, the methylol derivatives are compounded with fillers, plasticizers, pigments and catalyst and then cured by applying heat and pressure. During curing, three dimensional cross-linked polymer containing long –C–N–C–N– chains are formed.

Urea-formaldehyde resins give clear, water white products of good tensile strength, good electrical insulation, good electrical resistance, great hardness, great light-stability and good abrasion resistance.

(iii) Polyester resins

Polyester resins are the condensation products of dicarboxylic acid and dihydroxy alcohols. For example, terylene or decron is a saturated polyester, formed by condensation of ethylene glycol (a diol) and terephthalic acid (a saturated diacid).

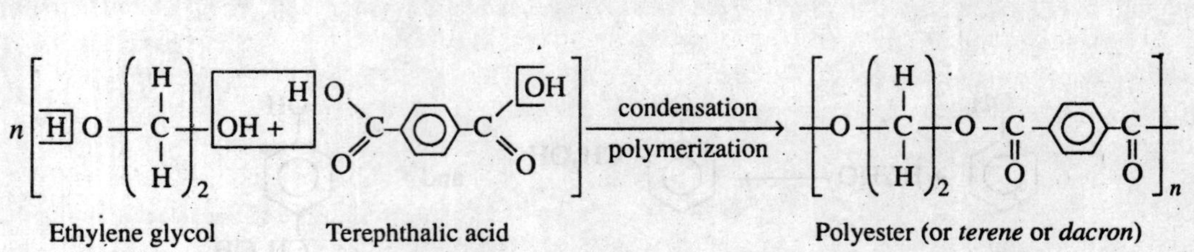

Because of its relatively symmetrical structure and presence of numerous polar groups, the polyester is a good fibre forming material and is converted into commercial fibres. Such fibres have high stretch resistance, high crease and wrinkle resistance. Polyethylene terephthalate is highly resistant to mineral and organic acids, but is less resistant to alkalis.

Uses

(i) It is mostly used for making synthetic fibres like Terylene, Dacron, etc.

(ii) For blending with wool and cotton to provide better crease and wrinkle resistance.

(iii) As glass reinforcing material in safety helmets, aeroplane tires, parasite cloth and wire insulators.

Advantages of blending with wool and cotton

(i) Blends do not shrink on ordinary washings and become better crease and wrinkle resistant.

(ii) Blends are stronger and more durable and are not ordinarily attacked by moth.

(iv) Alkyd resins

Alkyd resins are produced by reacting polybasic acid (like phthalic anhydride) with polyhydric alcohol (like glycerol) in correct proportions, in presence of heat and catalyst (CO_2 gas).

It is used for making good insulators, aircraft and automobile parts, sheets, rods, tubes, switches, gears, etc.

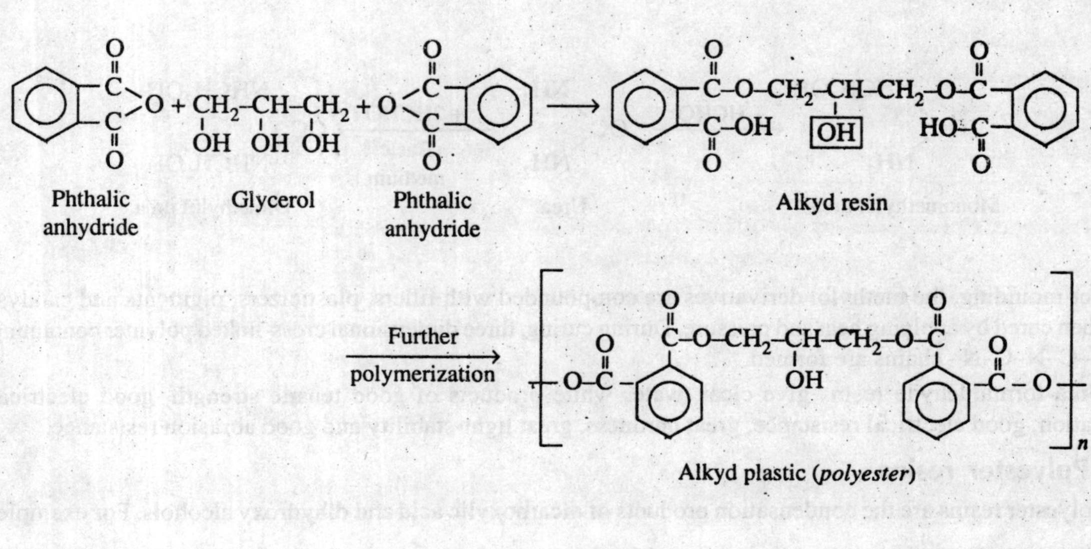

(v) Epoxy resins (Araldite)

Epoxy resins are produced by condensation of epichlorohydrin with bisphenol.

The reactive epoxide of epichlorohydrin and the hydroxyl group of bisphenol produce a three dimensional cross-linked structure. The value of n varies from 1 to 20. The molecular weight ranges from as low as 350 (a mobile liquid) to as high as 8000 (a solid melting at 145-155°C).

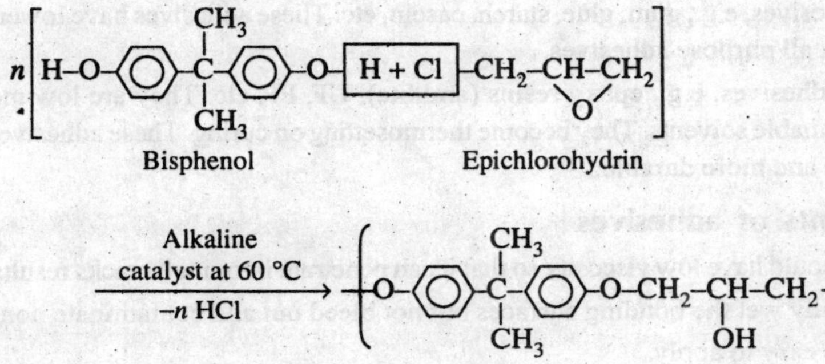

Bisphenol　　Epichlorohydrin

Alkaline catalyst at 60 °C, $-n$ HCl

Epoxy resins, due to presence of very stable ether linkages, have high chemical resistance to water, various solvents, acids, alkalis and other chemicals. Also the reactive groups (epoxide and hydroxyl groups) are comparatively widely spaced, thereby resulting in high flexibility, but at the same time, the presence of cross-links accounts for toughness and heat resistance of the cured polymers. Moreover, the polar nature of the molecules produces excellent adhesion quality.

Uses

1. As surface coatings, adhesives like araldite (for glass, metals, etc), glass-fibre-reinforced plastics.
2. For skid-resistance surfaces for high ways.
3. Moulds made from epoxy resins are employed for the production of components for aircrafts and automobiles.
4. Epoxy resins are applied over cotton, rayon, and bleached fabrics to impart crease-resistance and shrinkage control.
5. As laminating materials used in electrical equipments.

(vi) Polyurethanes

Polyurethanes are obtained, commercially by treating diisocyanate and diol. For example, Perlon-U (a crystalline polymer) is obtained by the reaction of 1, 4-butane diol with 1, 6-hexane diisocyanate.

$$\underset{\text{1, 6-Hexamethylene diisocyanate}}{OCN{-}(CH_2)_6{-}NCO} + \underset{\text{1, 4-dibutane diol}}{HO{-}(CH_2)_4{-}OH} \xrightarrow{\text{polymerization}}$$

$$\longrightarrow \underset{\text{Polyurethane (Perlon-U)}}{\left[OC{-}NH{-}(CH_2)_6{-}NH{-}CO{-}O{-}(CH_2)_4{-}O \right]_n}$$

Polyurethanes are less stable than polyamides (nylons) at elevated temperature but they are characterized by excellent resistance to abrasion and solvents.

Polyurethanes are used as coatings, films, foams, adhesives, and elastomers. Resilient polyurethane fibres (spandex) are used for foundation of garments and swim-suits. They also find use as a leather substitute (corfoam).

ADHESIVE

Adhesive is a non-metallic compound which can firmly hold two materials together by surface attachment. Adhesives are mainly used to join the surfaces of different substances such as metals, glasses, plastics, papers, etc.

Advantages of adhesive over welding, soldering, screwing, rivetting, etc. are many. For instance, adhesion may be carried out quickly, economically and easily. There are mainly two types of adhesives:

(i) Natural adhesives, e.g., gum, glue, starch, casein, etc. These adhesives have lower strength; moreover, they are not all purpose adhesives.

(ii) Synthetic adhesives, e.g., epoxy resins (araldite), UF, PF, etc. They are low molecular weight and soluble in suitable solvents. They become thermosetting on curing. These adhesives have high strength, tailor-made and more durable.

Basic requirements of adhesives

(i) Adhesive should have low viscosity so that it can penetrate into small cracks resulting in high cohesion.

(ii) It must readily wet the bonding surfaces but not bleed out and contaminate non-bonding areas.

(iii) It should be easy to apply.

(iv) It should be economically and easily available.

(v) Thermo curing adhesives should be cured at low temperatures.

(vi) It should have high strength.

(vii) It should be durable.

MOLECULAR WEIGHTS OF POLYMERS

Molecular weights of polymers are of great importance in both industry and research. The important properties of polymers are influenced by the molecular weights of polymers. As the intermolecular forces of attraction between different polymeric chains, increase with the length of the polymeric chains, hence, higher molecular weight polymers are tougher and more heat resistant. Their solution viscosities and softening temperature are also higher compared to the low molecular weight polymers, which are found to be soft and gum-like. Process ability begins to decrease past some molecular weight as the viscosity becomes too high and melt flow too difficult. Thus, polymers of specified average molecular weights are often required for particular purposes. In industrial polymerization, control of molecular weight is very essential.

All the molecules of the polymer sample do not have identical molecular weight, hence, polymers are poly disperse. This is a consequence of the fact that in a polymerization reaction, the length of chain formed is determined entirely by random events. In a condensation reaction, it depends on the availability of a suitable reactive functional group, and in addition polymerization reaction, it depends on the lifetime of the chain carriers. Inevitable, because of the random nature of the growth process, the product of a polymerization process is a mixture of chains of different lengths. As there is spread in the molecular weight of the product; hence the molecular weights are average. Different methods of determination of molecular weight provide different average molecular weights.

The methods for determining molecular weights of polymers are of two types:

(i) Primary or absolute methods, which are capable of determining molecular weights from first principles like colligate property measurements (which is denoted by ($\overline{M}_n$), light scattering measurements (which is denoted by $\overline{M}_w$).

(ii) Secondary or relative method, which requires calibration with samples of known molecular mass; examples, viscosity measurement (which is denoted by $\overline{M}_v$).

Number average molecular weight

The molecular weight obtained by colligative property measurement is known as *number average molecular weight* ($\overline{M}_n$), It is defined as the total weight (w) of all the molecules present in the polymer sample divided by the total number of moles present. Let a polymer sample contain $n_1, n_2, n_3, \cdots, n_i$ moles of polymer molecule having molecular weight $M_1, M_2, M_3, \cdots, M_i$ respectively, then by definition

$$\overline{M}_n = \frac{n_1 M_1 + n_2 M_2 + n_3 M_3 + \cdots + n_j M_j}{N_1 + n_2 + n_3 + \cdots + n_j} = \frac{\Sigma n_j M_j}{\Sigma n_j} = \frac{w}{\Sigma n_j},$$

where n_i is the number of molecules of molecular weight M_i

Thus, it is evident that M_n is simply the arithmetic mean of the molecular weight of all molecules in the sample.

Weight average molecular weight

The molecular weight determined by light scattering measurement and sedimentation method is known as weight average molecular weight ($\overline{M}_w$), Both the methods depend on the weights of the molecules present.

The weight average molecular weight ($\overline{M}_w$) is defined as the total contribution of various weight fraction s of polymeric species towards the total molecular mass of the polymeric sample.

$$\therefore \quad \overline{M}_w = w_1 M_1 + w_2 M_2 + w_3 M_3 + \cdots + w_j M_j = \Sigma w_j M_j$$

$$w_j = \frac{\text{mass of species}}{\text{total mass of polymer sample}} = \frac{n_j M_j}{n_1 M_1 + n_2 M_2 + \cdots + n_j M_j + \cdots} = \frac{n_j M_j}{\Sigma n_j M_j}$$

$$\therefore \quad \overline{M}_w = w_1 M_1 + w_2 M_2 + w_3 M_3 + \cdots + w_j M_j + \cdots = \Sigma w_j M_j$$

$$= \frac{M_1 n_1 M_1}{\Sigma n_j M_j} + \frac{M_2 n_2 M_2}{\Sigma n_j M_j} + \frac{M_3 n_3 M_3}{\Sigma n_j M_j} + \cdots$$

$$= \frac{1}{\Sigma n_j M_j}\left(n_1 M_1^2 + n_2 M_2^2 + \cdots + n_j M_j^2 + \cdots\right) = \frac{\Sigma n_j M_j^2}{\Sigma n_j M_j},$$

where n_i is the number of molecules of molecular weight M_i:

For a mono disperse polymer, $\overline{M}_n = \overline{M}_w$ i.e., $\overline{M}_n / \overline{M}_w = 1$.

But for a poly disperse sample of polymer, $\overline{M}_w$ is generally greater than $\overline{M}_n$.

Degree of polymerization (DP)

It represents the average number of mer units in a chain and is an alternate way of expressing average chain size of a polymer. Both number-average ($\overline{D}P_n$) and weight-average ($\overline{D}P_w$) degree of polymerization are possible. They are defined as:

$$\overline{D}P_n = \frac{\overline{M}_n}{M_0} \quad \text{and} \quad \overline{D}P_w = \frac{\overline{M}_w}{M_0},$$

where M_n, M_w, and M_0 are number average, weight average and monomer's molecular weights.

Molecular weight distribution

Index of poly dispersity or Poly Dispersity Index (PDI) is used as a measure of molecular weight distribution and is defined as:

$$\text{PDI} = \frac{\overline{M}_w}{\overline{M}_n}.$$

In the case of mono dispersity systems (Natural polymers and synthetic polymers made by anionic polymerization, PDI = 1 (unity), since, $\overline{M}_n = \overline{M}_w$.

In most other cases, PDI > 1 or $\overline{M}_w > \overline{M}_n$.

Problem

1. In a polymer, there are 100 molecules of molecular weight 100; 200 molecules of molecular weight 1000; and 300 molecules of molecular weight 10000. Find M_n, M_w, and PDI.

Solution

According to the problem, the given data are

$n_1 = 100$ $\quad n_2 = 200$ $\quad n_3 = 300$

$M_1 = 100$ $\quad M_2 = 100$ $\quad M_3 = 10000$

$$\overline{M}_n = \frac{\Sigma n_j M_j}{\Sigma n_j} = \frac{n_1 M_1 + n_2 M_2 + n_3 M_3}{n_1 + n_2 + n_3} = \frac{100 \times 100 + 200 \times 1000 + 300 \times 10000}{100 + 200 + 300}$$

$$= \frac{1 \times 10^4 + 2 \times 10^5 + 3 \times 10^6}{600} = \frac{3.21 \times 10^6}{600} = 5.35 \times 10^3.$$

$$\overline{M}_w = \frac{\Sigma n_i M_1^2}{\Sigma n_i M_i} = \frac{100 \times (100)^2 + 200 \times (1000)^2 + 300 \times (10000)^2}{100 \times 100 + 200 \times 1000 + 300 \times 10{,}000} = 9.4 \times 10^3.$$

Poly Dispersity Index = $\overline{M}_w / \overline{M}_n = 9.4 \times 10^3 / 5.35 \times 10^3 = 1.757$.

CONDUCTING POLYMERS

A polymer which can conduct electricity is termed *conducting polymer*. Conducting polymers can be classified into following types:

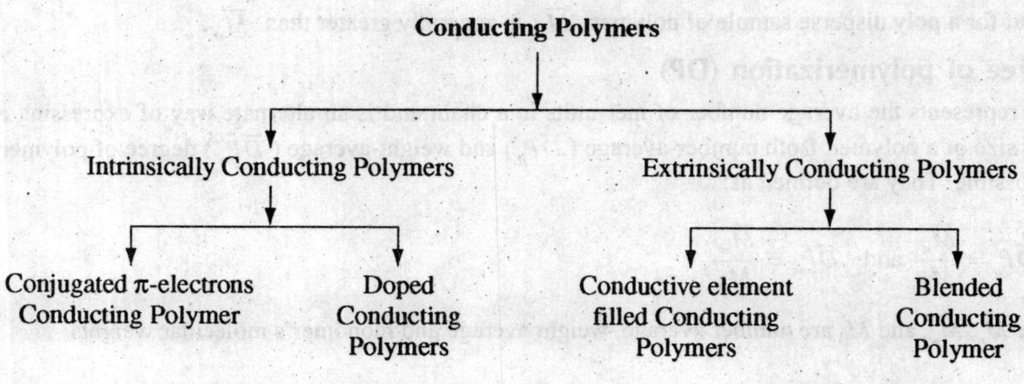

Intrinsically conducting polymers

(i) Conjugated π-electrons conducting polymers

Such polymers contain conjugated π-electrons in the backbone, which increases their conductivity to a large extent. This is because overlapping of conjugated π-electrons over the entire back bone results in the formation of valence bands as well as conduction bands, which extends over the entire polymer molecule. The valence bands and the conduction bands are separated by a significant band gap. Thus, electrical conduction could occur only after thermal or photolytic activation of electrons to give them sufficient energy to jump the gap and reach the lower levels of the conduction band. Presence of conjugated π-electrons in a polymer increases its conductivity to a larger extent. Important commercially produced conducting polymers are:

(i) Polyacetylene polymers; e.g., polyquinoline, poly-*p*-phenylene, etc.

(ii) Condensed aromatic polymers; e.g., polyaniline, polyanthrylene, etc.

(iii) Aromatic heteroatomamic and conjugated aliphatic unit polymers e.g., polypyrrole, polythiophene.

Conductivity of the above polymers having conjugated π-electrons in the backbone is not sufficient for their use in different applications.

(ii) Doped conducting polymers

Polymer is obtained by exposing a polymer to a charge transfer agent in either gas phase or in solution. Intrinsically conducting polymers possess low conductivity ($\sim 10^{-10}$ ohm^{-1} cm^{-1}), but these possess low ionization potential and high electron affinities, so these can be easily oxidized or reduced. Consequently, the conductivity of ICP can be increased by creating either positive or negative charges on the polymer backbone by oxidation or reduction. By analogy with semiconductor technology, it is referred to as doping and is of following two types

(i) **P-doping** involves treating an intrinsically conductive polymer with a Lewis acid, thereby oxidation process takes place and positive charges on the polymer backbone are created. Some of the common p-dopant used is $FeCl_3$, I_2, AsF_5, BF_3, PF_5, etc. For example,

$$\underset{\text{Polyacetylene}}{(C_2H_2)_n} + \underset{\text{(Lewis acid)}}{2\,FeCl_3} \longrightarrow (C_2H_2)_n^+ + FeCl_4^- + FeCl_2$$

$$2\,(C_2H_2)_n + 3\,I_3 \longrightarrow 2\,[(C_2H_2)_n^+ + I_3^-]$$

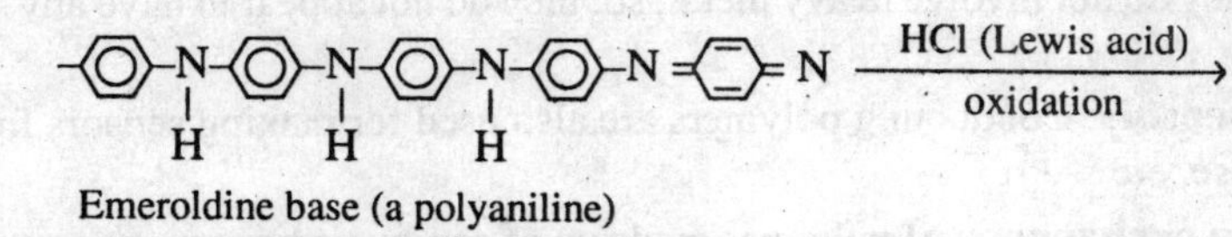

Emeroldine base (a polyaniline)
(Conductivity = ~ -10^{-10} Ω^{-1} cm^{-1})

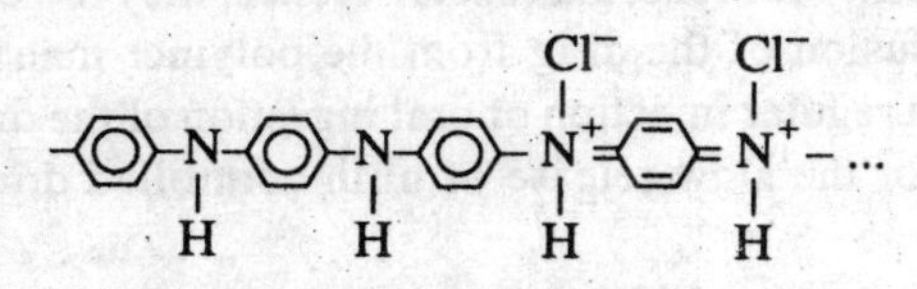

Emerolidine hydrochloride (a salt known as *synthetic metal*)
(Conductivity = ~ 10^3 Ω^{-1} cm^{-1})

(ii) **N-doping** involves treating an intrinsically conductive polymer with a Lewis base and thereby reduction process takes place and negative charges on the polymer backbone are created. Some of the common *n*-dopants used are alkali metals, alkaline earth metals, naphthylamine, sodium naphthalide, etc. For example:

$$\underset{\text{Polyacetylene}}{\cdots - CH = CH - CH = CH - \cdots} + \underset{(\textit{Lewis base})}{B} \xrightarrow{\text{reduction}} \cdots - CH = CH - \underset{\underset{B^+}{|}}{\bar{C}H} = CH - \cdots$$

$$\underset{\text{Polyacetylene}}{\cdots - CH = CH - CH = CH - \cdots} + \underset{\substack{\text{Naphthylamine}\\(\textit{Lewis base})}}{C_{10}H_7NH_2} \xrightarrow{\text{reduction}} \cdots - CH = CH - \underset{\underset{^+NH}{|}}{\bar{C}H} = CH - \cdots + \underset{\text{Naphthalene}}{C_{10}H_8}$$

Extrinsically conducting polymers

Extrinsically conducting polymers are those polymers whose conductivity is due to the presence of externally added ingredient in them. These are of the following two types:

(i) **Conductive element-filled polymer** is a resin or polymer filled with conducing elements such as carbon black, metallic fibres, metal oxides, etc. In this, the polymer acts as the binder to hold the conducting elements together in the solid entity. These polymers possess reasonably good bulk conductivity; and are, generally, low in cost, light in weight, mechanically durable and strong, and easily processable in different forms, shapes and sizes.

Minimum concentration of conductive filler which should be added so that polymer starts conducting is known as *percolation threshold*. At this concentration of filler, a conducting path is formed in polymeric material. Generally, special conducting grade carbon black is used as filler which has very high surface area (1000 m^2/gm), more porosity and more filamentous properties.

Such compounds are important, for example, in hospital operation theatres where it is essential that static charges do not build up, leading to explosion involving anaesthetics.

(ii) **Blended conducting polymer** is obtained by blending a conventional polymer with a conducting polymer either by physical or chemical change. Such polymers possess better physical, chemical mechanical and electrical properties and they can be easily processed.

Applications of conducting polymers

(i) **In rechargeable batteries**: These batteries are small in size (button type), longer lasting and can produce current density up to 50 mA/cm^2. Moreover, these rechargeable batteries have ecological advantage as they do not involve heavy metal; so, they do not appear to have any serious toxicological problems.

(ii) **In analytical sensors**: Conducting polymers are also used for making sensors for pH, O_2, NO_x, SO_2, NH_3 and glucose, etc.

(iii) **For making ion exchangers**: Membranes made up of conducting polymers can show boundary layer effects with selective permeability for ions, gases, etc.. Hence, they are useful for ion exchangers and control release of drugs. Diffusion of the drug from the polymer matrix permits a continuous and controlled release. With this, a regular injection or oral ingestion of the drug is no longer needed. This technique has been utilized for the slow release of birth controlled drug and it has potentiality for cancer chemotherapy.

(iv) **In electrochromic displays and optical filters**: ICP can absorb visible light to give coloured products; so, can be useful for electrochromic displays and optical filters. Thus, conducting polymers can be **used as electrochromic materials.**

(v) In wiring in aircrafts and aerospace components.

(vi) In telecommunication systems.

(vii) In antistatic coatings for clothing.

(viii) In electromagnetic screening materials.

(ix) In electronic devices such as transistors and diodes.

(x) In solar cells.

(xi) In photo voltaic devices, e.g., Al/Polymer/Au photovoltaic cells.

(xii) In molecular wires and molecular switches.

ELASTOMERS

In 1770, Joseph Priestley discovered that an exudates sample from the tree Hevea brasiliensis could rub pencil markings and he named it as rubber. Chemically speaking, natural rubber is *cis*-1, 4-polyisoprene.

In 1839, Charles Goodyear discovered that by heating natural rubber with sulphur (vulcanization) converts it into a more useful form, which is thermally stable.

In 1888, Dunlop's invention of pneumatic tire gave a boost to the demand for rubber. Subsequently, during the World Wars I and II, it was necessary to find the synthetic substitutes for natural rubber. All these synthetic substitutes which have rubber-like properties are called *elastomers*.

Elastomer is any rubber-like elastic polymer, which can be stretched to at least three times its length, but it returns to its original shape and dimensions as soon as stretching force is released. The elastic deformation in an elastomer arises from the fact that in an unstressed condition, an elastomer molecular chain is in the form of a coil and consequently, it can be stressed like a spring. This is due to absence of highly polar groups or sites for hydrogen bonding. Molecules in elastomers have weak intermolecular forces. For ready extension and recovery, the long chains of elastomers must be connected to each other by occasional cross links. These cross links prevent the slipping of molecules past each other but at the same time, these occasional cross links do not deprive the chains of their flexibility.

In a nut shell, long flexible chains, weak intermolecular forces and occasional cross links are the three essential structural requirements of an elastomer.

NATURAL RUBBER

By making small incisions on the barks of rubber trees, like Hevea brasiliensis and guayule, the rubber latex can be collected into small vessel, as it oozes out. Rubber latex contains 25-45% of rubber in the form of milky colloidal emulsion; the remainder of which is made mainly of water and small amounts of protein and resinous material. With time, the flow of latex from the cut decreases. Thus, at regular intervals (about six months), tapping (removal of another thin layer of bark) is necessary throughout the life of the tree, which is about 40 years. Latex is diluted to contain 15-20% of rubber. After dilution it is filtered to eliminate any dirt that may be present. It is then coagulated in a tank fitted with irregular partitions by adding about 1 kg acetic acid or formic acid per 200 kg of rubber, to a soft white mass. Sometimes ammonium or potassium alum is also used as coagulants. After washing and drying, the coagulum is subjected to any of the following processes.

(i) Crepe rubber

(ii) Smoked rubber

(i) *Crepe rubber*: It is prepared by adding little sodium bisulphite to bleach the rubber and then it is passed through a creping machine so that the coagulum is rolled out into sheets of about 1mm thickness. The sheet possesses an even rough surface resembling crepe paper. The sheet is then dried at 50°C in air.

(ii) *Smoked rubber*: It is made by eliminating the bleaching with sodium bisulphite and rolling the coagulum into somewhat thicker sheets having ribbed pattern on its surface. Ribbed surface pattern on the sheet prevents them from adhering together on stacking. It also facilitates consequent drying as it exposes greater surface area of the sheet. The sheets are then dried in smoke houses at about 50 °C in the smoke from burning wood or coconut shell. The rubber sheet thus obtained is translucent and amber in colour.

Drawbacks of Raw Rubber

(i) Raw or crude rubber becomes soft and sticky in hot summer, while in winter it becomes hard and brittle. Its usage temperature range is very limited.

(ii) It is weak, because of its low tensile strength (200 kg/cm^2)

(iii) It is attacked by oxidizing agents (HNO_3, H_2SO_4, etc)

(iv) In organic solvents it undergoes swelling and gradual disintegration.

(v) It possesses tackiness, which means that under pressure two fresh raw rubber surfaces coalesce together to form a single piece.

(vi) When stretched to a great extent, some molecular chains undergo sliding or slippage over each other, hence it suffers from permanent deformation.

(vii) Due to its oxidation in air, it is not durable.

Vulcanization

To improve the properties of raw rubber it is compounded with various ingredients, and the rubber mixture is prepared for vulcanization. In order to give it strength and elasticity it is vulcanized. In the vulcanization process, raw rubber is heated at 100-140°C with small amount of sulphur. The sulphur combines chemically with the double bonds of different rubber molecules and provides cross-linking between the chains. The cross-linking during vulcanization brings about a stiffening of the rubber by anchoring, and consequently preventing intermolecular movement of rubber springs. The amount of sulphur added determines the extent of stiffness of vulcanized rubber. For example, ordinary soft rubber (for tires) may contain 3-5% sulphur, but hard rubber (say, for a battery case) may contain as much as 30% sulphur. Sometimes, fillers such as carbon black and zinc oxide are usually added to the crude rubber before vulcanization in order to improve the wearing characteristics.

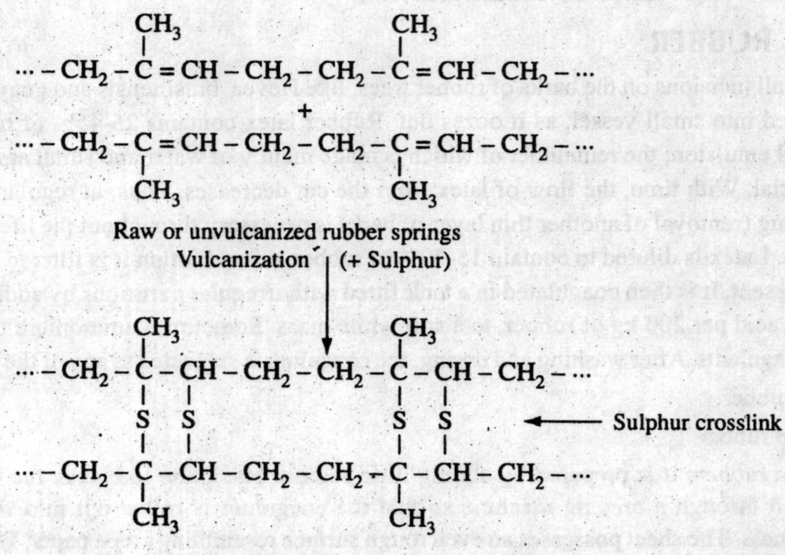

Vulcanization of raw rubber with sulphur as vulcanizing agent.

TABLE 9.5 : *Comparison of vulcanized rubber versus raw rubber*

Property	*Raw rubber*	*Vulcanized rubber*
1. Tensile strength	200 kg/cm^2	2,000 kg/cm^2
2. Elongation-break (%)	1200	800
3. Rapidity of retraction	Good	Very good
4. Water absorption	Large	Small
5. Swelling in organic solvents	Infinite (soluble)	Large, but limited
6. Tackiness	Marked	Slight
7. Useful temperature range	10 to 60°C	– 40 to +100°C
8. Chemical resistance	Very poor	Much better
9. Elasticity	Very high (300 to 1000%)	Low, depending on degree of vulcanization

SYNTHETIC RUBBER

Neoprene

It is the first synthetic rubber manufactured on large scale. Neoprene is prepared by the free radical polymerization of chloroprene at ordinary temperature in an emulsion system. It is also known as poly-chloroprene. The monomer, chloroprene is prepared from acetylene. Neoprene does not require sulphur for vulcanization; it can easily be vulcanized to a considerable extent by treatment of heat alone to have excellent tensile strength. However, its general physical properties are enhanced by compounding it with metallic oxides such as ZnO or MgO. Neoprene is particularly resistant to chemical action and is, therefore, used in making hoses for petrol oil and containers for corrosive chemicals. It is also used for making insulation coating on wires and cables.

$$2CH \equiv CH \xrightarrow[NH_4Cl]{Cu_2Cl_2} \underset{\text{Vinylacetylene}}{CH_2=CH-C\equiv CH} \xrightarrow{HCl} \underset{\text{Chloroprene}}{CH_2=CH-\overset{\overset{\displaystyle Cl}{|}}{C}=CH_2}$$

$$n\,\underset{\text{Chloroprene}}{[CH_2=CH-\overset{\overset{\displaystyle Cl}{|}}{C}-CH_2]} \xrightarrow[\text{1, 4-Addition}]{\text{polymerization}} \underset{\text{Neoprene}}{\left[CH_2-CH=\overset{\overset{\displaystyle Cl}{|}}{C}-CH_2\right]_n}$$

Buna S / Styrene Butadiene Rubber (SBR)

It is prepared by emulsion copolymerization of butadiene and styrene in the ratio 3:1 at 50°C in the presence of cumene hydro peroxide as catalyst. Since SBR contains double bonds, so vulcanization can be done in the same way as natural rubbers either by sulphur or sulphur monochloride (S_2Cl_2). Carbon black as reinforcing filler is essential to achieve good physical properties.

$$n\,\underset{\text{Butadiene}}{[CH_2=CH-CH=CH_2]} + n\,\underset{\text{Styrene}}{[C_6H_5-CH=CH_2]} \xrightarrow[\text{1, 4-Addition}]{\text{polymerization}}$$

$$\longrightarrow \underset{\text{Styrene-Butadiene Rubber (SBR)}}{\left[CH_2-CH=CH-CH_2-CH_2\overset{\overset{\displaystyle C_6H_5}{|}}{C}H\right]_n}$$

It is used for making foot wear component, insulation of wire and cables, carpet backing, gaskets, adhesives, etc.

Buna–N (Nitrile rubber/NBR)

It is a copolymer of 1, 3-butadiene and acrylonitrile and is made by emulsion copolymerization of them.

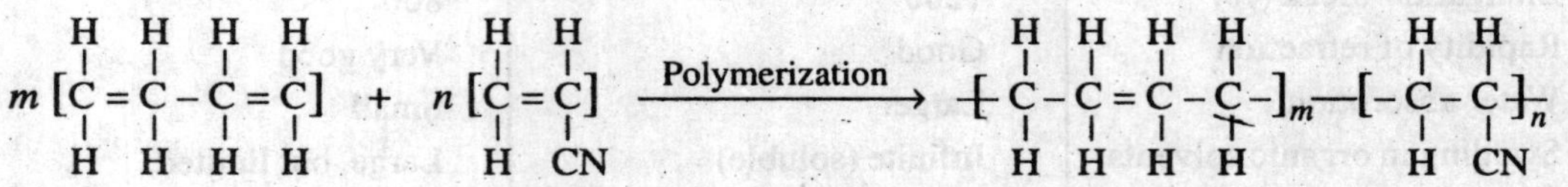

1, 3-Butadiene Acrylonitrile Polybutadiene-co-acrylonitrile
Buna-N (Nitrile rubber)

It possesses excellent resistance to heat, sunlight, oils, acids and salts but it is less resistant to alkalis than natural rubber because the presence of cyano groups –CN.

It is used for making conveyer belts, high altitude aircraft components, tank linings, hoses, gaskets, printing millers, adhesives, oil resistance foams and automobile parts.

Ebonite or vulcanite or hard rubber

It contains about 30% sulphur. The cross-linking of the isoprene spirals exists to such an extent that ebonite has practically no elasticity. Thus, ebonite is tough, high abrasion –resistant can be moulded, drilled and polished. It has extremely good electrical insulation properties.

Compounding of rubber

Compounding is the mixing of raw rubber, both synthetic and natural, with other substances so as to impart the product specific properties suitable for particular job. Besides rubber the following materials may be incorporated.

1. **Vulcanizing agents:** The main substance added is sulphur. Depending on the nature of the product required, the percentage of sulphur added varies between 0.15 to 32 percent. Many other vulcanizing agents are sulphur monochloride, hydrogen sulphide, benzoyl chloride, etc.
2. **Accelerators:** These materials drastically shorten the time required for vulcanization. The most usual accelerators used are 2-marcaptol, zinc alkyl zanthate and benzothiazole.
3. **Antioxidants:** Natural rubber has a tendency to perish due to oxidation. For this reason, antioxidant materials such as complex amines like phenyl naphthyl amine and phosphides are added.
4. **Reinforcing fillers:** These are added to give strength and rigidity to the rubber products. Common reinforcing fillers are carbon black, zinc oxide, calcium carbonate and magnesium carbonate.
5. **Colouring matters:** These are added to give the rubber product the desired colour. For white products the usual pigment is titanium dioxide; while for coloured products, pigments (like chromium oxide-green, ferric oxide-red, antimony sulphide-crimson, lead chromate-yellow) and organic lake colours are added.
6. **Softeners and plasticizers:** These are added to give the rubber the greater tenacity and adhesion. Important materials are vegetable oil, wax, stearic acid, rosin, etc.

APPLICATIONS OF RUBBER

Different varieties of rubber, natural as well as synthetic, find extensive use in the manufacture of following items:

1. Due to remarkable resistance to electricity, it is widely used as insulating materials coatings on wires and cables.

2. Due to its abrasion resistance (i.e., resistance to wearing away surface by friction) it is used for making bicycle, automobiles and aeroplane tires, shoe heads and soles, conveyor belts (used for conveying grains, coal, rocks, sand, and ores, etc., shock absorber, mounting for heavy machinery and automobile parts.
3. Due to its chemical resistance it is used for making rubber line metal tanks and vessels, used in chemical industry, where corrosive chemicals are to be processed.
4. Due to its chemical resistance to petrol, mineral oils, some solvents, etc. rubber hoses are used for their transmission. Rubber hoses are also used for water transmission for gardening, fire fighting, etc.
5. Due to its elasticity, it is used for making rubber bands, mechanical rubber goods, golf balls, printing rollers, tubes for automobiles, aeroplanes, motor cycles and bicycles.
6. Due to high abrasion resistance and good electrical insulation properties, hard rubber or ebonite is used for making storage battery cases, electrical switchboard panels, plugs and sockets, telephone receivers, etc. Hard rubber pumps, pipes, valves and fittings are used in chemical industry, where resistance to chemical corrosion is needed.
7. Sponge rubber is used for shock absorption, thermal and sound insulation purposes.
8. Foam rubber is used for making bed pillows, mattress, and cushion, and undercoat padding, etc.
9. Rubber gaskets are used for sealing various types of equipments such as cookers, refrigerator doors, autoclaves, oven doors, etc.
10. In medicine for making heart valves, blood filters, transfusion tubings, sheathe ring for nerve repairs and padding for plastic surgery.
11. As an excellent adhesive.
12. It is also used for making toys, sport goods, head bands for helmet and goggles, etc.

RECLAIMED RUBBER

Reclaimed Rubber is obtained from waste rubber articles like worn out tires, tubes, gaskets, hoses, foot-wears, etc. The process of reclamation of rubber is carried out as follows:

The waste is cut into small pieces and powdered by using a "cracker", which exerts powerful grinding and tearing action. Then, ferrous impurities, if any, are removed by the electromagnetic separator. The purified waste powdered rubber is then digested with caustic soda solution at about 200°C under pressure for 8-15 hours in "steam jacked autoclave". By this process the fibres are hydrolyzed. After the removal of fibres, reclaiming agents like petroleum and coal tar based oils and softeners are added. Sulphur gets removed as sodium sulphide and rubber becomes devocalized. The rubber is then thoroughly washed with water sprays and dried in hot air driers. Finally, the reclaimed rubber is mixed with small proportions of reinforcing agents like clay, carbon black, etc.

The reclaimed rubber is of less tensile strength, lower in elasticity, and possesses lesser water resistance than natural rubber. However, it is much cheaper; uniform in composition and has better ageing properties. Moreover, it is quite easy for fabrication.

It is used for manufacturing tires, tubes, automobile floor mats, belts, hoses, battery containers, shoes and heals etc.

POLYMERS IN MEDICINE AND SURGERY AS BIOMATERIALS

Biomaterials are materials that can be implanted in the body to provide prosthetic functions or used in diagnostic, surgical and therapeutic applications without causing adverse effect on blood and other tissues. Biomaterials have been developed from among metals, ceramics and polymers. However, use of biomaterials made from polymers is increasing day by day with the expansion of polymer industry. This is because a wide spectrum of polymers, having diverse properties, more similar to the body, have been made available during the past few decades, their appeal and acceptability is mainly due to their versatility and the fact they can be "tailor-made" or modified at will to suit specific body functions.

The most widely used polymer in medical applications is silicone rubber due to its desirable flexibility, inertness to body fluids, non-toxicity, easy sterilization, etc. Another polymer is polyurethane.

BIOPOLYMERS

Biopolymers are naturally occurring long chain molecules which are involved in biological changes important for our lives. These polymers are present in living matter (animals and plants). Broadly speaking, biopolymers are of four types, viz., Carbohydrates, Proteins, Lipids and Nucleic acids.

1. Carbohydrates provide necessary energy for functioning of the livening body and doing day's work. In the living organism, polysaccharides (such as starch and glycogen) first undergo hydrolysis in presence of enzymes to yield the monosaccharide, glucose. Glucose so produced is transported from cell to cell by the blood in animals or by the saps in plants. Slow oxidation of glucose (by a series of steps) provides energy for the functioning of living organism. Carbohydrates are digested, assimilated, and converted into energy quite speedily; and no other food can act so quickly as sugars to supply energy, when required.
2. Proteins are complex nitrogenous organic compounds of high molecular masses, synthesized by plants and animals, which on hydrolysis by enzymes yield amino acids. Proteins are among the most important types of substances present in living cells. They are found in all cells of living organisms; and they are ranked first among the biomolecule, essential for the growth and maintenance of life. Proteins (like meat, egg, fish, cheese, etc.) act as food; proteins (like wood, and silk) are used in textiles; amino acids (the hydrolyzed products of proteins) find medicinal use.
3. Lipids are a heterogeneous group of organic compounds, which are essential constituents of all plants and animals cells. They serve as the energy reservoir of body and facilitate the absorption of the fat–soluble vitamins. Their presence in tissues of animals serves as good heat insulators and shock absorbers.
4. Nucleic acids are high polymers of nucleotide (a nucleotide consists of a sugar, a phosphate group and a nitrogen base residue). Depending upon the type of sugar residues, there are two types of nucleic acids:

   (a) Deoxyribonucleic acid (DNA) and (b) Ribonucleic acid (RNA).

DNA is a polynucleotide consisting of: (i) deoxyribose (a pentose) as sugar and (ii) adenine, guanine, cytosine or thymine as the base in its structure. It is 'master molecule' contained in the chromosomes of nucleus of a cell. It is responsible for the inheritance of genetic character of a particular species of a living cell. In 1962 Watson, Crick and Wilkins won 'Nobel Prize' in Physiology and Medicine for their work on the DNA structure, which is considered by many to be the most significant development in biology in the twentieth century.

10

INSTRUMENTAL METHODS OF ANALYSIS

INTRODUCTION

Prior to about 1950, the determination of structure of a new compound was based almost entirely on its molecular formula and the chemical reactions. The compound was subjected to various tests and reactions, which give information about its structure and functional groups. This way of finding structure was time consuming and not always100% accurate. Today, such methods have been supplemented, and to some extent even replaced by modern instrumental methods of structure determination, called *spectroscopy*.

Spectroscopy is the study of interaction of matter with electromagnetic radiation with consequent exchange of energy leading to absorption or emission of radiation. The nature of this absorption or emission is characterized by molecular structure. The examination of this pattern of absorption or emission leads to unfolding of molecular structure, i.e., evaluation of parameters that decide the molecular structure, viz., bond length, bond angle, electron distribution force constant, bond dissociation energy, etc. The energy changes within a molecule during emission or absorption of electromagnetic radiation are quantized. The energy changes in a molecule are specified in terms of frequency (ν), wavelength (λ) and wave number ($\bar{\nu}$) as:

Energy change, $\Delta E = h\nu = hc/\lambda = hc\bar{\nu}$, where h is Planck's constant, c is the velocity of light.

ADVANTAGES OF SPECTROSCOPY

1. Spectroscopic methods are much easier and faster than most chemical reactions or tests.
2. It provides more information about molecular structure—detection of functional group and structural features can be done.
3. A very small amount of the sample is sufficient, and being non-destructive in nature, the entire sample can be recovered.

TYPES OF SPECTROSCOPY

There are many spectroscopic methods but the followings are discussed in short:

1. Ultra-violet-visible (UV Spectroscopy)
2. Infrared (IR Spectroscopy)
3. Nuclear magnetic resonance (NMR Spectroscopy)
4. Mass spectroscopy

In these methods, the molecule is made to absorb electromagnetic radiation, and then examined to see how it responds to absorption of energy. When a substance absorbs energy, it gets excited. The excited substance then emits radiation while returning back to ground state energy level. On recording the emitted radiation, an emission spectrum will be obtained.

There are three types of emission spectra:

1. **Continuous spectra:** If any solid is heated, it becomes incandescent emitting visible light which when examined by spectrometer (an instrument) gives a continuous spectrum. It is also observed that there is continuous gradation from violet to red throughout the spectrum and because of this continuity it is called *continuous spectrum*.
2. **Discontinuous spectra:** When incandescent gases and vapours are heated, they emit energy in the form of light. When this light is examined by spectrometer, it is found that the spectrum is non-continuous. Non-continuous spectra are of two types:
   (i) ***Line spectra***: Line spectra exhibit a series of sharp lines. Each line corresponds to a specific wavelength. The line spectrum is also called atomic spectrum because it originates from the atom of the element. Actually it is a characteristic of an atom.
   (ii) ***Band spectra***: Band spectra are characteristic of molecules and consist of broad luminous bands. Under high resolving spectrometer each band is found to be made up of large number of closely packed lines. This spectrum is generally called *molecular spectrum*.
3. **Electromagnetic radiation:** Electromagnetic radiation, of which visible light forms an obvious but very small part can be dispersed by a prism into a spectrum of colours. Electromagnetic radiation can be described as a stream of energetic particles called *photons* or as a wave. The fundamental equations of electromagnetic radiation are $E = h\nu$, where $\nu = c/\lambda$.

Here, E is the energy of radiation, h is Planck's constant, ν is the frequency of radiation and c is the velocity of light. Since ν and λ are interrelated, the energy of radiation can be expressed either by frequency or by wavelength. Higher the energy, lower is the wavelength of radiation.

REGIONS OF THE SPECTRUM

The various regions into which electromagnetic radiation has been divided can be summarized as given below:

1. Radio frequency region: $3 \times 10^6 - 3 \times 10^{10}$ Hz; Wavelength: 10 m – 1cm. The energy change involved is due to reversal of spin of a nucleus or electron. Nuclear magnetic resonance (NMR) and electron spin resonance (ESR) spectra are measured in this range.

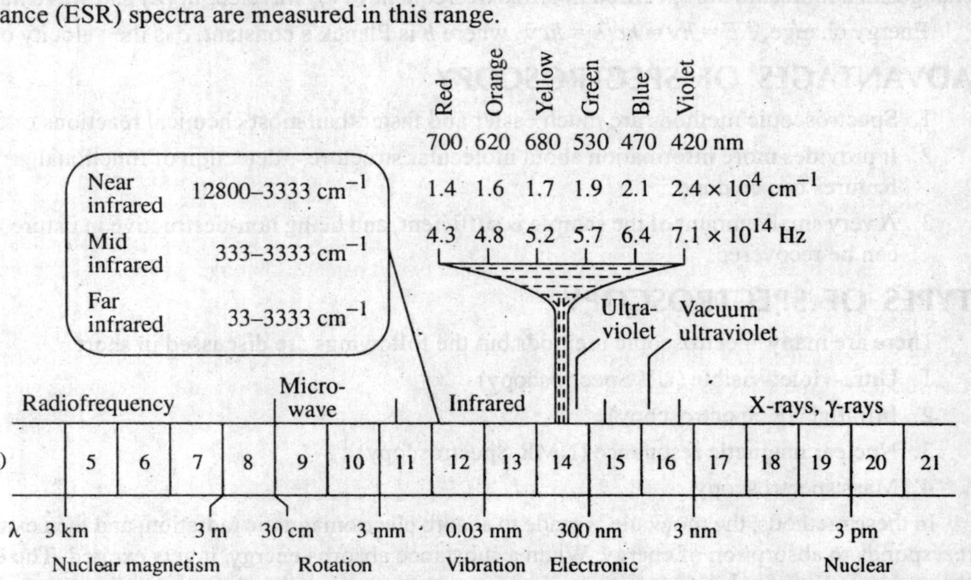

Fig. 10.1 : The electromagnetic spectrum specifying different spectral regions. The bands at the bottom indicate the type of transitions that absorbs (or emit) in the various regions.

2. Microwave region: $3 \times 10^{10} - 3 \times 10^{12}$ Hz; Wavelength: 1cm – 100μ. This region is suitable for rotational spectroscopy.
3. Infrared region: $(3 \times 10^{12} - 3 \times 10^{14})$ Hz; Wavelength: 100μ – 1μ. This region gives the results of vibrational spectroscopy.
4. Visible and ultraviolet region: $(3 \times 10^{14} - 3 \times 10^{16})$ Hz; Wavelength: 1μm – 10nm. In this region electronic spectroscopy is studied.
5. X-ray region: $(3 \times 10^{16} - 3 \times 10^{18})$ Hz; Wavelength: 10 nm – 100 pm. Energy changes involving the inner electrons of an atom or a molecule.

Parameters commonly used in molecules spectroscopy.

Parameter	wavelength (λ)	Frequency (ν)	Wavenumber ($\bar{\nu}$)
unit	nm or cm	Hz or S^{-1}	cm^{-1}

PRINCIPLE OF SPECTROSCOPY

A molecule in space can have

(i) rotational energy by virtue of bodily rotation about its centre of gravity
(ii) vibrational energy due to the periodic displacement of its atom from their equilibrium positions
(iii) electronic energy due to constant motion of the electron associated with each atom in bond.

All these energies are quantized and a particular molecule can exist in a variety of rotational, vibrational, etc. energy levels and can move from one level to another only by a sudden jump by absorbing a finite amount of energy. The absorption of wavelengths of radiation is selective for a particular transition, which depends on the structure of the molecule. By measuring the absorption spectra of known compounds, we can correlate the wavelengths of energy absorbed with characteristic structural features. It is then used to determine the structure of unknown compounds.

UV-VIS

In ultraviolet-visible (UV-VIS) spectroscopy, the 200-750 nm region of the spectrum is used. These include both the visible region (400-750 nm) and near ultraviolet region (200-400 nm). Radiation of this wavelength is sufficiently energetic to cause the promotion of loosely held electrons such as non-bonding electrons or electrons involved in a π-bond to higher energy levels. For absorption in this particular region of ultraviolet spectrum, the molecule must contain conjugated double bonds. If the conjugation is extensive, the molecule will absorb in the visible region.

Characterization of UV-VIS spectra

This is usually carried out by two parameters:

(i) λ_{max} value–λ_{max} value is the wavelength at which absorption maximum occurs. λ_{max} values are different for different molecules. For example, λ_{max} values for acetone and benzene are 280 nm and 255 nm respectively.
(ii) ε value–The extent of absorption for a given concentration, *C*, of a compound at any given wavelength is defined by molar absorptivity or molar extinction coefficient, which is indicated by ε value and is related to the height of the absorption band.

The parameters λ_{max} (the position of the absorption band) and ε (the extent of absorption) depend upon the concentration and structure of the molecule and hence, are regarded as characteristic properties of compounds. Consequently, these are extensively used in UV and Visible spectra in characterization as well as in quantitative estimation. The value of λ_{max} for compounds in UV-VIS spectra also depends upon the solvent employed but the variation is generally quite small and may be ignored.

Origin of absorption bands in UV and visible spectra

According to molecular orbital theory, the interaction of atomic orbital leads to the formation of bonding and anti bonding molecular orbitals. Depending upon the nature of the atomic orbitals involved the bonding orbitals may be of

(i) sigma (σ) type, in which the electron density is concentrated along the intermolecular axis, or
(ii) pi (π) type, in which the electron density is concentrated above and below inter nuclear axis.

Fig(a). depicts the electron density probability contours for electrons occupying (i) σ and π bonding, and (ii) σ* and π* anti bonding orbitals while Fig.(b) depicts the relative energies of bonding, anti bonding and nonbonding(*n*) molecular orbitals.

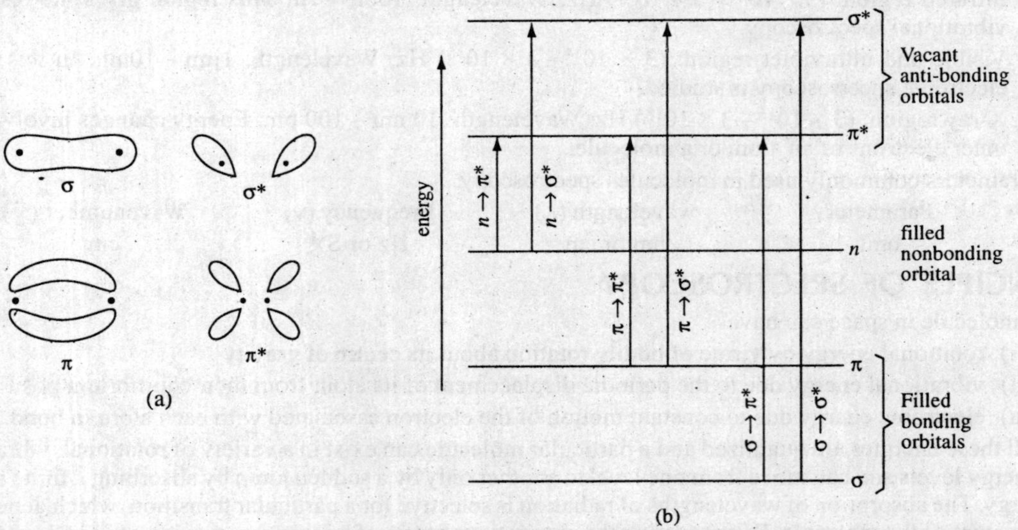

Fig. 10.2 : Shapes and relative energies of molecular orbitals: (a) Shapes of bonding and anti bonding orbitals, (b) Relative energies of orbitals and possible transitions between them.

When excitation occurs, an electron from one of the filled (σ), (π) or *n* molecular orbitals get excited to vacant σ* and π* (anti bonding) molecular orbitals. Corresponding to possible excitations, there are various absorptions:

$$\sigma \to \pi^* \qquad \pi \to \pi^* \qquad n \to \pi^*$$

$$\sigma \to \sigma^* \qquad \pi \to \sigma^* \qquad n \to \sigma^*$$

Since σ electrons require a high energy for excitation to σ*, so order of decreasing energy for the absorption is

$$\sigma \to \sigma^* > \sigma \to \pi^* > \pi \to \sigma^* > \pi \to \pi^* \sim n \to \sigma^* > n \to \pi^*$$

Out of the above possible transitions the last three ones account for the absorption in 200-800 nm region of the electromagnetic radiation. On the other hand, first three transitions require much higher energy; so only molecules with *n* and π electrons give rise to characteristic spectra in the region 200-800 nm of electromagnetic radiation. It may be pointed out that alkanes absorb in the region below 200 nm. It may be added here that the part of the molecule having *n* or π electrons is essentially responsible for absorption, and these fragments are known as chromophores.

The simple chromophores like >C = C <, –C ≡ C–, –C ≡ N–, –N = N–, and > C = O undergo $\pi \to \pi^*$ transitions in the short wavelength regions of ultraviolet radiations.

Polar groups such as –OH, –OR, $-NH_2$, –SH, and –X (halogen) having unshared pair of electrons may show absorption above 190 nm. Such grouping are called auxochromes, when conjugated with chromophores they modify the position of absorption band relative to that of the parent chromophore alone.

Compounds containing carbon oxygen double bond absorb light in the UV region. For example, acetone has a broad absorption peak at 280 nm, corresponding to excitation of an electron from nonbonding, *n* to π* of the carbon oxygen double bond. If the C = O double bond is conjugated with C = C double bond, then the absorption maxima correspond to $n \to \pi^*$ excitation and $\pi \to \pi^*$ occurs at a longer wavelength, but is much weaker. For example, the compound $CH_2 = CH–CO–CH_3$ gives $n \to \pi^*$ λ_{max} at 324 nm and $\pi \to \pi^*$ λ_{max} at 219 nm.

It may be pointed out that energy gap between nonbonding and anti bonding MO of ethene is greater than that between the corresponding MO in 1, 3-butadiene. Consequently, 1, 3-butadiene has its λ_{max} at 171 nm (of higher energy) than that of ethene at λ_{max} at 217 nm (of lower energy). Thus, in general the greater the number of double bonds present in a given compound, the lower is the wavelength at which it absorb UV region.

Theory of spectrophotometry

When light (monochromatic or heterogeneous) falls upon a homogeneous medium, a portion of the incident light is reflected, a portion is absorbed within the medium, and the remainder is transmitted. If the intensity of the incident light is expressed by I_0, that of the absorbed light by I_a, that of the transmitted light by I_t, and that of the reflected light I_r, then $I_0 = I_a + I_t + I_r$. For air-glass interfaces arising from the use of glass cells, it may be stated that about 4% of the incident light is reflected. I_r is usually eliminated by the use of a control, such as a comparison cell, hence $I_0 = I_a + I_t$. Credit for investigating the change of absorption of light with the thickness of the medium is frequently given to Lambert and concentration of solution to Beer. The two separate laws governing absorption are usually known as Lambert's Law and Beer's Law. In the combined form they are referred to as the *Lambert-Beer Law*. The mathematical form of Lambert-Beer law is $\log I_0/I_t = acl$, where I_0 is the intensity of incident light falling upon an absorbing medium of thickness, 1 and concentration c and I_t is the intensity of the transmitted light. If concentration, c is expressed in mole L^{-1} and length, 1 in centimeter then 'a' is given by the symbol, ε called the molar absorptivity (formerly the molar extinction coefficient).

Instrumentation

Ultra Violet–Visible spectra are measured by an instrument called *spectrophotometer* which can be single or double beam type. In a double beam spectrophotometer, a beam of light splits into two equal halves. One half of the beam (the sample beam) is directed through a transparent cell containing a solution of the compound being analyzed and the other half (the reference beam) is directed through an identical cell that contains only the solvent. The instrument is so designed that it can compare the intensities of the two beams at each wavelength of the region. If the compound absorbs light at a particular wavelength, then intensity of the sample beam (I) will be less than that of the reference beam (I_0). The instrument gives output graph – a plot of the wavelength of the entire region versus absorbance (A) of the light at each wavelength, where absorbance, $A = \log(I_0/I)$. Such a graph is known as an absorption graph.

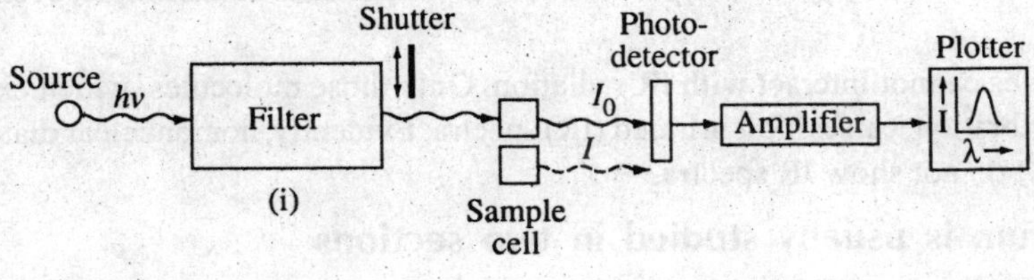

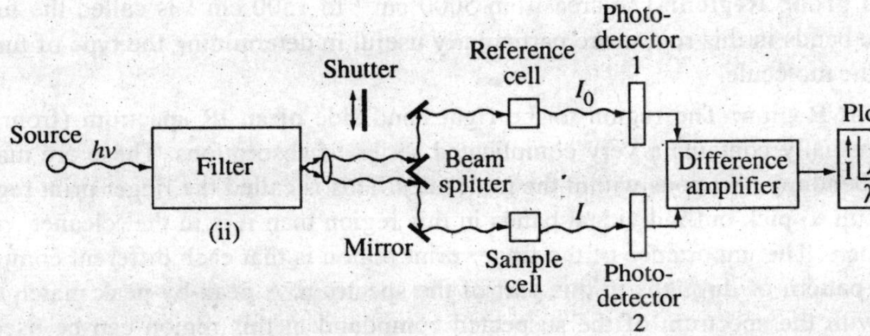

Fig. 10.3 : (i) Single beam & (ii) Double beam spectrophotometers.

The wavelength of maximum absorbance is referred to as λ_{max}.

The ultra violet –visible spectrum is composed of only a few broad bands of absorption.

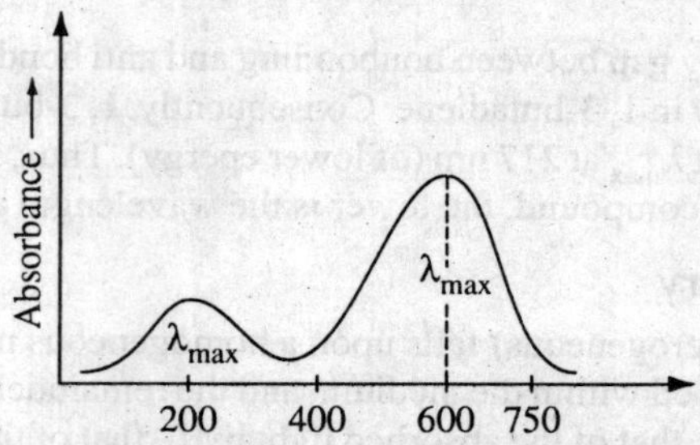

Fig. 10.4 : A sample UV-Vis spectra.

Applications

It is generally applied in predicting the relation between different groups, especially with respect to conjugation existing between

(i) two or more C–C multiple (double or triple) bonds.

(ii) C–C and C–O double bonds or

(iii) C–C double bonds and aromatic benzene ring.

INFRARED (IR) SPECTRA

Electromagnetic radiation in the infrared region of the spectrum (3×10^{12} to 3×10^{14} Hz or 100 cm^{-1} to 10^4 cm^{-1}) results in changes of the vibrational energy of molecules. The energy change required for molecular vibration is of the order 1 to100 $kJmol^{-1}$. The changes in vibrational energy levels are accompanied by transitions between rotational levels also. Consequently, each line in vibrational structure is accompanied by the rotational fine structure. Evidently, such a spectrum is known as vibrational-rotational spectrum. However, if the rotational fine structure is not considered, then absorption is known as vibrational spectrum. But, the complication can be avoided by taking liquid or solid sample as well as sample in solution . Under such conditions, the rotational fine structure becomes buried and consequently a broad peak is observed for a vibrational change only. Consequently, the interpretation of the IR spectra is made simple, even for polyatomic molecules.

All types of molecules cannot interact with IR radiation. Only those molecules, which exhibit change in dipole moment during vibration, can exhibit infrared (IR) spectra. Evidently, homonuclear diatomic molecules (e.g., H_2, O_2, N_2, or Cl_2) do not show IR spectra.

An infrared spectrum is usually studied in two sections

1. **Functional group Region:** The area from 5000 cm^{-1} to 1500 cm^{-1} is called the functional group region. The bands in this region are particularly useful in determining the type of functional groups present in the molecule.
2. **Finger print Region:** The region to the right hand side of an IR spectrum (from about 1500 to 500 cm^{-1}) usually contains a very complicated series of absorptions. These are mainly due to all manner of bending vibrations within the molecules. This is called the finger print region. It is much more difficult to pick out individual bands in this region than it is in the 'cleaner' region at higher wave numbers. The importance of the finger print region is that each different compound produces a different pattern of throughs in this part of the spectrum. A peak-by-peak match of an unknown spectrum with the spectrum of the suspected compound in this region can be used, much like a fingerprint, to confirm its identity.

Some characteristic IR absorption bands

TABLE 10.1: *Characteristic infrared absorption frequencies (in* cm^{-1}*) for different groups*

Bond	Type of compound	Frequency (cm^{-1})	Bond	Type of compound	Frequency (cm^{-1})
C–H	Aliphatic	2700-3000	O–H	Monomeric alcohols	3200-3600
C–H	Aromatic	3000-3100	O–H	Carboxy acids	3300-3370
C–H	Cycloalkane	2920-3100	O–H	Phenolic	3600-3700
C–H	Alkanes	1350-1480	O–H	Phenolic (H-bonding)	3200-3600
C–H	Alkenes	3000-3100	S–H	Marcaptan	2570-2600
C–H	Alkynes	3300	N–H	Amines, amides	3300-3500
C–H	Open chain aliphatic	750-1100	C–N	Amines (primary)	1050-1400
C=C	Aromatic rings	1460-1600	C=N	Amines (secondary)	1590-1660
C=C	Alkenes	1620-1670	C≡N	–Nitriles	2200-2250
C≡C	Alkynes	2100-2250	$-NO_2$	–Nitro compounds	1515-1560
C=O	Aldehydes (aliphatic)	1720-1740	C–F		1000-1400
C=O	Aldehydes (aromatic)	1495-1715	C–Cl		600-800
C=O	Ketones (aliphatic)	1705-1725	C–Br		500-600
C=O	Ketones (aromatic)	1720-1740	C–I		500
C=O	Esters (aliphatic)	1700-1750			
C=O	Esters (aromatic)	1230-1650			
C=O	Carboxylic acids	1650-1750			
C=O	Amides	1630-1690			
C–O	Alcohals, ethers, esters	1080–1300			

Applications of Infrared spectroscopy

1. Different functional groups absorb infrared radiation at different wavelengths and the presence or absence in a molecule can be determined by examination of an IR spectrum. No two compounds have exactly identical infrared spectra. This is generally done by fingerprint technique, which involves matching the spectrum (IR) of unknown compound with that of a known compound.
2. *Identification of functional groups in organic molecules*: As different functional groups in organic compounds have the characteristic value of μ and *K*, hence, they will absorb at characteristic frequencies in the IR spectrum. Thus, from the IR spectrum of a compound, the presence and absence of certain functional groups in the compound is confirmed.
3. *Detection of impurities in a sample*: IR spectra of impure sample will show some extra absorption bands. By comparison with IR spectra of pure compound, presence of impurity can be detected.
4. *To ascertain hydrogen bonding in molecules*: Generally, it is not possible to distinguish between intra- and intermolecular hydrogen bonding. This can be ascertained by taking a series of IR spectra of the compound at different dilutions. As the dilution is increased, the absorption band due to intermolecular hydrogen bonding diminishes, while that due to intra molecular hydrogen bonding remains unchanged.
5. IR spectra may also be employed for rapid quantitative analysis of a mixture of compounds, in pollution detection, in milk analysis, etc.

6. *Elucidation of structure*: Structure elucidation is possible by IR spectroscopy because it provides valuable information regarding molecular symmetry, dipole moments, bond lengths, bond strength, etc.
7. Distinguishing positional isomers of a compound.
8. Determination of force constant from vibrational spectrum.

NUCLEAR MAGNETIC RESONANCE

E.M. Purcell (1912-1997), an American physicist, was awarded Nobel Prize in 1952 for his independent discovery (1946) of nuclear magnetic resonance (NMR). Nuclear magnetic resonance (NMR) has been widely used to study the molecular structure of pure materials and the composition of mixtures.

It has been found experimentally that any nucleus consisting of either an odd number of protons or odd number of neutrons or both has the property of nuclear spin. For example, ^{1}H, ^{13}C, and ^{19}F nuclei possess spin, but not ^{12}C and ^{16}O. We are particularly concerned almost exclusively with the nuclear spin associated with proton (^{1}H). Since the spinning proton generates a circulating electric current, which in turn produces a magnetic field, so a spinning proton behaves like a tiny magnet. Now, when a proton is placed in a magnetic field, it can align itself in one of the two possible orientations with respect to the applied field:

(i) Low energy alignment of the nucleus in which the magnetic field of the nucleus is in the same direction as the applied magnetic field, and

(ii) High-energy orientation in which the two magnetic fields oppose each other as shown in Fig. 10.5.

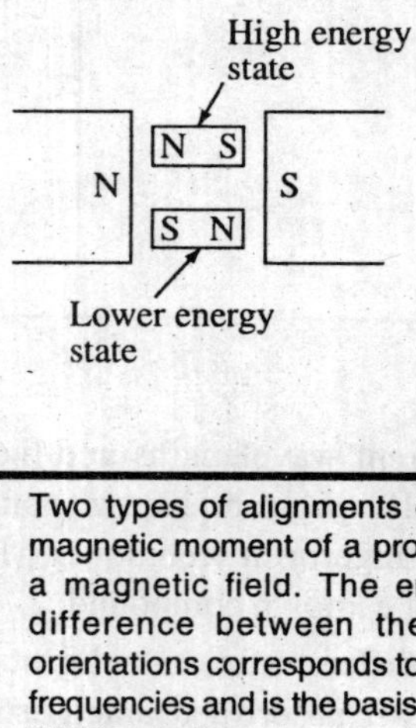

Fig. 10.5 : Two types of alignments of the magnetic moment of a proton in a magnetic field. The energy difference between the two orientations corresponds to radio frequencies and is the basis of the NMR technique.

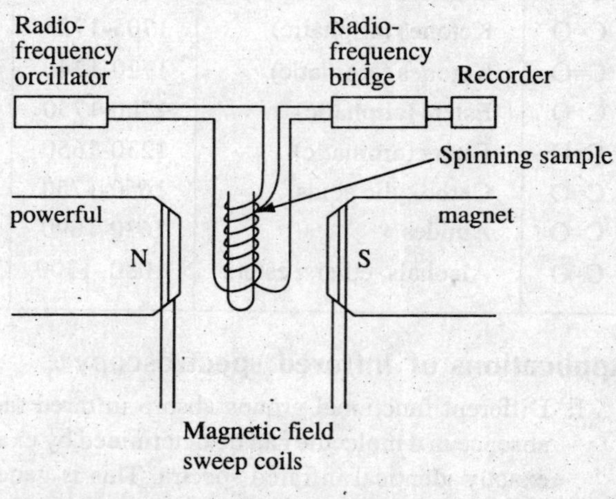

Fig. 10.6 : Basic arrangement for NMR.

A transition from the low energy state to the high-energy state of the nucleus can be obtained by providing energy equal to the difference in energy between the two states, ΔE. This energy can be supplied by electromagnetic radiation. Also energies are dependent on the magnitude of the applied field, so, it is possible to choose the external magnetic field to give absorption of electromagnetic radiation in a suitable part of spectrum. Usually, a field of approximately 2.3 T (Tesla) or 23000 Gauss results in absorption in radio frequencies of 10MHz. Consequently, nuclear magnetic resonance (NMR) is normally observed in this region.

It may be pointed here, that the character of NMR spectra is not solely determined by nuclear properties, but also by the environment. In other words, the frequency of radiation absorbed by a given nucleus is strongly influenced by the kind of environment in which it exists. For example, CH_3CH_2OH contains 6 protons. So, its NMR spectra contain 3 peaks for 3 protons of CH_3, 2 protons of CH_2 and 1 proton of OH respectively.

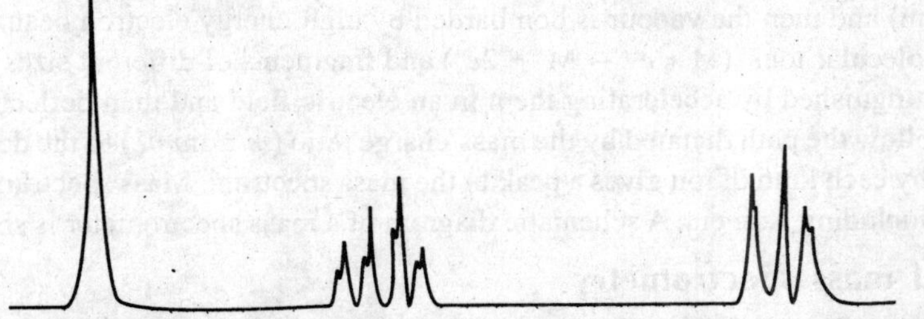

Fig. 10.7 : Proton resonance spectrum of ethyl alcohol at 40 MHz. The signal from the radiofrequency receiver is plotted vertically, and the magnetic field strength is plotted horizontally. The field increases linearly from left to right.

Applications of NMR spectroscopy

Information obtained from NMR spectroscopy:

1. The number of signals in the spectrum indicates types of proton in the compound.
2. Integration of spectrum gives the relative number of protons of different types.
3. Chemical shifts give the type of structure associated with particular protons.
4. Positions of the peaks indicate the type of electronic environment of each kind of proton.
5. The splitting patterns give the number of protons on atoms adjacent to the group, which contains protons, whose resonance is being measured.
6. The intensities (relative height) of signal give the number protons of each kind. For example, if three signals recorded on NMR spectrum are of relative heights 3:2:1, then there are 3 protons of one type, 2 protons of other type and 1 proton of another type.

MASS SPECTROMETRY

Mass spectrometry is an important analytical tool, which provides qualitative and quantitative information regarding the atomic and molecular composition of inorganic and organic materials. F. W. Aston (1877-1945) received Nobel Prize in 1922 for the discovery of mass spectrograph, a device that separates atoms or molecular

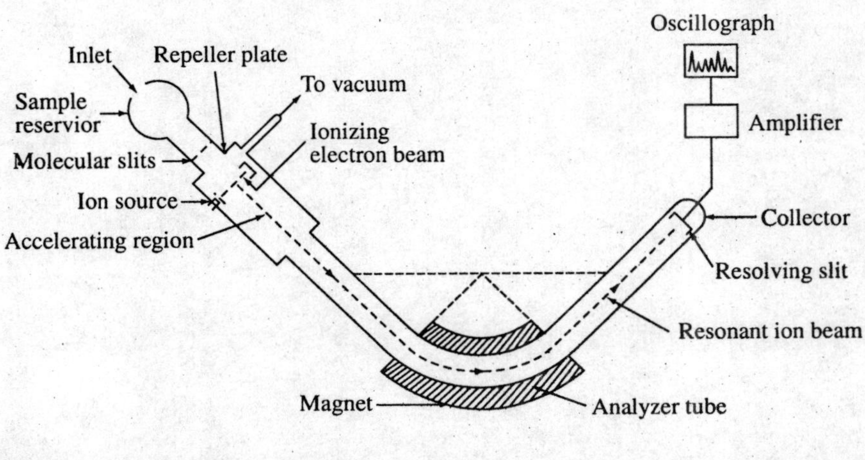

Fig. 10.8 : Schematic diagram of a mass spectrometer.

fragments of mass, and measures with remarkable accuracy. The mass spectrograph is widely used in chemistry, biology, geology and nuclear physics.

In mass spectrometry, the material (inorganic or organic or biological) under investigation is vapourized (using high vacuum) and then the vapour is bombarded by high-energy electron beam. Many of the vapour molecules form molecular ions ($M + e^- \rightarrow M^+ + 2e^-$) and fragments of different sizes called *daughter ions*. These ions are distinguished by accelerating them in an electric field and then deflecting them in magnetic field. Here, they follow the path dictated by the mass/charge ratio (m/z or m/e) to the detection and recording equipment, whereby each kind of ion gives a peak in the mass spectrum. Mass spectrum is a characteristic of every compound, including isomers. A schematic diagram of a mass spectrometer is shown in Fig.10.8.

Applications of mass spectrometry

1. Determination of molecular weight with high accuracy.
2. For analysis of gas and petroleum products.
3. Detection of trace impurities in semiconductors.
4. Establishing the structure of organic compounds
5. Investigation of reaction mechanism in radioactive tracer work.
6. Understanding of kinetics and mechanism of molecular reactions.
7. Identification of unknown compounds by comparing their mass spectra with literature.

Disadvantages of mass spectrometry

1. Samples employed cannot be recovered.
2. Difficulties in introducing sample into high vacuum system.
3. High cost of the instrument.

INDEX

UNIVERSITY QUESTION PAPERS WITH ANSWERS

2005

Time Allotted : 3 hours **Full marks : 70**

The figures in the margin indicate full marks. Candidates are required to give their answers in their own words as far as practicable.

Answer Question 1 and any five from the rest

1. (a) Answer any five of the following:
   (i) What does the abbreviation TEL stand for?
   (ii) What is the relation between C_p and C_v of an ideal gas?
   (iii) Write the name of a natural polymer.
   (iv) What is the relation between enthalpy and internal energy?
   (v) What is inversion temperature?
   (vi) Name the most commonly used vulcanising agent.
   (b) State True or False:
   (i) For an isothermal process $\Delta E = 0$.
   (ii) Homolytic fission leads to generation of free radicals.
   (iii) S_N2 reaction occurs with retention of configuration.
   (iv) Maximum available work is obtained from a process when it is carried out isothermally and reversibly.
   (v) Bakelite is a thermoplastic.
2. (a) Define the thermodynamic terms: Open system and closed system.
   (b) For an adiabatic process, prove that PV^{γ} = constant.
   (c) Calculate the amount of work done when 5 moles of an ideal gas expands isothermally and reversibly at 27 °C from 5 litres to 50 litres.
3. (a) State and explain the First Law of Thermodynamics.
   (b) Deduce the expression for the maximum work done when *n* moles of an ideal gas expand isothermally and reversibly.
   (c) Explain what is meant by Gibbs free energy.
   (d) Deduce the Gibbs-Helmholtz equation in any of its standard forms.
4. (a) Discuss the shapes of NH_3 and H_2O molecules on the basis of VSEPR theory.
   (b) Explain–H_2O is a liquid at ordinary temperature, whereas H_2S is a gas.
   (c) Discuss Schottky and Frenkel defects with the help of diagrams.
5. (a) Distinguish between 'order' and 'molecularity' of a reaction.
   (b) Deduce the expression for the rate constant of a first order reaction. Hence, show that the half-life period of a first order reaction is independent of the initial concentration of the reactant.
   (c) A first order reaction is 20% complete in 10 minutes. Calculate the time taken by the reaction to go to 80% completion.
6. (a) Define 'Ligand' and 'Coordination number'.
   (b) (i) Name the following complex in the IUPAC system:
   $[CO(en)_2ONOCl]Cl$

(ii) Write the formula of the following complex:
tetramminedicholoro cobalt (III) ion.

(c) Define linkage isomerism for coordination compounds with examples.

(d) What are S_N1 and S_N2 reactions? Give example.

(e) Write down examples of two electrophiles.

7. (a) Define addition and condensation polymerisations with examples.

(b) Distinguish between thermosetting and thermoplastic resins. Give examples.

(c) Classify polymers on the basis of tacticity.

8. (a) Distinguish between Gross and Net Calorific Values of a fuel.

(b) What do you understand by HTC and LTC of coal? Write down the usefulness of each process.

(c) What are (i) straight run gasoline and (ii) jet fuels?

9. (a) Explain the variation of equivalent conductances of strong and weak electrolytes with concentration?

(b) State and explain Kohlrausch's law of independent migration of ions.

(c) Differentiate between electrolytic cells and galvanic cells.

(d) Calculate the EMF of the following cell:
$Ni\,|\,Ni^{2+}(1M)\,||\,Pb^{2+}(1M)\,|\,Pb$ at 25 °C
Given: standard electrode potentials of Ni and Pb are –0.24 V and –0.13 respectively at 25 °C.

SOLUTION

1. (a) (i) TEL–Tetraethyl lead $(C_2H_5)_4Pb$

(ii) $C_P – C_V = R$ for one mole of gas
R is universal gas constant

(iii) Natural polymer–Starch, silk, wool, natural rubber, etc.

(iv) $H = U + PV$
$H \rightarrow$ Enthalpy of a system
$U \rightarrow$ Internal energy of the system
$PV \rightarrow$ Pressure-volume work

(v) For every gas there is a temperature where Joul-Thomson coefficient $\mu_{JT} = \left(\frac{\partial T}{\partial P}\right)_H \approx \left(\frac{\Delta T}{\Delta P}\right)_H = 0$
i.e., when neither heating nor cooling of the gas due to passage of gas through fine orifices would occur. This is called the inversion temperature of the gas. Inversion temperature of N_2 is 621 K, for H_2 is 202 K.

(vi) Most commonly used vulcanising agent is sulphur.

(b) (i) For an isothermal process, $\Delta E = 0$ —True

(ii) Homolytic fission leads to generation of free radicals —True

(iii) S_N2 reaction occurs with retention of configuration—False

(iv) Maximum available work is obtained from a process when it is carried out isothermally and reversibly—True

(v) Bakelite is a thermoplastic—False.

2. (a) **Open System:** In an open system, the transfer of both energy and mass between system and surroundings can take place. For example, a glass of water; water can be poured out of the glass or added to the glass, again the water in the glass may be heated or cooled thus enabling the transfer of both mass and energy.

Closed System: In a closed system, no transfer of mass is possible, only energy transfer may take place. For example, if we seal the mouth of the glass containing the water, no mass transfer can take place. However, the water in the glass may still be heated or cooled.

(b) We know that for adiabatic process, $đq = 0$.

Therefore, $dU = đw$ (from first law);

Now $dU = nC_V dT$ and $đw = -PdV$

Therefore, $nC_V dT = -PdV$

$= -nRT\, dV/V$

or, $C_V\, dT/T = -RdV/V$.

Integrating between limits T_1 and T_2 and limits V_1 and V_2 respectively, we get

$C_V \ln (T_2/T_1) = -R\ln (V_2/V_1)$

or, $T_2/T_1 = (V_1/V_2)^{R/C_V}$

or, $T_2/T_1 = (V_1/V_2)^{\gamma-1}$.

or, $T_2 (V_2)^{\gamma-1} = T_1(V_1)^{\gamma-1}$

i.e., $T(V)^{\gamma-1} = \text{constant}$.

Replacing T in the above equation, we get

$PV(V)^{\gamma-1}/nR = \text{constant}$

$\therefore PV^{\gamma} = \text{constant}$.

(c) We know, for an isothermal and reversible expansion of 5 moles of ideal gas from 5 litres to 50 litres at 27°C,

$$(-W) = nRT \ln \frac{V_2}{V_1}$$

$$= 5 \times 0.0821\ (273 + 27) \ln \frac{50}{5}$$

$$= 283.61 \text{ litre–atmosphere}$$

or, $$(-W) = nRT \ln \frac{V_2}{V_1}$$

$$= 5 \times 8.314 \times (273 + 27) \ln \frac{50}{5} \text{ Joule}$$

$$= 28720.7 \text{ Joule.}$$

3. (a) **The First Law of Thermodynamics:**

The total energy of the Universe is constant. Energy can neither be created nor be destroyed. It can only be changed from one form to another. Supposing that a system in its initial state is associated with energy U_i, and heat q is given to it, the heat absorbed by the system may be used to increase its internal energy or perform work or both. If the energy of the system in the final state is U_f, then $(U_f - U_i) = \Delta U$ is the increase in internal energy. The law of conservation of energy demands that:

Heat absorbed = increase in energy + work done by the system, i.e.,

$q = \Delta U + (-w) = \Delta U - w$.

The minus sign appears because the work done by the system is taken as $(-w)$ since it will decrease the internal energy of the system.

Thus, $\Delta U = q + w$.

or, in the differential form $dU = đq + đw$.

(b) For a **reversible isothermal expansion** from volume V_1 to volume V_2 the work done by the system is

$$(-w) = \int_{V_1}^{V_2} P\mathrm{d}V.$$

Now we know that

for an ideal gas, $PV = nRT$.

Therefore, $P = nRT/V$.

$$\text{Therefore,} \quad (-w) = \int_{V_1}^{V_2} (nRT/V)\mathrm{d}V$$

$$= nRT \ln (V_2/V_1).$$

(c) Gibbs free energy may be defined as $G = H\text{–}TS$ where H = enthalpy

T = absolute temperature

S = entropy

The decrease in Gibbs free energy is equal to the maximum amount of work, which a system can do isothermally and isobarically over and above the mechanical work.

$\therefore -\Delta G = (-W_{net})$.

(d) Gibbs-Helmholtz equation relates the free energy change to the enthaply change and the rate of change of free energy with temperatures.

We know that

$G = H - TS \quad (1)$

$= U + PV - TS$

For an infinitesimal change,

$dG = dU + PdV + VdP - TdS - SdT$ combining the first and second law of thermodynamics and considering that the external work is due to volume change only it can be shown that

$dG = VdP - SdT \quad (2)$

At constant pressure, $dP = 0$, therefore equation (2) becomes

$$\left(\frac{\partial G}{\partial T}\right)_P = -S \quad (3)$$

Substitution of equation (3) in equation (1) gives.

$$G = H + T\left(\frac{\partial G}{\partial T}\right)_P$$

$$\text{or,} \quad G - T\left(\frac{\partial G}{\partial T}\right)_P = H \quad (4)$$

This is referred to as Gibbs-Helmholtz equation.

4. (a) According to VSEPR theory, the repulsion is of the order $lp - lp > lp - bp > bp - bp$.

(i) H_2O: Oxygen has $2s^2\ 2p^4$ valence electrons. It undergoes sp^3 hybridization, two hybrid orbitals contain 2 paired δ electrons and the other two have single unpaired electrons. These, then, overlap with the 1*s* orbital of hydrogen forming two single σ bonds. The two paired or non-bonding electrons repel each other, and so due to $lp - lp$ interaction the H-O-H angle decreases slightly and the value is 104.5° instead of 109°28′.

(ii) NH_3: Nitrogen has $2s^2 2p^3$ electrons. It undergoes sp^3 hybridization with one lone pair in one of the hybrid orbital and due $lp - lp$ interaction H-N-H bond angle is 107.3° instead of 109°28′.

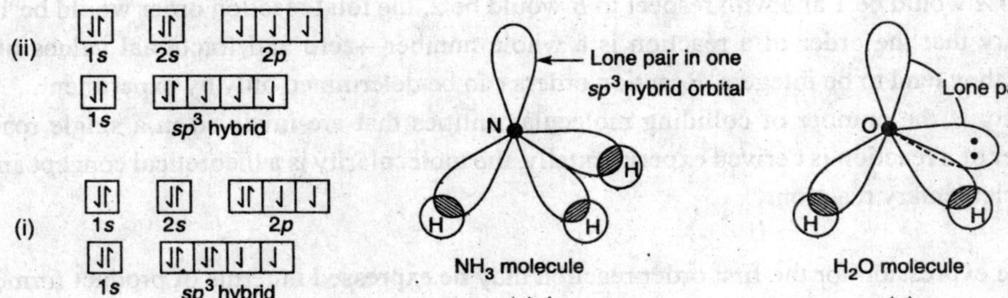

NH$_3$ molecule (c) H$_2$O molecule (a)

(b) Electronegetivities of O, S, H are 3.5, 2.5, 2.1 respectively. Hydrogen can form hydrogen bond with oxygen easily but not with sulphur. Hence, water is an associated molecule of large effective molecular weight as shown in the figure, but *S* forms isolated H_2S molecule of lower effective molecular weight. Hence water has high boiling and melting points and is liquid in nature; H_2S is gas.

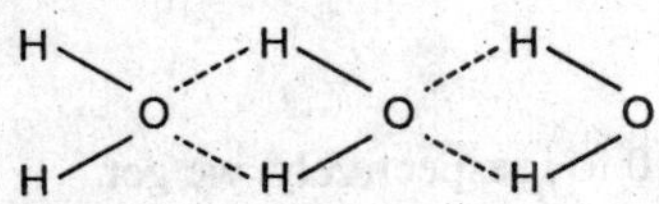

(a)

(c) In stoichiometric crystal defects the ratio of cations and anions represent their chemical formula. These defects are to two types:

Schottky defect: Here the ratio of $\frac{A^+}{B^-}$ is same but same number of cations and anions are missing from the crystal lattice creating equal number of vacancies or holes. This type of defect is generally found in highly ionic compounds of high coordination number, where size of cation and anion are equal, e.g., NaCl, CsCl, etc.

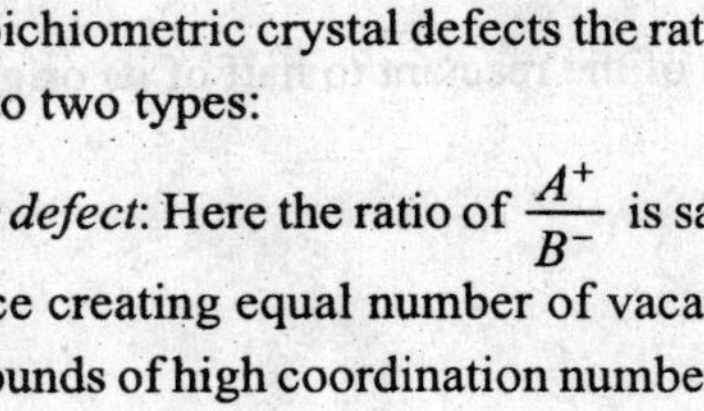

Schottky defect

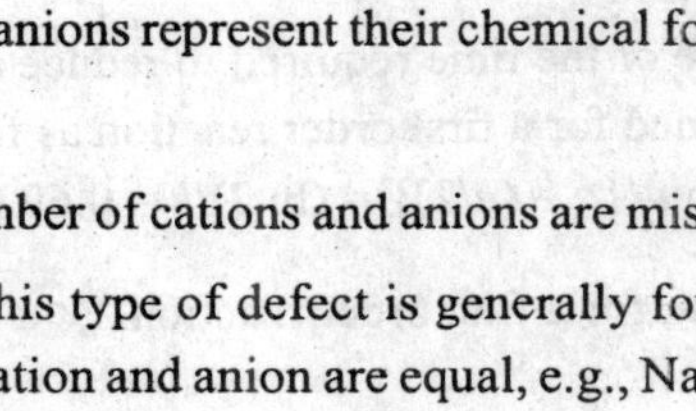

Frenkel defect

Frenkel defect: In this case an ion is shifted from its normal lattice position and occupies an interstitial position. This occurs in compounds, where A^+ and B^- differs widely in size, and coordination number is low, e.g., AgBr, ZnS, etc. Here the cations are smaller, and so take up the interstices of bigger anions. Such compounds are polar or covalent due to greater polarization. These compounds are feeble conductor of electricity as the cation moves from one lattice to the other leaving a hole, which is occupied by another, creating a new hole and in this way conduction occurs.

5. (a) The order of reaction with respect to a certain reactant is defined, in chemical kinetics, as the power to which its concentration term in the rate equation is raised.

For example, a given chemical reaction $A + 2B \rightarrow C$, with a rate equation, $r = k[A]^1[B]^2$, the reaction order with respect to A would be 1 and with respect to B would be 2, the total reaction order would be $1 + 2 = 3$. It is not necessary that the order of a reaction is a whole number—zero and fractional values of order are possible—but they tend to be integers. Reaction orders can be determined only by experiment.

Molecularity is the number of colliding molecular entities that are involved in a single reaction step. While the order of a reaction is derived experimentally, the molecularity is a theoretical concept and can only be applied to elementary reactions.

(b) The rate expression for the first order reaction may be expressed in terms of product formed.

	A	$\rightarrow$	B
Time $t = 0$	a		0
Time $t = t$	$a - x$		x

If 'a' moles per litre is the initial concentration of A and 'x' moles per litre be the concentration of the product formed after time 't', then the rate expression is given by

$-d(a - x)/dt = k(a - x)$

or, $dx/dt = k(a - x)$

or, $dx/(a - x) = kdt$.

Integrating between limits 0 to x and 0 to t respectively, we get

$\ln a/(a - x) = kt$,

the unit of k is time^{-1}.

The half life or the time required to reduce the concentration of the reactant to half of its original value, can be determined for a first order reaction as follows:

$t_{½} = (1/k) \ln a/\{a - (a/2)\} = (\ln 2)/k = 0.693/k$.

(c) We know, for a first order reaction, $K = \frac{1}{t} \ln \frac{a}{a - x}$.

Here $x = 20\% = \frac{a}{5}$.

$$K = \frac{1}{10} \ln \frac{a}{a - \frac{a}{5}} = \frac{1}{10} \ln \frac{5}{4} = \frac{1}{10} \ln 1{\cdot}5 = 0{\cdot}1 \times 0{\cdot}223$$

$= 0{\cdot}0223$/minute.

When the reaction is 80% complete, $x = 80\% = \frac{4a}{5}$

$$\therefore t = \frac{1}{K} \ln \frac{a}{a - \frac{4a}{5}} = \frac{1}{0{\cdot}0223} \ln 5 = \frac{1{\cdot}61}{0{\cdot}0223} = 72{\cdot}17 \text{ minute}$$

So time taken for 80% completion is 72.17 minutes.

6. (a) **Ligand:** Ligands are atoms or group of atoms containing at least a pair of non bonded electrons which they can donate to form a coordinate bond with a metal ion or atom with deficiency of electrons

Ligands are of the following types, e.g.,

Anionic mono dentate ligands: Cl^-, Br^-, I^-, CN^-, SCN^-, NO_2^-, NH_2^-, etc.

Anionic bi dentate ligand : Oxalate ion, $^-OOC\text{-}COO^-$

Neutral mono dentate ligands: NH_3, H_2O, CO, NO, etc

Neutral bi dentate ligand: Ethylenediamine, $H_2N - CH_2 - CH_2 - NH_2$

COORDINATION NUMBER

The coordination number of a metal ion in a complex moiety is the number of ligands that surround the metal ion and held by coordinate valencies or secondary valencies. Here, the ligands are donor, i.e., Lewis base and metal ion is the acceptor, i.e., Lewis acids. The number of the secondary valencies of a metal ion can be determined by Effective Atomic Number Rule, i.e., E. A. N. Rule. The coordination number,

$$\text{C.N.} = \frac{\text{number of extra nuclear electron in next inert gas} - \text{number of extra nuclear electron in metal ion}}{2}.$$

As for example, Fe^{2+} has 24 extra nuclear electrons in the orbits. The next higher inert gas Krypton has 36 extra nuclear electrons, so C.N. of $Fe^{+2} = \frac{36-24}{2} = 6$. Similarly, Cu^{2+} has 27 extra nuclear electrons, so, C.N. of Cu^{+2} will be $\frac{36-27}{2} = \frac{9}{2} = 4.5$. As coordination number cannot be a fraction, it will be 4. It is evident that through coordination, the metal ion will assume the electronic configuration of next higher inert gas.

But there are some compounds which do not obey the E. A. N. rule, e.g., $[Cr^{3+}(NH_3)_6]^{2+}$; $[Ni^{3+}(NH_3)_6]^{2+}$, $[Cu^{1+}Cl_2]^-$, etc.

In $[Cr^{3+}(NH_3)_6]^{3+}$ the electron on the Cr^{3+} is $(24-3+2\times6) = 33$. In $[Ni^{2+}(NH_3)_6]^{2+}$, it is $(28-2+2\times6) = 38$; in $[Cu^{1+}Cl_2]^-$, it is $(29-1+2\times2) = 32$ and so on. Thus during complex formation, the trend of the metal ion is to attain the higher inert gas configuration, but it is not the necessary condition.

(b) (i) Chloro-bis (ethylenediamine) nitrito cobalt (III) chloride

(ii) $[Co(NH_3)_4Cl_2]^+$

(c) **LINKAGE ISOMERISM**

It deals with ligands that are capable of bonding through one type of donor atom in one situation but a different atom in another complex. Some authors refer to this type of isomerism as "structural isomerism" but inasmuch as all isomerism is basically 'structural', the term linkage isomerism is preferable. The first example of this type of isomerism was provided by Jorgensen, Werner's contemporary. His method of preparation was as follows:

$$[Co(NH_3)_5Cl]Cl_2 \xrightarrow{NH_3} \xrightarrow{HCl} \xrightarrow{NaNO_2} \text{Solution A}$$

$$\text{Solution A} \xrightarrow{\text{Let stand in cold}} [(NH_3)_5\,Co\,ONO]Cl_2 \ \text{red}$$

$$\text{Solution A} \xrightarrow{\text{Heat}} \xrightarrow[\text{Conc HCl}]{\text{Cool}} [(NH_3)_5\,CoNO_2]Cl_2 \ \text{yellow}$$

According to Jorgensen and Werner, the difference between the two isomers resides in the linkage of the NO_2 group to the cobalt. The N-bonded (for nitro) structure was assigned to the yellow isomer and the O-bonded (for nitrito) structure to the red isomer on the basis of the colour of similar compounds.

In general the other types of isomerism for coordination compounds are less interesting than those discussed previously, but will be listed briefly to show the variety of possibilities.

(d) S_N1 reaction

S_N1 is the abbreviation used for unimolecular nucleophilic substitution reaction. Here, the rate of the reaction is first order, i.e., depends only on the concentration of the substrate, i.e., the alkyl halide. Generally, tertiary alkyl halides undergo this type of reaction and the reaction proceeds in two steps.

(a) Formation of carbocation intermediate by the heterolytic cleavage of substrate molecule

$R-X \rightarrow [R\cdots X] \rightarrow R^{(+)} + X^{(-)}$, (slow). Hence, it is rate-determining step. Rate α [alkali halide].

(b) Combination of carbocation intermediate with the nucleophile to yield product.

$R^{(+)} + OH \rightarrow R-OH$ (fast)

The term unimolecular is used here since the rate-determining step involves only one molecule.

S_N2 reaction

S_N2 is the abbreviation used for bimolecular nucleophilic substitution reaction. Here, the rate of the reaction depends on the concentration of the alkyl halide as well as the base. So, the order of the reaction is two.

Rate α [alkali halide] [base]

Generally primary alkyl halides undergo this type of substitution reaction and it is a single step process.

(e) Examples of electrophiles:

(i) Charged electrophiles: NO_2^+, Br^+, Cl^+, etc.

(ii) Neutral electrophiles (Lewis acids): BF_3, $ZnCl_2$, $FeCl_2$, etc.

7. (a) Addition polymerization:

The process in which the polymers are formed from the corresponding monomer without losing any materials and the polymer formed is an exact multiple of monomer, is called *addition polymerization*.

$$n\ CH_2=\underset{\displaystyle X}{\underset{|}{CH}} \longrightarrow \left[CH_2-\underset{\displaystyle X}{\underset{|}{CH}} \right]_n$$

where X may be H, CH_3, Cl, CN, C_6H_5, etc.

Peroxides or acids catalyze these reactions. The reaction requires high pressures and temperatures. Much lower temperature and pressure can be used with so-called Ziegler-Natta catalysts, which consist of a trialkyl aluminium and titanium tetrachloride in an inert solvent.

Condensation polymerization

This type of polymerization takes place between two bi- or poly-functional monomers with elimination of small molecules like water, HCl, etc. For example, hexamethylenediamine and adipic acid condense to form a polymer, nylon 6:6.

$$\underset{\text{Hexamethylenedimine Diamine}}{H_2N-(CH_2)_6-NH_2} + \underset{\text{Adipic acid Diacid}}{HOOC-(CH_2)_4-COOH} \xrightarrow{-2n\,H_2O}$$

$$\underset{\text{Polyhexamethylene adipate (Polyamide) (Nylon 6:6)}}{\left[HN-(CH_2)_6-NH-CO-(CH_2)_4-CO\right]_n}$$

(b) According to thermal behaviour polymer is divided into two types: *thermoplastic* and *thermosetting polymers* (thermosets). Thermoplastic polymers are those, which softens readily on heating but on cooling they stiffen again reversibly, that is why they can be moulded, reshaped and reclaimed from wastes. Thus, they can be processed again and again. These polymers are usually soluble in suitable solvents. Most addition polymers and some of condensation polymers are thermoplastic (e.g., polyethylene, polypropylene, polystyrene, polyamides, nylons, PVC, polyesters, PET, PMMA, PTEE, etc). Thermoplastics are amorphous or semi-crystalline and have either liner or branched structures.

Thermosetting polymers are those polymers, which during moulding (by heat) become hardened and once they are solidified, cannot be softened, i.e., they are permanent setting polymer. Such polymers are prepared in two steps. The first step is the formation of long chain molecules, which are capable of further reaction with each other. The second step is the application of heat, which causes a reaction to occur between the chains; thus producing a complex three-dimensional cross-linked network structure; so thermosetting polymers cannot be remoulded, reused or reclaimed. These polymers are hard, strong and brittle, and are generally insoluble in solvent. Examples: PF, UF, MF, vulcanized rubber, epoxy, etc.

(c) According to the stereochemistry (orientation) of monomeric, units in a macromolecule can be classified into the following categories:

(i) Atactic polymers in which the side groups are oriented at random around the backbone chain, e.g., polypropylene.

(ii) Isotactic polymers in which the side groups are all on the same side of the main chain, e.g., natural rubber.

(iii) Syndiotactic polymers in which the side groups are arranged in alternating fashion around the main chain. Polystyrene made by metallocene catalysis polymerisation is crystalline with melting point 270°C is an example of syndiotactic polymer.

8. (a) **Higher or gross calorific value(HCV):**

Generally fuels contain hydrogen. During combustion hydrogen is converted into steam. If the net heat liberated is measured at room temperature, this steam is condensed to water giving out latent heat of condensation. This heat is included in the measurement and gives higher value. This is called higher or gross calorific value (HCV). Thus, HCV is the total heat produced on complete combustion of unit mass/

volume of fuel and the products are cooled to room temperature 15 °C or 59 °F. But, if the products of combustion are not cooled to room temperature the gaseous products escape with hot combustion gases. So, less amount of heat is available. This is called lower or net calorific value (LCV). Thus, LCV is the amount of heat liberated on complete combustion of unit mass/volume of fuel and the products are permitted to escape.

LCV = HCV – latent heat of water vapour formed. Since 1 part by mass of hydrogen produces 9 parts by mass of water, hence LCV = HCV – mass of $H_2 \times 9 \times$ latent heat of vaporization.

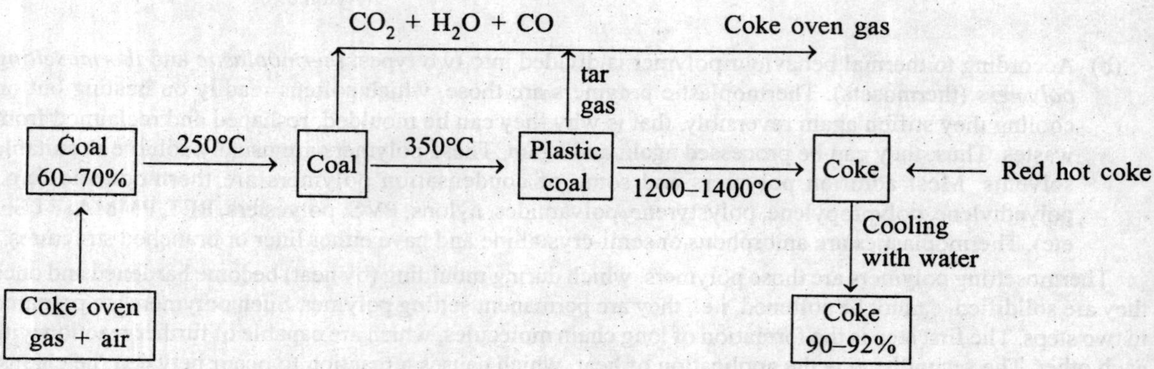

(b) CARBONIZATION OF COAL:

Carbonization is the process by which undesirable and volatile matters are removed from coal by heating in absence of air; as a result percentage of carbon increases. This is destructive distillation.

Conversion of coal to coke is done by two methods

1. High temperature carbonization using 900-1200°C and
2. Low temperature carbonization using 500-750°C.

In high temperature process, carbonization yields coke having properties like

i) high porosity, strength, purity, etc
ii) suitable for metallurgy process
iii) carbon content is 85-90% and
iv) volatile matter is 1-3%.

The low temperature process yields coke having properties like

i) not mechanically strong, so not used in metallurgical purpose
ii) burns with smokeless hot and radiant fire and hence used for domestic purpose
iii) the gaseous fuel has high calorific value.

A comparative chart of high temperature and low temperature carbonization is given below:

TABLE

Low temperature carbonization	High temperature carbonization
1. Temperature used 500–750 °C. 2. 75–80% coke is produced. 3. Good for domestic purpose.	1. Temperature used 900–1200°C. 2. 65–70% coke is produced. 3. Good for metallurgical purpose.

cont.

Low temperature carbonization	High temperature carbonization
4. Not mechanically strong.	4. Hard and good mechanical strength.
5. Volatile matter is 5–10%.	5. Volatile matter is 1–3%.
6. Coal gas produced is 130–150 m^3/ton.	6. Coal gas is produced is 300–390 m^3/ton.
7. Calorific value of gas is 6500–9500 kcal/m^3.	7. Calorific value gas is 5400–6000 kcal/m^3.
8. Produces soft coke.	8. Produces hard coke.
9. 20–40 gallon/ton light oil and tar is produced.	9. 10–17 gallons/ton light oil and tar is produced.
10. 12-15 lbs/ton ammonium sulphate is produced.	10. 24–26 lbs/ton ammonium sulphate is produced.

(c) (i) **Straight run gasoline:** It is the fractional distillation product of crude petroleum oil having boiling range 40-120°C, used as fuel in internal combustion engines of automobiles and aeroplanes. Its composition is C = 84%, H = 15%, N + S + O_2 = 1%, calorific value is 11250 kcal/kg. Main constituents are C_5H_{12} to C_8H_{16} saturated hydrocarbons. This gasoline obtained from fractional distillation is of inferior quality with only 20% yield. Gasoline of better antiknock property is obtained from cracking of higher hydrocarbon to lower one.

(ii) **Jet fuel:** Jet fuel is a type of aviation fuel designed for use in jet-engine aircraft. It is a clear to straw colored fuel, based on either an unleaded paraffin oil (Jet A-1), or a naphtha-kerosene blend (Jet B). It is similar to diesel fuel, and can be used in either compression ignition engines or turbine engines. Jet fuel is a mixture of a large number of different hydrocarbons, possibly as many as a thousand or more. The range of their sizes (molecular weights or carbon numbers) is restricted by the requirements for the product, for example, freezing point or smoke point. Kerosene-type jet fuel (including Jet A and Jet A-1) has a carbon number distribution between about 8 and 16 carbon numbers; wide-cut or naphtha-type jet fuel (including Jet B), between about 5 and 15 carbon numbers.

9. (a) Equivalent conductance has been found to increase with increase in dilution of the electrolytic solution. However, the trend is different for strong and weak electrolytes.

Strong electrolyte: The change in equivalent conductance of strong electrolytes with dilution is relatively small and ultimately tends to reach a limiting value. The accepted idea is that strong electrolytes are completely ionized in solution even in ordinary concentrations, so the number of ions will remain the same in all concentrations. With dilution, the speed of the ions increases. This is because the density of ions in strong electrolytes is very high; each ion is surrounded by oppositely charged ions resulting in interionic attraction. The ions may even give rise to ion pairs of the type ($\overset{+}{A}\ \overset{-}{B}$). Such interionic attraction effectively reduces the speed of the ions and hence equivalent conductance of the solution. With increasing dilution the ions move further apart thereby reducing the interionic attraction and thus move more freely. At infinite dilution, the interionic attraction practically disappears and the limiting value of equivalent conductance is reached.

Weak electrolyte: The equivalent conductance of weak electrolytes is low at ordinary concentration owing to its small extent of ionization. With increase in dilution the ionization increases considerably and with the increase in number of ions the equivalent conductance also increases. In case of weak electrolyte interionic attraction is not of any importance.

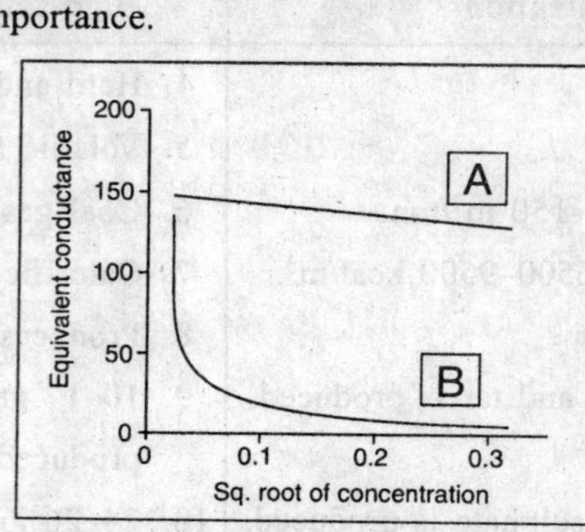

A — Curve for KCl
B — Curve for CH_3COOH

(b) KOHLRAUSCH'S LAW OF INDEPENDENT MIGRATION OF IONS

The law states that at infinite dilution, where all electrolytes are completely ionized and where all interionic effects are absent, each ion migrates independently of its co-ion and contributes a definite share to the total equivalent conductance of the electrolyte.

$$\Lambda_0 = \lambda_0^+ + \lambda_0^-,$$

where λ_0^+ and λ_0^- are the conductance of cation and anion respectively.

In case of weak electrolytes where direct determination of Λ_0 is not possible Kohlrausch's law may be applied. For example, we can calculate Λ_0 for acetic acid in the following manner—

$$\Lambda_0^{\text{Hac}} = \Lambda_0^{\text{NaAc}} + \Lambda_0^{\text{HCl}} - \Lambda_0^{\text{NaCl}}$$

$$= \lambda_0^{\text{Na}^+} + \lambda_0^{\text{Ac}^-} + \lambda_0^{\text{H}^+} + \lambda_0^{\text{Cl}^-} - \lambda_0^{\text{Na}^+} - \lambda_0^{\text{Cl}^-} = \lambda_0^{\text{Ac}^-} + \lambda_0^{\text{H}^+}.$$

(c) A galvanic cell or electrochemical cell is a device in which the free energy of a chemical process is converted into electrical energy. In an electrolytic cell the electrical energy from an external source is used to perform a chemical change.

(d) The Cell is $\text{Ni}|\text{Ni}^{2+}(\text{eq})||\text{Pb}^{2+}(\text{eq})|\text{Pb}$

$$E_{\text{Cell}} = E_{\text{R. H. S.}} - E_{\text{L. H. S.}}$$

$$= E_{\text{Pb}^{2+}/\text{Pb}} - E_{\text{Ni}^{2+}/\text{Ni}}$$

$$= E^{\circ}_{\text{Pb}^{2+}/\text{Pb}} - \frac{RT}{2F}\ln\frac{a_{\text{Pb}}}{a_{\text{Pb}^{2+}}} - E^{\circ}_{\text{Ni}^{2+}/\text{Ni}} + \frac{RT}{2F}\ln\frac{a_{\text{Ni}}}{a_{\text{Ni}^{2+}}}$$

$$= \left(E^{\circ}_{\text{Pb}^{2+}/\text{Pb}} - E^{\circ}_{\text{Ni}^{2+}/\text{Ni}}\right) - \frac{RT}{2F}\ln\frac{1}{1} + \frac{RT}{2F}\ln\frac{1}{1}$$

$$= -0.13 - (-0.24) \text{ volt}$$

$$= 0.11 \text{ volt.}$$

2006

Time Allotted : 3 hours **Full marks : 70**

Group–A
(Multiple Choice Questions)

1. Choose the correct alternatives for any ten of the following: (10×1=10)

(i) $[CO(NH_3)_5Br]SO_4$ and $[CO(NH_3)_5SO_4]Br$ are examples of

(a) Geometrical (b) Optical

(c) Linkage isomers. (d) Ionisation

(ii) Free radicals take part in

(a) Addition (b) Substitution

(c) Elimination (d) Disproportionate reactions

(iii) The time required to reduce the concentration of a reactant from 0.02 M to 0.01 M is 15 sec and from 0.01 M to 0.005 M is also 15 sec. The order of the reaction is

(a) Zero (b) One

(c) Two (d) Three

(iv) Cellulose acetate is a

(a) Natural polymer (b) Synthetic polymer

(c) Semisynthetic polymer (d) Thermosetting polymer

(v) The polydispersity index of a commercial polymer is

(a) 1.0 (b) 5.0

(c) 50.0 (d) 100

(vi) During mixing, entropy

(a) Increases (b) Decreases

(c) Remains unchanged (d) Cannot be predicted

(vii) The unit of equivalent conductance is

(a) ohm^{-1} cm^2/gm-eqv (b) ohm^{-1} m^2/gm-eqv

(c) ohm^{-1} cm^{-1} (d) ohm^{-1} m^{-1}

(viii) The monomer unit present in natural rubber is

(a) 1, 3-butadiene (b) 2-methyl-1, 3-butadiene

(c) 2-ethyl-1, 3-butadiene (d) 2-chloro-1, 3-butadiene

(ix) Hydrolysis of ethyl acetate follows the

(a) Second order reaction kinetics (b) First order reaction kinetics

(c) Zero order reaction kinetics (d) Pseudo first order reaction

(x) By proximate analysis we can determine

(a) The moisture content of coal only.

(b) The moisture co: tent and ash content only.

(c) The moisture content, volatile matter, ash and fixed carbon contents of coal.

(d) The moisture content, volatile matter and fixed carbon of coal.

(xi) The relation between HCV and LCV will be

(a) HCV = LCV
(b) HCV > LCV
(c) HCV < LCV
(d) no relation exists

(xii) The presence of hydrogen atom attached to an aromatic ring may be identified by which of the following spectroscopies?

(a) UV-vis
(b) IR
(c) NMR
(d) EPR

Group–B
(Short Answer Questions)

Note : (i) Each answer should not exceed 50 words.
(ii) Write short notes on any three of the following: (3×5=15)

2. (a) Prove that for an adiabatic reversible process. PV^{γ} = constant.
(b) Show that for an ideal gas, $C_P - C_V = R$.
3. The standard reduction potential for the electrode Sn^{4+}/Sn^{2+} is 0.16 volt and that for Fe^{3+}/Fe^{2+} electrode is 0.77 volt. Write the cell reaction and calculate the value of the equilibrium constant of the reaction at 25°C.
4. State Kohlrausch's law of independent migration of ions and show how this law is utilised to determine the equivalent conductance of acetic acid at infinite dilution.
5. What is entropy? What is its physical significance? Explain what you understand by internal energy of a system.
6. What is LPG? Why is it used as domestic fuel? Define octane number of a fuel and explain how the octane number can be improved.

Group–C
(Long Answer Questions)

Answer any three question

7. Discuss the physicochemical principle involved in the measurement of pH of an aqueous solution by using a hydrogen electrode.
8. Define the term Gibbs free energy and derive the Gibbs-Helmholtz equation relating the change in Gibbs free energy (ΔG) and the temperature coefficient of ΔG.
9. What is meant by inversion temperature?,Define the expression for inversion temperature of a gas obeying van der Waals equation.
10. (a) What are the important products formed from the atmospheric distillation of crude oil?
(b) How do gasoline and diesel differ in chemical composition?
11. What are water gas and semi-water gas? How is water gas produced from coke?

SOLUTION

GROUP–A

1. (i) (d) ionization isomerism
(ii) (a) addition reaction
(iii) (b) one
(iv) (c) semisynthetic polymer
(v) (a) 1.0
(vi) (a) increases
(vii) (a) $ohm^{-1}\ cm^2/gm\text{-}eqv$
(viii) (b) 2-methyl-1, 3-butadiene
(ix) (d) pseudo first order reaction
(x) (c) The moisture content, volatile matter, ash and fixed carbon contents of coal.
(xi) (b) HCV > LCV
(xii) (c) NMR

GROUP–B

2. (a) See Q. 2(b) of 2005

(b) Heat capacity at constant volume may be defined as the heat required to increase the temperature of the system by 1°C at constant volume.

Therefore, $C_V = (\partial q/\partial T)_V = (\partial U/\partial T)_V$.

Again, the heat required to increase the temperature of the system by 1°C at constant pressure is the heat capacity at constant pressure.

Therefore, $C_P = (\partial q/\partial T)_P = (\partial H/\partial T)_P$.

$C_P - C_V = (\partial H/\partial T)_P - (\partial U/\partial T)_V$

$= (\partial U/\partial T)_P + P(\partial V/\partial T)_P - (\partial U/\partial T)_V$. (1)

We know that $U = f(V, T)$.

Therefore, $dU = (\partial U/\partial T)_V\, dT + (\partial U/\partial V)_T dV$.

Dividing throughout by dT and imposing the condition of constant pressure, we get

$(\partial U/\partial T)_P = (\partial U/\partial T)_V + (\partial U/\partial V)_T(\partial V/\partial T)_P$. (2)

or, $(\partial U/\partial T)_P - (\partial U/\partial T)_V = (\partial U/\partial V)_T(\partial V/\partial T)_P$.

Inserting the value of equation (1) in equation (2), we get

$C_P - C_V = (\partial U/\partial V)_T(\partial V/\partial T)_P + P(\partial V/\partial T)_P$.

$= [(\partial U/\partial V)_T + P](\partial V/\partial T)_P$.

For ideal gas, $(\partial U/\partial V)_T = 0$ and $(\partial V/\partial T)_P = nR$.

Therefore, for 1 mole of ideal gas, $C_P - C_V = R$.

3. **Standard reduction potential E_0 of Fe^{3+}/Fe^{2+} is 0.77 volt and for Sn^{4+}/Sn^{2+} system is 0.16 volt, in combination of two systems in the cell, Sn^{2+} will reduce Fe^{3+} and the cell reaction will be**

$$Sn^{2+} - 2e^- \rightarrow Sn^{4+}[ox]$$
$$2\left(Fe^{3+} + e^- \rightarrow Fe^{2+}[Red]\right)$$
$$\overline{Sn^{2+} + 2Fe^{3+} \rightarrow Sn^{4+} + 2Fe^{2+}}$$

The cell will be $Sn^{2+} | Sn^{4+} || Fe^{3+} | Fe^{2+}$

The potential of the cell will be

$E^o = 0.77 - 0.16 = 0.61$ volt

Now $\Delta G^0 = -nFE^0$

Again $\Delta G^0 = -RT \ln K_{eq}$.

$\therefore -nFE^0 = -RT \ln K_{eq}$.

or, $\log K_{eq} = \dfrac{nFE^o}{RT}$

$$\log K_{eq} \text{ at } 293°\text{K} = \frac{n(E_2 \sim E_1)}{0.059} = \frac{2 \times \cdot 61}{0.059} = 20 \cdot 6$$

$\therefore K_{eq} = 4 \cdot 8 \times 10^{20}$.

4. (b) See Q. 9(b) of 2005.

5. Entropy change has often been defined as a change to a more disordered state at a molecular level. In recent years, entropy has been interpreted in terms of the "dispersal" of energy. Entropy is an extensive state function that accounts for the effects of irreversibility in thermodynamic systems. When a system's energy is defined as the sum of its "useful" energy, (e.g., the energy which is used to push a piston), and "useless energy", i.e., the energy which cannot be used for external work, then entropy may be (most concretely) visualized as the "scrap" or "useless" energy. When heat is absorbed by a gas, the kinetic energy of the gas molecules will be increased and they will move faster and collide with (i) each other (ii) the container and (iii) the face of the piston, thereby the piston will be pushed. Only the third type of collision will bring about the expansion of the gas. Thus a fraction of the total amount of heat absorbed is wasted. This is the unavailable work, and entropy is a measure of this unavailable work. Under an ideal situation, if all the gas molecules were striking the piston perpendicularly, then whole of the heat absorbed would have been converted into work. This can happen if there is a regular and only one possible arrangement. In other situations when the molecules are in random motion, a fraction of work is converted into such a factor which is a measure of the randomness of the molecular system. Entropy is thus a measure of the randomness or disorderness of the molecular system. More the randomness of the system, more is its entropy. Quantitatively, entropy is defined by the differential quantity,

$dS = đq_{rev}/T$.

INTERNAL ENERGY (*U*)

We are going to be talking about our chemical system losing and gaining energy, so it is logical to introduce a quantity, which represents the total energy of the system. This is known as the internal energy, *U*. As with all types of energy, the unit of *U* is the joule. Now, the internal energy is defined as the total energy of our system–

chemical, nuclear, heat, gravitational, any other type of energy one can think of, the sum of all the system's energy. Although we cannot measure the absolute value of the internal energy of a system, we can measure changes in the internal energy. Internal energy is a state function, i.e., its change is not governed by the process of change and only depends on the initial and final state of the system.

$$\Delta U = U_f - U_i,$$

where U_f = internal energy of the final state

U_i = internal energy of the initial state.

6. **LPG:** The first fraction of the distillation product of petroleum is the LPG. It is also obtained from the cracking of heavy oil and from natural gas. Main constituents of Liquefied Petroleum Gas, LPG are *n*-butane, isobutane, butylene and propane. Under ordinary pressure, they are gas but can be liquefied and bottled easily under pressure. Trade name of LPG is Indane, HP, etc. The calorific value is around 27800 kcal/m^3, which are 3 times that of natural gas and 7 times that of coal gas. The use of this gas is safe, as it does not contain carbon monoxide. Some amount of organic sulphides (mercaptans) is mixed with the gas to recognize the gas leak. Fire hazards are as great as any other gaseous fuel. Its use as motor fuel is advantageous over gasoline as it is cheaper, highly knock resistant and also load sensitivity is high.

KNOCKING: Gasoline obtained from the fractional distillation of crude petroleum is not always suitable for the internal combustion engine. Gasoline for this purpose must be such that burning would be smooth and the flame must spread rapidly. It has been found that if the gasoline contains normal straight chain hydrocarbon, its combustion becomes irregular and ignition becomes very rapid. The rate of oxidation becomes so great that the last portion of fuel-air mixture ignited instantaneously producing an explosive violence to the engine known as KNOCKING. This causes the loss of efficiency of the engine. For example, if the gasoline contains *n*-heptane, it is the most inferior fuel and knocking is high or antiknock value is zero. If it contains 2:2:4-trimethyl pentane, i.e., isooctane, its knocking property is very little, i.e., antiknock value is very high, assumed to be 100. Octane number or octane rating grades quality of gasoline. Thus, octane number of fuel of an internal combustion engine is defined as the percentage by volume of isooctane in a mixture of isooctane and *n*-heptane, which matches with the knocking characteristics of the fuel under test. Say, a fuel has "Octane 80" means the fuel has the combustion characteristics that matches with the characteristics of the fuel containing mixture of 80 volume isooctane and 20 volume *n*-heptane. The knocking tendency of hydrocarbons increases of the order *n*-paraffin > branched chain paraffins > olefins > cyclo paraffins > aromatics. The reasons of knocking tendency of normal hydrocarbon and branched chain compound can be understood in this way. The normal chain compounds have greater surface area, and intramolecular atomic interactions are less, whereas branched compounds have less surface area, and intramolecular atomic interactions are higher. So, surface energy of the branched chain compounds are higher and hence, rupture of bonds are easier and regular, and knocking tendency is less.

The octane number of a fuel is determined by combusting the fuel under test in an internal combustion engine at 600 rpm at 52 °C. The value thus obtained is known as Research Octane Number (RON). Then the fuel is burnt in the same engine at 900 rpm at 150 °C. The value is known as Motor Octane Number (MON). The average of the two is a measure of Antiknock Index.

IMPROVEMENT OF ANTIKNOCK PROPERTY: The antiknock property of fuel can be improved by addition of some chemical compounds – mostly used lead tetraethyl, TEL. For motor fuel 0.5ml TEL and for aviation fuel 1-1.5 ml TEL per litre of petrol is added. For motor fuel, with TEL, some amount of 1, 2-dibromo ethane or 1, 2- dichloro ethane is added. Otherwise, deposition of finely divided PbO particle occurs that is harmful for engine. Addition of the chloro and bromo compounds forms volatile $PbBr_2$ or $PbCl_2$ and this escapes with exhaust gas. Diethyl telluride is also used for the purpose.

Alternate method is to add isopentene, isooctane, ethylbenzene, isopropylbenzene, methyl tertiary butyl ether (MTBE) to increase the octane rating. MTBE is more suitable as the oxygen in ether group helps better combustion of the fuel. Thus, unleaded petrol has high octane rating without addition of lead compound. For use of this unleaded fuel the engine is provided with a catalytic converter containing Rhodium metal that converts toxic gasses CO and NO to harmless CO_2 and N_2 and also converts unburnt hydrocarbon to CO_2 and H_2O.

GROUP–C

7. The solution whose pH is to be determined is taken in a vessel and a Pt-wire carrying Pt-foil coated with platinum black is immersed in the solution. Pure H_2-gas at one atmospheric pressure is continuously bubbled so that the Pt-black and the solution are saturated and at equilibrium. This half-cell is then coupled with a reference electrode (a SHE or SCE) through a KCl salt bridge.

For SHE

$$\text{Cell } Pt|H_2O(p=1\,atm.)|H^+(a_{H^+})||H^+(a=1)|H_2(p=1\,atm.)|Pt$$

$$E_{Cell} = E_{SHE} - E^0_{H^+/\frac{1}{2}H_2} + RT/F \ln p^{1/2}_{H_2}/a_{H^+}$$

$$= 0 - 0 - RT/F \ln a_{H^+} = -0.0591 \log a_{H^+}$$

$$\text{Now } pH = -\log a_{H^+} \quad \therefore E_{Cell}/0.0591 = pH$$

For SCE,

$$\text{Cell } Pt|H_2(p=1\,atm.)/H^+(a_{H^+})||Cl^-|Hg_2Cl_2(S)|Hg$$

$$E_{Cell} = E_{SHE} - E^0_{H^+/\frac{1}{2}H_2} + RT/F \ln p^{1/2}_{H_2}/a_{H^+}.$$

$$= 0.2415V - 0 - RT/F \ln a_{H^+}$$

$$\therefore (E_{Cell} - 0.2415)/0.0591 = pH.$$

8. See Q. 3(c) & (d) of 2005.

9. For every gas there is a temperature where Joule-Thomson coefficient $\mu_{JT} = \left(\frac{\partial T}{\partial P}\right)_H \approx \left(\frac{\Delta T}{\Delta P}\right)_H = 0$. i.e., when neither heating nor cooling of the gas due to passage of gas through fine orifices would occur. This is called the inversion temperature of the gas. Inversion temperature of N_2 is 621K, for H_2 is 202K.

For van der Waals' gas

$$\mu_{JT} = \left(\frac{\partial T}{\partial P}\right)_H = \frac{1}{C_P}\left[\frac{2a}{RT} - b\right],$$

where a and b are van der Waals' constants.

$\therefore$ at inversion temperature $\mu_{JT} = 0$.

$$\therefore \frac{2a}{RT_i} = b \quad \text{or, } T_i = \frac{2a}{Rb}.$$

[T_i = inversion temperature]

10. (a)

TABLE

Name of fraction	Boiling range	Approx. composition in terms of number of carbon atoms	Uses
1. Uncondensed gas	Below 30°C	C_1–C_4, C_2H_6, C_3H_6, iso-C_4H_{10}	L.P.G. as domestic and industrial fuel
2. Petroleum ether	30-70°C	C_5–C_7	As a solvent
3. Gasoline or petrol or motor spirit	40-120°C	C_5–C_9, C_5H_{12}–C_8H_{18} C = 84%, H = 15%; N + S + O = 1%, Calorific value is 11250 kcal/kg	Used as fuel for internal combustion engines of automobiles and aeroplanes
4. Solvent Naphtha or solvent spirit	120-180°C	C_{10}–C_{16} calorific value = 11000 kcal/kg	As solvent and in dry cleaning
5. Kerosene oil	180-250°C	C_{10}–C_{16}, such as $C_{10}H_{22}$ to $C_{16}H_{34}$ C = 84%, H = 16% Density is 0.86–0.95 Calorific value = 11000 kcal/kg	Used as domestic fuel and jet engine fuel, as illuminant and for preparing laboratory gas.
6. Diesel oil or fuel oil	250-320°C	C_{15}–C_{18}, i.e., $C_{15}H_{32}$–$C_{18}H_{38}$ Density is 0.86–0.95 Calorific value 11000 kcal/kg	Used as diesel engine fuel
7. Heavy oil from which lubricating oil, petroleum jelly, grease, and paraffin wax are obtained	320-400°C	C_{17}–C_{30}	Used to get gasoline after cracking. These are used as lubricant, in cosmetics and in medicines, Wax is used as candle, boot polish, etc.
8. Solid residue is asphalt petroleum coke.	>400°C	>C_{30}	Used as waterproof for roofs, fuel, arc light rods, etc.

(b) Gasoline contains the hydrocarbons C_5H_{12} to C_8H_{18}. Analysis of gasoline shows that it has C = 84%, H = 15%, N + S + O = 1%. Gasoline obtained from fractional distillation of crude oil is of inferior quality, but cracked one is of better quality. Hydrocarbons here must have branch or side chains and then knocking will be low. Boiling range of gasoline is 40–120 °C. Diesel is of higher boiling range 250–300°C containing $C_{15}H_{32}$ to $C_{18}H_{38}$. The hydrocarbon must not contain any side chain and should not contain aromatics. The ignition quality of hydrocarbon is of the order

n–alkanes > naphthalene > alkenes > branched > alkenes > aromatics.

11. WATER GAS

It is a mixture of nearly equal volume of CO and H_2 produced by passing steam through white-hot carbon at 1300 °C. If the temperature falls to 800-900 °C, then CO_2 proportion increases. So, the condition is to be maintained such that temperature of bed does not fall below 1000 °C. Its composition is CO = 43%, H_2 = 51%, CO_2 = 3% N_2 = 3%. As the gas burns with blue flame, it is called blue water gas or blue gas. But composition varies with temperature of the bed as given below:

Temperature	% Decomposition of steam	H_2	CO	CO_2
900°C	42	62	15	23
1010°C	94	49	50	1
400°C	99.4	51	48	1

Its calorific value is 2800 kcal/m^3. It is used as a source of hydrogen, as illuminating gas, and as reducing agent in metallurgy. The reaction which is taking place is shown below.

Supplied steam reacts with red-hot coke at 1100-1300°C.

$$C + H_2O = CO + H_2 - 29 \text{ kcal}$$

This being an endothermic reaction, temperature begins to fall and CO_2 will be produced. Hence, to increase the temperature of the bed, steam line is put off and air is passed through when exothermic reaction, $C + O_2 = CO_2 + 97$ and $2C + O_2 = 2CO + 59$ Kcal takes place and temperature of the bed rises to desired value. Thus, initially steam is passed for 8 min and then air for 2 min alternately. The gases produced during second operation are vents out separately.

The reactor is a 4 m high 3 m wide steel vessel lined inside with fire bricks and refractory bricks. It is provided with a cup and cone feeder at the top, and outlet for the gas is near the top. Near the base, inlet pipes for air and steam and at the base an opening for taking out the ash are provided.

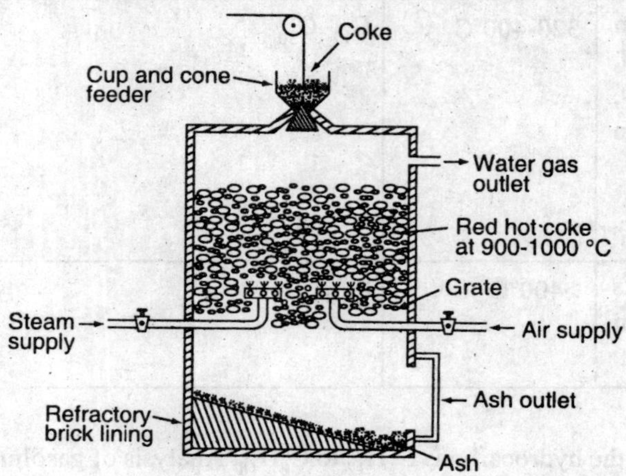

SEMI WATER GAS

It is prepared by passing the mixture of steam and air – the proportion adjusted in such a way that the heat change in exothermic and endothermic reaction is adjusted and the temperature of the bed remains constant. Its average composition is H_2 = 12%, CH_4 = 3%, CO = 30%, CO_2 = 2%, N_2 = 53% and calorific value is 1700 kcal/m^3. It is used as a source of N_2 and H_2 for the manufacture of NH_3, as fuel, etc.

2007

Time Allotted : 3 hours **Full marks : 70**

Group–A
(Multiple Choice Questions)

1. Choose the correct alternatives for any **ten** of the following: (10×1=10)

(i) The change in internal energy of a system depends on
(a) Initial and final states of the system
(b) The reversible path
(c) The irreversibility in the process
(d) The initial state of the system only

(ii) The Joule-Thomson expansion is
(a) Isobaric
(b) Isentropic
(c) Isothermal
(d) Isenthalpic

(iii) In normal hydrogen electrode, the activity of H^+ ion is
(a) 0.1
(b) 0.2
(c) 2
(d) 1

(iv) Which of the following has the least bond angle?
(a) NH_3
(b) H_2O
(c) CH_4
(d) BeF_2

(v) The half-life period of a reaction is found to be directly proportional to the initial concentration. The order of the reaction is
(a) Zero
(b) One
(c) Two
(d) Three

(vi) A conducting polymer is
(a) Polyethylene
(b) Polypropylene
(c) Polyaniline
(d) Bakelite

(vii) The highest ranking coal is
(a) Anthracite
(b) Bituminuous
(c) Lignite
(d) Peat

(viii) Which of the following is true always for a spontaneous change at all temperatures?
(a) $\Delta H > 0$ and $\Delta S < 0$
(b) $\Delta H < 0$ and $\Delta S < 0$
(c) $\Delta H < 0$ and $\Delta S > 0$
(d) $\Delta H > 0$ and $\Delta S > 0$

(ix) If the rate of a reaction becomes twice for every 10°C rise in temperature, by what factor will the rate of the reaction increases when temperature is raised from 30°C to 80°C?

(x) NMR spectra are normally carried out on
(a) Liquids or solutions
(b) Gases
(c) Solids
(d) Suspensions

(xi) The calorific value should be the highest for
(a) Water gas
(b) LPG
(c) Producer gas
(d) Carburetted water gas

(xii) Most commonly used vulcanising agent is
(a) Graphite
(b) Sulphur
(c) Lampblack
(d) Soft soap

Group–B
(Short Answer Questions)

Answer any three questions

(3×5=15)

2. State and explain the First Law of Thermodynamics. Deduce the expression for the maximum work done when an ideal gas expands isothermally and reversibly.
3. Explain the principle of Ultraviolet spectroscopy in determining the constitution of an organic compound. Write down the mathematical form of Lambert-Beer's law and explain each term.
4. What do you understand by polymerisation? What is the degree of polymerisation? Write down the structure and use of Nylon-666 and PVC.
5. Deduce the expression for the rate constant of a second order reaction where the initial concentrations of the two reactants are same.
6. Discuss about S_N1 and S_N2 reactions with examples.

Group–C
(Long Answer Questions)

Answer any three question

(3×15=45)

7. Tabulate the products available from the distillation of crude petroleum against the range of temperature. Define octane and cetane numbers. What is biofuel? (7+6+2)
8. State and explain the basic features of high and low temperature carbonisations of coal. Write the usefulness of both the processes. Define gross and net calorific values of fuel. What are proximate and ultimate analysis of coal? (7+2+2+4)
9. (a) Explain addition and condensation polymerisation with example. (4)
   (b) Distinguish between thermoplastic and thermosetting resins citing examples. (4)
   (c) Deduce Gibbs-Helmholtz equation in any of its standard forms. (2)
   (d) What is the difference between *p*-type semiconductor and *n*-type semiconductor? (2)
   (e) What is the 'Finger Print Region' in IR-spectroscopy? (2)
10. (a) Draw the conductometric titration curve for the titration of HCl vs. NaOH and explain the salient features in the curve. (5)
    (b) Define specific conductance and equivalent conductance. Show how they are related. (6)
    (c) State and explain Hess law with examples. (4)
11. Write explanatory notes on any three of the following: (3+5)
    (a) Schottky and Frenkel defects
    (b) Order and molecularity of a reaction
    (c) Gibbs-Duhem equation for a two-component system
    (d) Nernst's electrode potential
    (e) Arrhenius equation
    (f) Ligand.

SOLUTION

GROUP–A

1. (i) (a) initial and final states of the system
   (ii) (d) isoenthalpic
   (iii) (d) 1
   (iv) (b) H_2O
   (v) (a) zero
   (vi) (c) polyaniline
   (vii) (a) anthracite
   (viii) (c) $\Delta H < 0$ and $\Delta S > 0$
   (ix) (b) 32
   (x) (a) liquids or solutions
   (xi) (b) LPG
   (xii) (b) sulphur.

GROUP–B

2. See Q. 3(a) & (b) of 2005.

3. **U–V spectroscopy:** According to molecular orbital theory, the interaction of atomic orbital leads to the formation of bonding and anti bonding molecular orbitals. Depending upon the nature of the atomic orbitals involved the bonding orbitals may be of

(i) sigma (σ) type, in which the electron density is concentrated along the intermolecular axis, or

(ii) pi (π) type, in which the electron density is concentrated above and below inter nuclear axis.

Fig(a). depicts the electron density probability contours for electrons occupying (i) σ and π bonding, and (ii) σ* and π* anti bonding orbitals while Fig.(b) depicts the relative energies of bonding, anti bonding and nonbonding(*n*) molecular orbitals.

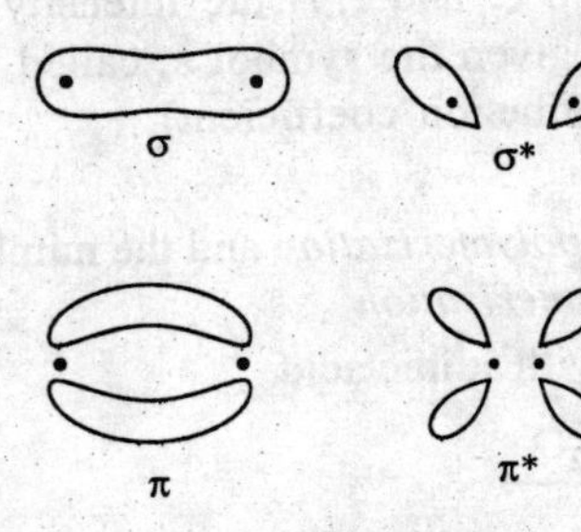

(a)

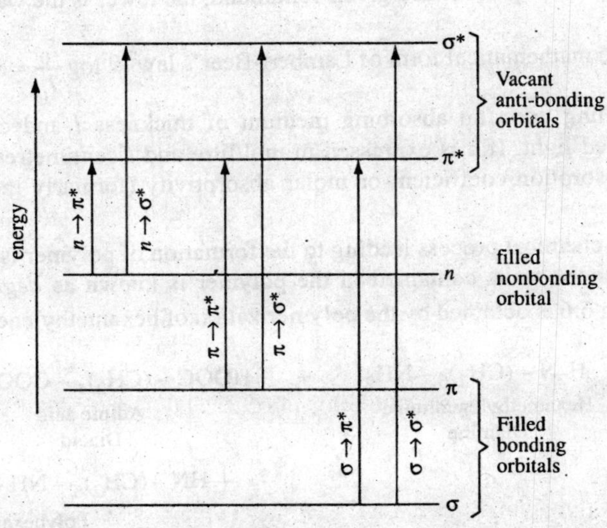

(b)

When excitation occurs, an electron from one of the filled (σ), (π) or n molecular orbitals get excited to vacant σ^* and π^* (anti bonding) molecular orbitals. Corresponding to possible excitations, there are various absorptions:

$$\sigma \rightarrow \pi^* \qquad \pi \rightarrow \pi^* \qquad n \rightarrow \pi^*$$

$$\sigma \rightarrow \sigma^* \qquad \pi \rightarrow \sigma^* \qquad n \rightarrow \sigma^*$$

Since σ electrons require a high energy for excitation to σ^*, so order of decreasing energy for the absorption is

$$\sigma \rightarrow \sigma^* > \sigma \rightarrow \pi^* > \pi \rightarrow \sigma^* > \pi \rightarrow \pi^* \sim n \rightarrow \sigma^* > n \rightarrow \pi^*$$

Out of the above possible transitions the last three ones account for the absorption in 200-800 nm region of the electromagnetic radiation. On the other hand, first three transitions require much higher energy; so only molecules with n and π electrons give rise to characteristic spectra in the region 200-800 nm of electromagnetic radiation. It may be pointed out that alkanes absorb in the region below 200 nm. It may be added here that the part of the molecule having n or π electrons is essentially responsible for absorption, and these fragments are known as chromophores.

The simple chromophores like >C = C <, –C ≡ C–, –C ≡ N–, –N = N–, and > C = O undergo $\pi \rightarrow \pi^*$ transitions in the short wavelength regions of ultraviolet radiations.

Polar groups such as –OH, –OR, $-NH_2$, –SH, and –X (halogen) having unshared pair of electrons may show absorption above 190 nm. Such grouping are called auxochromes, when conjugated with chromophores they modify the position of absorption band relative to that of the parent chromophore alone.

Compounds containing carbon oxygen double bond absorb light in the UV region. For example, acetone has a broad absorption peak at 280 nm, corresponding to excitation of an electron from nonbonding, n to π^* of the carbon oxygen double bond. If the C = O double bond is conjugated with C = C double bond, then the absorption maxima correspond to $n \rightarrow \pi^*$ excitation and $\pi \rightarrow \pi^*$ occurs at a longer wavelength, but is much weaker. For example, the compound $CH_2 = CH\text{–}CO\text{–}CH_3$ gives $n \rightarrow \pi^*$ λ_{max} at 324 nm and $\pi \rightarrow \pi^*$ λ_{max} at 219 nm.

It may be pointed out that energy gap between nonbonding and anti bonding MO of ethene is greater than that between the corresponding MO in 1, 3-butadiene. Consequently, 1, 3-butadiene has its λ_{max} at 171 nm (of higher energy) than that of ethene at λ_{max} at 217 nm (of lower energy). Thus, in general the greater the number of double bonds present in a given compound, the lower is the wavelength at which it absorb UV region.

The mathematical form of Lambert-Beer's law is $\log \frac{I_0}{I_t} = acl$ where I_0 is the intensity of the incident light, falling upon an absorbing medium of thickness l and concentration c, and I_t is the intensity of transmitted light. If c is expressed in mol/litre and l centimetres then a is given the symbol ε, called the molar absorption coefficient or molar absorptivity (formarly the molar extinction coefficient).

4. The chemical process leading to the formation of polymer is known as *polymerization* and the number of monomeric units contained in the polymer is known as *degree of polymerization.*

Nylon 6:6 is obtained by the polymerization of hexamethylene diamine with adipic acid.

$$\underset{\substack{\text{Hexamethylenedimine} \\ \text{Diamine}}}{H_2N-(CH_2)_6-NH_2} + \underset{\substack{\text{Adipic acid} \\ \text{Diacid}}}{HOOC-(CH_2)_4-COOH} \xrightarrow{-2n\,H_2O}$$

$$\underset{\substack{\text{Polyhexamethylene adipate (Polyamide)} \\ \text{(Nylon 6:6)}}}{\left[HN-(CH_2)_6-NH-CO-(CH_2)_4-CO \right]_n}$$

Polyvinyl Chloride (PVC)

It is obtained by heating a water emulsion of vinyl chloride containing water, in presence of a small amount of benzoyl peroxide or hydrogen peroxide in an autoclave maintained at 40-45 °C under pressure. It is used in the manufacture of imitation leather sheets, floor covering corrugated roofing materials, moulding articles, insulators and gramophone records, etc.

$$n\,(CH_2=\underset{|}{\overset{Cl}{C}}H) \xrightarrow{\text{polymerization}} \left[CH_2-\overset{Cl}{\overset{|}{C}}H \right]_n$$

Vinyl chloride → Polyvinyl chloride

5.

	$2A$	→	Product	or,	A	+	B	→	Product
Time $t = 0$	a		0	Time $t = 0$	a		a		0
Time $t = t$	$a - x$		x	Time $t = t$	$a - x$		$a - x$		x

If 'a' moles per litre is the initial concentration of A and 'x' moles per litre be the concentration of the product formed after time 't' then the rate expression is given by

$-d\,(a-x)\,/dt = k\,(a-x)^2$

or, $dx/dt = k(a-x)$

or, $dx/(a-x)^2 = kdt$.

Integrating between limits 0 to x and 0 to t respectively, we get

$x\,/a\,(a-x) = kt$

Half life, $t_{1/2} = 1/ak$.

6. See Q. 6(d) of 2005.

GROUP–C

7. See Q. 10(a) of 2006.

Octane number:

It has been found that if the gasoline contains *n*-heptane, it is the most inferior fuel and knocking is high or antiknock value is zero. If it contains 2:2:4-trimethyl pentane, i.e., isooctane, its knocking property is very little, i.e., antiknock value is very high, assumed to be 100. Octane number or octane rating grades quality of gasoline. Thus, octane number of fuel of an internal combustion engine is defined as the percentage by volume of isooctane in a mixture of isooctane and *n*-heptane, which matches with the knocking characteristics of the fuel under test. Say, a fuel has "Octane 80" means the fuel has the combustion characteristics that matches with the characteristics of the fuel containing mixture of 80 volume isooctane and 20 volume *n*-heptane. The knocking tendency of hydrocarbons increases of the order *n*-paraffin > branched chain paraffins > olefins > cyclo paraffins > aromatics. The reasons of knocking tendency of normal hydrocarbon and branched chain compound can be understood in this way. The normal chain compounds have greater surface area, and intramolecular atomic interactions are less, whereas branched compounds have less surface area, and intramolecular atomic interactions are higher. So, surface energy of the branched chain compounds are higher and hence, rupture of bonds are easier and regular, and knocking tendency is less.

The octane number of a fuel is determined by combusting the fuel under test in an internal combustion engine at 600 rpm at 52 °C.The value thus obtained is known as Research Octane Number (RON). Then the fuel is burnt in the same engine at 900 rpm at 150 °C. The value is known as Motor Octane Number (MON). The average of the two is a measure of Antiknock Index.

Catane number:

Diesel oil is exploded in the diesel engine by application of heat and pressure. Ignition quality of the hydrocarbon in diesel oil is of the order *n*-alkanes > naphthalene > alkenes > branched alkenes > aromatics. The quality of diesel oil or fuel efficiency is expressed in terms of **"Cetane number"**. Cetane is the *n*-hexadecane, $C_{16}H_{34}$. It has good ignition value taken as 100 and 1-methyl naphthalene has least ignition value taken as zero. So, cetane number of diesel oil is the percentage by volume of cetane in a mixture of cetane and 1-methyl naphthalene that has the same ignition characteristics as the diesel oil in question. Thus, if the cetane number of diesel oil is 60, then the diesel oil in question has the same ignition characteristic as the mixture of 60% by volume of cetane and 40% 1-methyl naphthalene. Depending on the cetane number, diesel oil is high speed (cetane number is higher than 45 and maximum 60. It requires $1/500^{th}$ second to heat the diesel drop to ignition point), low speed (cetane number is 25) and medium speed (cetane number is 35).

Biofuel: (if cultivated, then also called **agrofuel** or **agrifuel)** can be broadly defined as solid, liquid, or gas fuel consisting of/or derived from, recently dead biological material, most commonly plants. This distinguishes it from fossil fuel, which is derived from long dead biological material.

Biofuel can be theoretically produced from any (biological) carbon source. The most common by far is photosynthetic plants that capture solar energy. Many different plants and plant-derived materials are used for biofuel manufacture.

The most common use for biofuels is as liquid fuels for automobiles. The use of renewable biofuels provides increased independence from petroleum and enhances energy security.

One of the greatest technical challenges is to develop ways to convert biomass energy specifically to liquid fuels for transportation. To achieve this, the two most common strategies are:

1. To grow sugar crops (sugar cane, sugar beet, and sweet sorghum), or starch (corn/maize), and then use yeast fermentation to produce ethanol (ethyl alcohol).
2. To grow plants that (naturally) produces oils, such as oil palm, soybean, algae, or jatropha. When these oils are heated, their viscosity is reduced, and they can be burned directly in a diesel engine, or the oils can be chemically processed to produce fuels such as biodiesel.

Wood and its byproducts can also be converted into biofuels such as woodgas, methanol or ethanol fuel.

Biomass is material derived from recently living organisms. This includes plants, animals and their by-products. For example, manure, garden waste and crop residues are all sources of biomass. It is a renewable energy source based on the carbon cycle, unlike other natural resources such as petroleum, coal, and nuclear fuels.

Animal waste is a persistent and unavoidable pollutant produced primarily by the animals housed in industrial sized farms. It was found that microorganisms turn farm waste into alternative energy, providing farmers with a simple way to treat their waste and convert it into energy.

There are also agricultural products specifically grown for biofuel production which include corn, switchgrass, soybeans, rapeseed, wheat, sugar beet, sugar cane, palm oil, miscanthus, sorghum, cassava, and jatropha. Hemp has also been proven to work as a biofuel. Biodegradable outputs from industry, agriculture, forestry, and households can be used for biofuel production, either using anaerobic digestion to produce biogas as in the case of gobar gas, or using second generation biofuels; examples include straw, timber, manure, rice husks, sewage, and food waste. The use of biomass fuels can therefore contribute to waste management as well as fuel security and help to prevent climate change, though alone they are not a comprehensive solution to these problems.

Using waste biomass to produce energy can reduce the use of fossil fuels, reduce greenhouse gas emissions and reduce pollution and waste management problems.

8. See Q. 8(b) of 2005.

Gross and net calorific value: See Q. 8(a) of 2005.

Quality of coal is assessed depending on the analysis of coal that is done in two ways:

(A) Proximate analysis: Here percentage of moisture, volatile matter, ash and carbon is determined.

(a) **Moisture:** Moisture in coal lowers the effective calorific value of coal as it takes latent heat of evaporation from the fuel. Coal must not contain more than 10% moisture.

About 1g of air dried finely powdered coal is taken in a constant weighed crucible and placed inside an electrically hot air oven and heated at 105-110 °C for about one hour and then cooled in a desiccator and weighed. Loss in weight is due to the amount of moisture.

$$\% \text{ of moisture} = \frac{\text{loss in weight}}{\text{weight of coal taken}} \times 100.$$

(b) **Volatile matter:** During burning of coal, volatile matter escapes away unburned, lowering the calorific value. The volatile matter burns with long flame and produces smoke. Volatile matter may be combustible (such as methane, hydrogen, carbon monoxide and other hydrocarbon) and non-combustible (such as carbon dioxide nitrogen, etc.) and these are not dissociable. Coal having high volatile matter, i.e., low rank coal (50% volatile matter) is suitable in coal gas manufacturing but do not cake well. Coal having medium quantity of volatile matter (26-30%) produces hard and strong coke on carbonization. Low volatile matter content coal do not cake at all and unsuitable for coke making.

The dried sample of carbon left after removing of moisture is covered with the lid and heated to 925 ± 20 °C for seven minutes in a muffle furnace, cooled first in air, then in desiccator and then weighed. The second loss in weight of the sample is due to the volatile matter.

$$\% \text{ of volatile matter} = \frac{\text{second loss in weight}}{\text{weight of coal taken}} \times 100.$$

(c) **Ash:** It is the undesirable non-combustible constituent of coal. It causes lowering of calorific value, hindrance to the flow of air and heat. Fused ash, called clinkers, blocks the interspaces of grates on which coal is burnt causing obstruction of air supply and so, burning of coal becomes irregular. It causes wear of furnace wall, burning apparatus and feeding mechanism. Ash contains mainly oxides of inorganic compounds. The approximate compositions are

SiO_2	40-60%	MgO	0.5-4.0
Al_2O_3	20-35%	CaO	1-1.5%
Fe_2O_3	2-2.5%	$Na_2O + K_2O$	1-4%
SO_3	0.2-15%		

The residual coal after removing volatile matters is then heated without lid at (700± 20) °C for 30 min in the muffle furnace. The crucible is then cooled in air first and then in desiccator, and weighed finally after repeated heating and cooling until constant weight is obtained.

$$\% \text{ of ash} = \frac{\text{weight of ash left}}{\text{weight of coal taken}} \times 100.$$

(d) **Fixed carbon:**

% of carbon = (100–)% of (moisture + volatile matter + ash).

(B) Ultimate analysis of coal: It involves the determination of carbon, hydrogen, oxygen, nitrogen, sulphur and ash.

(a) **Carbon and hydrogen:** Percentage of carbon increases from lignite to anthracite. Higher the percentage of carbon and hydrogen better is the quality of coal.

About 1-2 g accurately weighed coal is burnt in a combustion chamber in a current of oxygen when CO_2 and H_2O produced are carried out and absorbed first in weighed fused $CaCl_2$ tube for H_2O and then into weighed KOH bulb for CO_2.

$$\underset{12}{C} + O_2 = \underset{44}{CO_2} \quad \text{and} \quad \underset{2}{H_2} + \tfrac{1}{2}O_2 = \underset{18}{H_2O}$$

$$2KOH + CO_2 = K_2CO_3 + H_2O \qquad CaCl_2 + 2H_2O = CaCl_2, 2H_2O$$

From the increase in weight of potash bulb and $CaCl_2$ tube, percentage of carbon and water is calculated.

$$\% \text{ of C} = \frac{\text{increase in weight of potash bulb} \times 12 \times 100}{\text{weight of coal taken} \times 44}$$

$$\% \text{ of } H_2 = \frac{\text{increased weight of } CaCl_2 \text{ tube} \times 2 \times 100}{\text{weight of coal taken} \times 18}$$

(b) **Nitrogen:** It has no calorific value and hence undesirable. Hence, good coal should contain nitrogen as low as possible.

Nitrogen is determined in Kjeldehl's flask taking about 1g of coal using a chemical balance together with concentrated H_2SO_4 and K_2SO_4 as catalyst and then heating until the solution becomes clear. At this stage, nitrogen is converted to ammonium sulphate. The solution is then boiled with excess KOH solution and the liberated NH_3 is absorbed in known volume of standard acid solution. The unconsumed acid is then back titrated against standard NaOH solution. From the volume of acid consumed by NH_3 and the strength of acid, nitrogen is estimated.

$$\% \text{ of N} = \frac{\text{volume of acid consumed} \times \text{normality} \times 1{\cdot}4 \times 100}{\text{weight of coal taken}}$$

(c) **Sulphur:** Its presence in coal is undesirable. During combustion it produces acid fumes of SO_2 and SO_3 that corrodes the equipment and also pollutes the atmosphere. Again, if this coal is used in the production of coke used for the metallurgy of iron, this sulphur is transferred to iron and induces the bad effects in the quality of steel.

(d) It is determined from the washings obtained from the known weight of coal used in bomb calorimeter for the determination of calorific value of coal. In this operation all sulphur is converted to sulphuric acid. So the washings are treated with acidified barium chloride solution when H_2SO_4 is precipitated as $BaSO_4$. It is filtered, washed and heated to a constant weight.

From the weight of $BaSO_4$, sulphur is determined.

$$\% \text{ of S} = \frac{\text{weight of } BaSO_4 \times 32 \times 100}{\text{weight of coal taken in bomb calorimeter} \times 233}$$

The sulphur content is usually 0·5-1·5%.

Ash: Ash is determined as in proximate analysis.

Oxygen: It is determined by the defference.

% of O_2 = 100 – percentage of (C + H + S + N + ash).

9. (a) See Q. 7(a) & (b) of 2005.

(c) See Q. 3(d) of 2005.

(d) *p*-type and *n*-type semiconductor:

The conductivity of semiconductors can be increased by adding impurities (i.e., doping), like atoms of neighbouring group elements of the periodic group. This type of semi-conductivity is known as *extrinsic semiconductivity*. These types of semiconductors are of two types: (1) *n*-type semiconductors and (2) *p*-type semiconductors.

1. *p-type Semiconductors*: Here the doped material has one electron less (Ga, In) than the host semiconductor Si or Ge. The empty band of Ga, called the acceptor band (as it accepts electron from the host) is localized near the filled energy band of host Si or Ge crystal lattice. On heating, some electrons from the energy band move to acceptor band creating electron holes in the energy band, then electron from the neighbouring energy band moves to occupy the hole making another positive hole. So, as positive holes move through the crystal lattice of the host Si or Ge lattice causing conduction, this is called the *p*-type semiconductor.

2. *n-type Semiconductors*: Si, Ge, etc. are semiconductors. If some As (arsenic) (having 5 valence electrons) is added ($1:10^9$ atoms ratio to Si) the As replacing the Si as host material has one electron in excess and is localized in a band called the donor band. This donor band is localized above the filled energy band and near the empty conduction band. If temperature is increased, some electron from the donor band will move to the empty conduction band and consequently some electron from the filled energy band will move to the donor band to fill the vacancy. As a result conduction continues and electron flow occurs throughout the crystal lattice of the host element. Since negative electron causes the conductivity it is called *n*-type semiconduction and the material as *n*-type semiconductor.

(e) Finger print region in IR spectra:

The region to the right hand side of an IR spectrum (from about 1500 to 500 cm^{-1}) usually contains a very complicated series of absorptions. These are mainly due to all manner of bonding vibrations within the molecule. This is called the finger print region.

It is much more difficult to pick out individual bands in this region than it is in the 'cleaner' region at higher wave numbers. The importance of the finger print region is that each different compound produces a different pattern of troughs in this part of the spectrum.

10. (a) The reaction

$$H^+Cl^- + Na^+OH^- \rightarrow H_2O + Na^+Cl^-$$

occurs. For each amount of NaOH added equivalent amount of hydrogen ions is removed. Effectively, the faster moving H^+ cation is replaced by the slower moving Na^+ ion, and the conductivity of the titrated solution as well as the measured conductance of the cell fall. This continues until the equivalence point is reached, at which we have a solution of sodium chloride, NaCl. If more base is added, an increase in conductivity or conductance is observed, since more ions are being added and the neutralization reaction no longer removes an appreciable number of any of them. Consequently, in the titration of a strong acid with a strong base, the conductance has a minimum at the equivalence point. This minimum can be used instead of an indicator dye to determine the end-point of the titration. Conductometric titration curve, that is a plot of the measured conductance or conductivity values against the number of millilitres of NaOH solution, is shown in Fig. in next page:

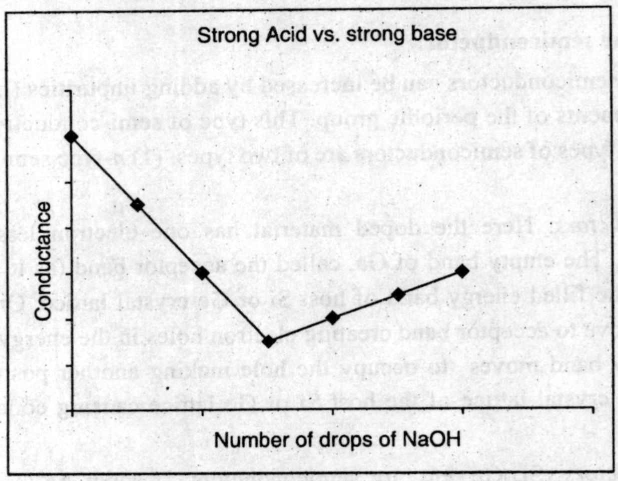

(b) Definition of specific conductance or conductivity may be stated as: the conductance of a solution enclosed between two electrodes of unit area of cross section, unit distance apart. Now the conductance of the solution is dependent on the number of ions present or the concentration of the solution. Hence, terms that can demonstrate this dependence have to be introduced.

Equivalent conductance (Λ): The conductance of a solution containing 1g-equivalent of dissolved electrolyte placed between two electrodes, unit distance apart, is called its *equivalent conductance*. Direct experimental determination of equivalent conductance entails the use of electrodes of enormous sizes so the specific conductance is first determined experimentally followed by the determination of equivalent conductance from it using the following formula:

$\Lambda = a/l \times \kappa$.

If we take a solution having concentration c g-equivalent per litre, then the volume of solution containing 1g-equivalent of dissolved electrolyte would be $1/c$ litre. If this volume of solution is placed between two electrodes, unit distance apart, then area of cross section, $a = 1/c$ sq m.

Therefore, $\Lambda = \kappa/c$.

(c) Hess's law (1840): The heat exchange accompanying a transformation is the same whether the process occurs in one or several steps.

Example: Let us consider the formation of CO_2.

(In one step) ⇒ $C(s) + O_2(g) \rightarrow CO_2(g)$ $\Delta H = -393.5$ kJ

(In two steps) ⇒ $C(s) + \frac{1}{2}O_2(g) \rightarrow CO(g)$ $\Delta H_1 = -110.5$ kJ

$CO(g) + \frac{1}{2}O_2(g) \rightarrow CO_2(g)$ $\Delta H_2 = -283$ kJ

Adding: $C(s) + O_2 \rightarrow CO_2(g)$ $\Delta H = -393.5$ kJ

11. (a) See Q. 4(c) of 2005.

(b) See Q. 5(a) of 2005.

(c) **THE GIBBS-DUHEM RELATION**

$$dG=(\partial X/\partial T)_{P,n_1,n_2,\cdots,n_j}dT+(\partial G/\partial P)_{T,n_1,n_2,\cdots,n_j}dP+(\partial G/\partial n_1)_{T,P,n_2,\cdots,n_j}dn_1+(\partial X/\partial n_2)_{T,P,n_1,n_3,\cdots,n_j}$$
$$dn_2+\cdots+(\partial G/\partial n_i)_{T,P,n_1,n_3,\cdots,n_j}dn_1+\cdots \quad (1)$$

Integrating we have

$$G_{P,T}=\mu_1n_1+\mu_2n_2+\mu_3n_3+\cdots\mu_in_i+\cdots \quad (2)$$

Now complete differentiation of equation (2) gives

$$dG_{P,T}=(\mu_1dn_1+\mu_2dn_2+\mu_3dn_3+\cdots+\mu_idn_i+\cdots)+(n_1d\,\mu_1+n_2d\,\mu_2+n_3d\,\mu_3+\cdots+n_id\,\mu_i+\cdots). \quad (3)$$

Subtracting equation (1) from equation (3),

$$(n_1d\,\mu_1+n_2d\,\mu_2+n_3d\,\mu_3+\cdots+n_id\,\mu_i+\cdots)=0. \quad (4)$$

These three relations for chemical potentials,

$G_{P,T}=\mu_1n_1+\mu_2n_2+\mu_3n_3+\cdots\mu_in_i+\cdots=\Sigma\,\mu_in_i$ [from equation (2)]

$dG_{P,T}=\mu_1dn_1+\mu_2dn_2+\mu_3dn_3+\cdots\mu_idn_i+\cdots=\Sigma\,\mu_idn_i$ [from equation (1)]

$0=n_1d\,\mu_1+n_2d\,\mu_2+n_3d\,\mu_3+\cdots+n_id\,\mu_i+\cdots=\Sigma\,n_id\,\mu_i$ [from equation (4)]

are commonly known as the Gibbs-Duhem relations.

(d) **Nernst Equation:**

Nernst equation gives the electrode potential *(E)*, relative to the standard electrode potential, (E^0), of the electrode couple or, equivalently, of the half-cells of a battery.

For the reaction,

$Ox + ne^- \rightarrow red$

$E=E^0+(RT/nF)\ln(a_{Ox}/a_{red})$.

The activities of pure solid or liquid phases are taken as unity. For a solution at room temperature (25 °C) replacing activities by concentration:

$E=E^0+(0.0591/n)\log([Ox]/[red])$.

(e) The Arrhenius equation is an expression that shows the dependence of the rate constant k of chemical reactions on the temperature T (in Kelvin) and activation energy E_a. He gave the empirical formulation,

$d\ln k/dt=E_a/RT^2$

or, $k=Ae^{-Ea/RT}$,

where A is the pre-exponential factor or simply the prefactor and R is the gas constant.

(f) See Q. 6(a) of 2005.

Some exciting gifts are up for grabs with this book!!

ENGINEERING CHEMISTRY

B. C. Ray ◆ S. N. Das ◆ S. Biswas

FREE Gifts Rs. 1000/-

DEAR READERS

By now you must have gone through this book in the entirety. So we invite your assessment on this title with absolute impartiality and candid observations. After all, we publishers blindly depend on your benchmarking. Friends, please fill up the open spaces below with CAPITAL LETTERS. You may add some extra sheets of paper if the space provided is not sufficient for you.

i. What are the important topics of your syllabus which are not adequately covered in this book?

ii. In which chapters and on what topics the treatment of the subject-matter by the author is not satisfactory and why?

iii. Have you come across the typos, factual inaccuracies, inconsistencies, and other forms of mistakes in the book? If yes, then please specify the mistakes and page numbers.

iv. Name top three books on the same subject, which, in your consideration, are better than this book. Could you provide some reasons?

Your 1st Preference: ....

Your 2nd Preference: ....

Your 3rd Preference: ....

v. Any suggestion to improve the quality of illustrations, diagrams etc.?

vi. How did you like the cover design? Any suggestions on colours?

vii. Suggestions for further improvement of production standard:

viii. Who recommended you this book:

☐ Teacher ☐ Friend ☐ Parents ☐ Bookseller ☐ Librarian

ix. Name of the person who has recommended:

Designation: ....

Address: ....

Phone No.: ....

e-mail: ....

x. Name and address of the bookseller from whom you have purchased this book:

....

....

....

xi. Name and address of the library where you would like to find the book:

....

....

xii. Name and address of your institution (Please mention the University or Board):

....

....

xiii. Your name, e-mail, and complete postal address:

....

....

....

xiv. A few other titles published by us that you would like to read:

....

....

The best assessment will be rewarded half-yearly. The award will be in the form of our publications, as decided by the Editorial and Marketing Operations Division, amounting to Rs. 1000/- (Rupees one thousand) only.

....

Please mail the filled up coupon at your earliest to :

Editorial & Marketing Operations Division

NEW CENTRAL BOOK AGENCY (P) Ltd

8/1 Chintamoni Das Lane, Kolkata 700 009

West Bengal, INDIA

Ph. No.: 033 + 22418411, 22418573, 22419094

Fax: 033 + 22419573, 22413852

e-mail: ncbapvtltd@eth.net, ncba@cal2.vsnl.net.in

Website: www.centralgroup.com